Brooklands Books

CW01667248

A·J·S
and
MATCHLESS
Gold Portfolio
1945-1966

Compiled by
R M Clarke

ISBN 9781855203525

Brooklands Books
BROOKLANDS BOOKS LTD.
P.O. BOX 146, COBHAM,
SURREY, KT11 1LG. UK
sales@brooklands-books.com

AJ45BGP

Brooklands Books

ROAD TEST SERIES

Abarth Road Test Portfolio 1950-1971
AC Cars 1904-2011
AC & Cobra 1962-2011
AC Ace Aceca Greyhound 1953-1962
Alfa Romeo A Brooklands Portfolio 1920-1940
Alfa Romeo Giulietta Gold Portfolio 1954-1965
Alfa Romeo Giulia Berlina Lim. Edit. Extra 1962-76
Alfa Romeo Giulia Coupes Lim. Edit. Ultra 1973-76
Alfa Romeo Alfasud 1972-1984
Alfa Romeo Spider Ultimate Portfolio 1966-1994
Alfa Romeo Spider & GTV Perf. Port. 1995-2005
Allard Limited Edition Ultra
Alpine Renault Ultimate Portfolio 1958-1995
Alvis Gold Portfolio 1919-1967
AMC Rambler Limited Edition Extra 1956-1969
AMX & Javelin Gold Portfolio 1968-1974
Armstrong Siddeley Gold Portfolio 1945-1960
Aston Martin Gold Portfolio 1921-1947
Aston Martin Ultimate Portfolio 1948-1968
Aston Martin Ultimate Portfolio 1968-1980
Aston Martin Ultimate Portfolio 1981-1993
Aston Martin Ultimate Portfolio 1994-2006
Auburn Cord Duesenberg - A Brooklands Portfolio
Audi Quattro Gold Portfolio 1980-1991
Audi TT Performance Portfolio 1998-2006
Austin Seven & Ten Road Test Portfolio
Austin-Healey 100 & 100/6 Gold Port. 1952-1959
Austin-Healey 3000 Ultimate Portfolio 1959-1967
Bentley & Rolls-Royce Portfolio 1990-2002
BMW 6 & 8 Cyl. Cars Limited Edition 1935-1960
BMW 700 Limited Edition 1959-1965
BMW 2002 Ultimate Portfolio 1968-1976
BMW 6 Cylinder Coupes & Saloons Gold P. 1969-1976
BMW 316, 318, 320 (4 cyl.) Gold Port. 1975-1990
BMW 320, 323, 325 (6 cyl.) Gold Port. 1977-1990
BMW 3 Series Gold Portfolio 1991-1997
BMW M3 Ultimate Portfolio 1986-2006
BMW M5 Gold Portfolio 1980-2003
BMW 5 Series Gold Portfolio 1988-1995
BMW 6 Series Ultimate Portfolio 1976-1989
BMW 7 Series Performance Portfolio 1977-1986
BMW 7 Series Performance Portfolio 1986-1993
BMW 8 Series Performance Portfolio
BMW X5 Limited Edition Extra 1999-2006
BMW Alpina Performance Portfolio 1967-1987
BMW Alpina Performance Portfolio 1988-1998
BMW Z3, M Coupe & M Roadster Gold Port. 1996-02
Bond Cars 1949-1974 Road Test Portfolio
Borgward Isabella Limited Edition
Bristol Cars Portfolio HARD COVER
Bugatti Type 10 to Type 40 Road Test Portfolio
Bugatti Type 41 to Type 55 Road Test Portfolio
Bugatti Type 57 to Type 251 Road Test Portfolio
Bugatti Type 10 to Type 251 Road Test Portfolio
Buick Performance Portfolio 1947-1962
Buick Muscle Portfolio 1963-1973
Buick Riviera Performance Portfolio 1963-1978
Cadillac Performance Portfolio 1948-1958
Cadillac Performance Portfolio 1959-1966
Cadillac Eldorado Performance Portfolio 1967-1978
Cadillac Allante Limited Edition Extra
Caterham Seven Road Test Portfolio 1974-1999
Caterham Seven Road Test Portfolio 2000-2010
Checker Automobiles Road Test Portfolio
Impala & SS Muscle Portfolio 1958-1972
Corvair Performance Portfolio 1959-1969
El Camino & SS Muscle Portfolio 1959-1987
Chevy II & Nova SS Gold Portfolio 1962-1974
Chevelle & SS Gold Portfolio 1964-1972
Camaro Muscle Portfolio 1967-1973
Blazer & Jimmy Limited Edition Extra 1969-1982
Blazer & Jimmy Limited Edition Extra 1983-1994
Camaro Performance Portfolio 1993-2000
Chevrolet Caprice Road Test Portfolio 1965-1990
Chevrolet Corvette Gold Portfolio 1953-1962
Chevrolet Corvette Sting Ray Gold Port. 1963-1967
Chevrolet Corvette Gold Portfolio 1968-1977
Chrysler Imperial Gold Portfolio 1951-1975
Citroen Traction Avant Limited Edition Premier
Citroen 2CV Ultimate Portfolio 1948-1990
Citroen DS & ID 1955-1975
Citroen DS & ID Gold Portfolio 1955-1975
Citroen CX Road Test Portfolio
Shelby Cobra Gold Portfolio 1962-1969
Crosley & Crosley Specials Limited Edition
Damlier SP250 Dart & V-8 2.5 litre 250 RTP 1959-69
Datsun Roadsters Performance Portfolio 1960-71
Datsun 240Z & 260Z Road Test Portfolio
Datsun 280Z & 280ZX Road Test Portfolio
Delahaye Road Test Portfolio
Delage Road Test Portfolio
DeLorean Gold Portfolio 1977-1995
De Soto Limited Edition 1952-1960
Dodge Limited Edition 1949-1959
Dodge Dart Limited Edition Extra 1960-1976
Dodge Muscle Portfolio 1964-1971
Charger Muscle Portfolio 1966-1974
Elfin Sports and Racing Cars
Elva Sports Racers Road Test Portfolio
ERA Gold Portfolio 1934-1994
Facel Vega Limited Edition Extra 1954-1964
Ferrari Road Cars 1946-1956 Road Test Portfolio
Ferrari Dino Limited Edition Extra 1965-1974
Ferrari 308 & Mondial Ultimate Portfolio 1975-85
Ferrari 328 348 Mondial Ultimate Portfolio 1986-94
Ferrari F355 & 360 Gold Portfolio 1995-2004
Fiat 500 1936-1972 Road Test Portfolio
Fiat Dino Road Test Portfolio
Fiat 124 Spider Performance Portfolio 1966-1985
Fiat Abarth 1972-1987 Road Test Portfolio
Fiat X1/9 Gold Portfolio 1973-1989
Fiat Barchetta Road Test Portfolio
Ford Zephyr, Zodiac, Executive Mk. III & IV 1962-1971
High Performance Capris Gold Portfolio 1969-1987
Capri Muscle Portfolio 1974-1987
High Performance Fiestas 1979-1991
Ford Escort RS & Mexico Limited Edition 1970-1979
High Performance Escorts 1980-1985
High Performance Escorts 1985-1990
Ford Thunderbird Performance Portfolio 1955-1957
Ford Thunderbird Performance Portfolio 1958-1963
Ford Thunderbird Performance Portfolio 1964-1976
Ford Fairlane Performance Portfolio 1955-1970
Ford Ranchero Muscle Portfolio 1957-1979
Ford Galaxie & LTD Gold Portfolio 1960-1976
Falcon Performance Portfolio 1960-1970
Ford GT40 & GT Ultimate Portfolio 1964-2006
Ford Torino Performance Portfolio 1968-1974
Ford Bronco 4x4 Performance Portfolio 1966-1977
Ford Bronco 1978-1988
Shelby Mustang Ultimate Portfolio 1965-1970
Mustang Muscle Portfolio 1967-1973
High Performance Mustang IIs 1974-1978
Mustang 5.0L Muscle Portfolio 1982-1993
Ginetta Cars Limited Edition Ultra 1958-2007
Goggomobil Limited Edition
Hispano-Suiza Road Test Portfolio
Honda - Acura NSX Ultimate Portfolio 1989-2005
Honda CRX 1983-1987
Honda S2000 Performance Portfolio 1999-2008
Hudson Performance Portfolio 1946-1957
International Scout Gold Portfolio 1961-1980
Isetta Gold Portfolio 1953-1964
ISO & Bizzarrini Limited Edition Ultra 1962-1974
Jaguar XK120 • 140 • 150 Gold Portfolio 1948-1960
Jaguar Mk 7, 8, 10 & 420 Road Test Portfolio
Jaguar Mk 1 & Mk 2 1955-1969 Road Test Portfolio
Jaguar C-Type & D-Type Gold Portfolio 1951-1960
Jaguar E-Type Ultimate Portfolio 1961-1975
Jaguar XJ6 Series I & II Gold Portfolio 1968-1979
Jaguar XJ6 Series III Perf. Portfolio 1979-1986
Jaguar XJS Gold Portfolio 1975-1988
Jaguar XJ-S V12 Ultimate Portfolio 1988-1996
Jaguar XK8 & XKR Performance Portfolio 1996-2005
Jeep CJ-5 Limited Edition 1960-1975
Jeep CJ-5 & CJ-7 4x4 Perf. Portfolio 1976-1986
Jeep Wagoneer Performance Portfolio 1963-1991
Jeep J-Series Pickups 1970-1982
Jeepster & Commando Limited Edition 1967-1973
Jeep Cherokee & Comanche Pickups P. P. 1984-91
Jeep Wrangler 4x4 Performance Portfolio 1987-99
Jeep Cherokee & Grand Cherokee 4x4 P. P. 1992-98
Jensen Road Test Portfolio 1934-1965
Jensen Interceptor Ultimate Portfolio 1966-1992
Jensen - Healey Road Test Portfolio 1972-1976
Jowett Road Test Portfolio
Kaiser - Frazer Limited Edition 1946-1955
Lagonda Gold Portfolio 1919-1964
Lancia Aurelia & Flaminia Gold Portfolio 1950-1970
Lancia Fulvia Gold Portfolio 1963-1976
Lancia Montecarlo & Scorpion 1975-1982 RTP
Lancia Stratos Limited Edition Extra
Lancia Delta & integrale Ultimate Portfolio
Land Rover Series I, II & IIA Gold Portfolio 1948-71
Land Rover Series III 4x4 Perf. Portfolio 1971-1985
Land Rover 90 110 Defender Gold Portfolio 1983-94
Land Rover Discovery Perf. Port. 1989-2000
Lamborghini Performance Portfolio 1964-1976
Lamborghini Performance Portfolio 1977-1989
Lamborghini Gold Portfolio 1990-2004
Lincoln Gold Portfolio 1949-1960
Lincoln Continental Performance Portfolio 1961-1969
Lincoln Continental 1969-1976
Lola Sports Racing Cars Lim. Ed. Premier 1958-1985
Lotus Sports Racers Portfolio - covering 1951-1965
Lotus Seven Gold Portfolio 1957-1973
Lotus Elite Limited Edition 1957-1964
Lotus Elan Ultimate Portfolio 1962-1974
Lotus Cortina Road Test Portfolio
Lotus Elite & Exlat Road Test Portfolio 1984-1812
Lotus Excel Road Test Portfolio
Lotus Europa Gold Portfolio 1966-1975
Lotus Elise & Exige Gold Portfolio 1995-2005
Marcos Coupés & Spyders Gold Portfolio 1960-1997
Maserati Cars Performance Portfolio 1957-1970
Maserati Cars Performance Portfolio 1971-1982
Maserati Cars Performance Portfolio 1982-1998
Maserati Cars Ultimate Portfolio 1999-2007
Matra Road Test Portfolio
Mazda Miata MX-5 Performance Portfolio 1989-1997
Mazda Miata MX-5 Performance Portfolio 1998-2005
Mazda Miata MX-5 Takes On The Competition
McLaren F1 • GTR • LM Sportscar Perf. Portfolio
Mercedes-Benz 1925-1939 - A Portfolio
Mercedes 190SL 300SL 300SLR - A Portfolio
Mercedes S & 600 Limited Edition Extra 1965-1972
Mercedes S Class 1972-1979
Mercedes S Class Limited Edition Extra 1980-1991
Mercedes 230SL 250SL 280SL Ultimate Port. 1963-71
Mercedes-Benz SLs & SLCs Ultimate Port. 1971-1989
Mercedes SLs Performance Portfolio 1989-1994
Mercedes E-Class W124 Road Test Portfolio 1986-1995
Mercedes G-Wagen Gold Portfolio 1981-2005
Mercedes AMG Gold Portfolio 1983-1999
Mercedes AMG Ultimate Portfolio 2000-2006
Mercury Gold Portfolio 1947-1966
Mercury Comet & Cyclone Lim. Edit. Extra 1960-1975
Cougar Muscle Portfolio 1967-1973
Messerschmitt Gold Portfolio 1954-1964
MG Gold Portfolio 1929-1939
MG TD & TF Gold Portfolio 1949-1955
MG Y Type & Magnette Road Test Portfolio
MGA & Twin Cam Gold Portfolio 1955-1962
MG Midget Road Test Portfolio 1961-1979
MGB Roadsters 1962-1980
MGB MGC & V8 Gold Portfolio 1962-1980
MGC & MGB GT V8 Limited Edition
MGF & TF Performance Portfolio 1995-2005
Mini Gold Portfolio 1959-1969
Mini Gold Portfolio 1969-1980
Mini Gold Portfolio 1981-1997
High Performance Minis Gold Portfolio 1960-1973
Mini Cooper Gold Portfolio 1961-1971
Mini Moke Ultimate Portfolio 1964-1994
Mini Performance Portfolio 2001-2006
Starion & Conquest Performance Portfolio 1982-1990
Mitsubishi 3000GT & Dodge Stealth P.P. 1990-1999
Morgan Three-Wheelers Ultimate Portfolio 1909-52
Morgan Four-Wheelers Ultimate Portfolio 1936-67
Morgan Ultimate Portfolio 1968-1990
Morgan Ultimate Portfolio 1991-2009
Morris Minor MM & Series II Road Test Portfolio
Morris Minor 1000 Road Test Portfolio
Nash Limited Edition Extra 1949-1957
Nash-Austin Metropolitan Gold Portfolio 1954-1962
Nissan 350Z Road Test Portfolio
Nissan Skyline GT-R Ultimate Portfolio 1969-2010
Noble Sports Cars Road Test Portfolio
NSU Ro80 Limited Edition
Oldsmobile 1948-1963 Limited Edition Premier
Oldsmobile Muscle Portfolio 1964-1971
Cutlass & 4-4-2 Muscle Portfolio 1964-1974
Opel GT Ultimate Portfolio 1968-1973
Opel Manta Limited Edition 1970-1975
Packard Automobiles 1920-1958 - A Portfolio
Pantera Ultimate Portfolio 1970-1995
Panther Gold Portfolio 1972-1990
Plymouth Limited Edition 1950-1960
Plymouth Fury Limited Edition Extra 1956-1976
Barracuda Muscle Portfolio 1964-1974
Plymouth Muscle Portfolio 1964-1971
Pontiac 1946- 1963 Limited Edition Premier
High Performance Firebirds 1982-1988
Firebird & Trans Am Performance Portfolio 1993-00
Pontiac Fiero Performance Portfolio 1984-1988
Porsche Sports Racing Cars UP 1952-1968
Porsche 917 • 935 • 956 • 962 Gold Portfolio
Porsche 912 Road Test Portfolio
Porsche 365 Ultimate Portfolio 1952-1965
Porsche 911 1965-1969
Porsche 911 SC & Turbo Gold Portfolio 1978-1983
Porsche 911 Ultimate Portfolio 1990-1997
Porsche 911 Takes On The Competition 1990-1997
Porsche 911 Ultimate Portfolio 1998-2004
Porsche 914 Ultimate Portfolio
Porsche 924 Gold Portfolio 1975-1988
Porsche 928 Gold Portfolio 1977-1995
Porsche 928 Takes On The Competition
Porsche 944 Ultimate Portfolio
Porsche 968 Limited Edition Extra
Porsche Boxster Ultimate Portfolio 1996-2004
Railton & Brough Superior Gold Portfolio 1933-1950
Range Rover Gold Portfolio 1970-1985
Range Rover Gold Portfolio 1985-1995
Range Rover Performance Portfolio 1995-2001
Range Rover Takes on the Competition
Riley RM Series Pathfinder 2.6 A Brooklands Port.
Rolls-Royce Silver Clould & Bentley S Ultimate Port.
Rolls-Royce Silver Shadow Ultimate Port. 1965-80
Rover P4 1949-1959
Rover P6 1963-1977
Rover 2000 & 2200 1963-1977
Studebaker Ultimate Portfolio 1946-1966
Studebaker Hawks & Larks Lim. Edit. Premier 1956-66
Avanti Limited Edition Extra 1962-1991
Subaru Impreza Turbo Limited Edition Extra 1994-00
Subaru Impreza WRX Performance Port. 2001-2005
Sunbeam Alpine Limited Edition Extra 1959-1968
Sunbeam Tiger Limited Edition Extra 1964-1967
Suzuki SJ Gold Portfolio 1971-1997
Vitara, Sidekick & Geo Tracker Perf. Port. 1988-1997
Tatra Cars Road Test Portfolio
Toyota Land Cruiser Gold Portfolio 1956-1987
Toyota Land Cruiser 1988-1997
Toyota Supra Performance Portfolio 1982-1998
Toyota MR2 Road Test Portfolio 1984-1989
Toyota MR2 Road Test Portfolio 1990-1999
Toyota MR2 Road Test Portfolio 2000-2007
Toyota MR2 Takes On The Competition
Triumph TR2 & TR3 Road Test Portfolio
Triumph TR4, TR5 & TR250 Road Test Portfolio
Triumph TR6 Road Test Portfolio
Triumph Herald 1959-1971
Triumph Vitesse 1962-1971
Triumph 2000, 2.5, 2500 1963-1977
Triumph GT6 Gold Portfolio 1966-1974
Triumph Spitfire Road Test Portfolio
Triumph Stag Road Test Portfolio
TVR Limited Edition Ultra 1958-1985
TVR Performance Portfolio 1986-1994
TVR Performance Portfolio 1995-2000
TVR Performance Portfolio 2000-2005
VW Beetle Gold Portfolio 1935-1967
VW Beetle Gold Portfolio 1968-1991
VW Bus Camper Van Perf. Portfolio 1954-1967
VW Bus Camper Van Perf. Portfolio 1979-1991
VW Karmann Ghia Gold Portfolio 1955-1974
VW Golf GTI Limited Edition Extra 1976-1991
VW Golf Cabriolet 1979-2002 Road Test Portfolio
VW Corrado Limited Edition Premier 1989-1995
Volvo PV444 & PV544 Perf. Portfolio 1945-1965
Volvo 120 Amazon Ultimate Portfolio
Volvo 1800 Ultimate Portfolio 1960-1973
Volvo 140 & 160 Series Gold Portfolio 1966-1975
Forty Years of Selling Volvo
Westfield Performance Portfolio 1982-2004

MILITARY VEHICLES

Complete WW2 Military Jeep Manual
Dodge WW2 Military Portfolio 1940-1945
German Military Equipment WW2
Hail To The Jeep
Combat Land Rover Portfolio No. 1
Land Rover Military Portfolio
Off Road Jeeps Civilian & Military 1944-1971
US Military Vehicles 1941-1945
Standard Military Motor Vehicles-TM9-2800 (WW2)
VW Kubelwagen Military Portfolio 1940-1990
WW2 Allied Vehicles Military Portfolio 1939-1945
WW2 Jeep Military Portfolio 1941-1945

RACING & THE LAND SPEED RECORD

The Land Speed Record 1898-1919
The Land Speed Record 1920-1929
The Land Speed Record 1930-1939
The Land Speed Record 1940-1962
The Land Speed Record 1963-1999
Can-Am Racing 1966-1969
Can-Am Racing 1970-1974
Can-Am Racing Cars 1966-1974
The Carrera Panamericana Mexico - 1950-1954
Le Mans - The Bentley & Alfa Years - 1923-1939
Le Mans - The Jaguar Years - 1949-1957
Le Mans - The Ferrari Years - 1958-1965
Le Mans - The Ford & Matra Years - 1966-1974
Le Mans - The Porsche Years - 1975-1982
Le Mans - The Porsche & Jaguar Years - 1983-91
Le Mans - The Porsche & Peugeot Years - 1992-99
Mille Miglia - The Alfa & Ferrari Years - 1927-1951
Mille Miglia - The Ferrari & Mercedes Years - 1952-57
Targa Florio - The Porsche & Ferrari Years - 1955-1964
Targa Florio - The Porsche Years - 1965-1973
Brabham Ralt Honda The Ron Tauranac

RESTORATION & GUIDE SERIES

BMW 2002 - A Comprehensive Guide
BMW 02 Restoration Guide
BMW E30 - 3 Series Restoration Bible
BMW E36 - 3 Series Restoration Tips & Techniques
BMW - 5 & 6 Series Restoration Tips & Techniques
Classic Camaro Restoration
Engine Swapping Tips & Techniques
Ferrari Life Buyer's Portfolio
Land Rover Restoration Portfolio
PC on Land Rover Series I Restoration
Lotus Elan Restoration Guide
MG T Series Restoration Guide
MGA Restoration Guide
PC on MGB Restoration
Mustang Restoration Tips & Techniques
Practical Gas Flow
Restoring Sprites & Midgets an Enthusiast's Guide
Solex Carburettors Tuning Tips and Techniques
SU Carburetters Tuning Tips & Techniques
The Great Classic Muscle Cars Compared
Weber Carburettors Tuning Tips and Techniques

ROAD & TRACK SERIES

Road & Track on Aston Martin 1962-1990
Road & Track on Austin Healey 1953-1970
Road & Track on BMW Cars 1966-1974
R & T BMW Z3, M Coupe & M Roadster Port. 96-02
R & T Camaro & Firebird Portfolio 1993-2002
Road & Track on Corvette 1968-1982
Road & Track on Corvette 1982-1986
Road & Track on Corvette 1986-1990
Road & Track Corvette Portfolio 1997-2002
Road & Track Dodge Viper Portfolio 1992-2002
Road & Track on Ferrari 1975-1981
Road & Track on Ferrari 1984-1988
Road & Track Ferrari V-12 Portfolio 1992-2002
Road & Track Ferrari F355 360 F430 Portfolio 95-06
Road & Track on Jaguar 1950-1960
Road & Track on Jaguar 1961-1968
Road & Track on Jaguar 1968-1974
R & T Jaguar XJ-S - XK8 - XKR Portfolio 1975-2003
Road & Track MX-5 Miata Portfolio 1989-2002
Road & Track on Mercedes 1952-1962
R & T Mercedes SL - SLK - CLK Portfolio 1990-2003
Road & Track on MG Sports Cars 1949-1961
Road & Track on MG Sports Cars 1962-1980
Road & Track Mustang Portfolio 1994-2002
Road & Track Nissan 300ZX & 350Z Portfolio 1984-03
Road & Track on Porsche 1951-1967
Road & Track on Porsche 1975-1978
Road & Track on Porsche 1979-1982
Road & Track Porsche 928 Portfolio 1977-1994
Road & Track Porsche 911 Portfolio 1990-1997
R & T on Rolls Royce & Bentley 1950-1965
R & T on Triumph Sports Cars 1967-1974
R & T on Triumph Sports Cars 1974-1982
Road & Track on Volkswagen 1951-1968
Road & Track on Volkswagen 1968-1978
Road & Track on Volvo 1957-1974
Road & Track on Volvo 1977-1994
Road & Track - Best of PS
Road & Track - Peter Egan Side Glances 1983-1992
Road & Track - Peter Egan Side Glances 1992-1997
Road & Track - Peter Egan Side Glances 1998-2002
Road & Track - Peter Egan Side Glances 2002-2006

CAR AND DRIVER SERIES

Car and Driver on BMW 1957-1977
Car and Driver on Ferrari 1955-1962
Car and Driver on Ferrari 1963-1975
Car and Driver on Porsche 1955-1962
Car and Driver on Porsche 1963-1970
Car and Driver on Porsche 1970-1976
Car and Driver on Porsche 1977-1981
Car and Driver on Porsche 1982-1986
Car and Driver on Volvo 1955-1976

HOT ROD 'ENGINE' SERIES

Chevy 265 & 283
Chevy 302 & 327
Chevy 348 & 409
Chevy 396 & 427
Chevy 454 thru 512
Chevy Monster Big Blocks
Chrysler Hemi
Chrysler 273, 318, 340 & 360
Chrysler 361, 383, 400, 413, 426 & 440
Ford 289, 302, Boss 302 & 351W
Ford 351C & Boss 351
Ford Small Block
Ford Big Block

MOTORCYCLES

To see our range of over 70 titles visit www.brooklands-books.com

Brooklands Books

Contents

5	Two AJS's Enter Production	*Motor Cycle*	June	28	1945
6	A Flip on a 1945 Model - 500cc AJS	*Motor Cycle*	June	28	1945
7	Two Matchless Models - 350 & 500 Clubman	*Motor Cycle*	July	5	1945
8	The AJS Racing Twin	*Motor Cycling*	May	29	1947
10	Assignment at Assen... - AJS Model 18 498cc	*Motor Cycling*	July	17	1947
12	The New AJS "Junior" Racer - 350cc	*Motor Cycling*	Feb	5	1948
14	For the Private Owner - 350cc Racing AJS	*Motor Cycle*	Feb	5	1948
16	348cc Racing Single AJS	*Motor Cycle*	May	27	1948
19	Gulley King - AJS Scrambler 500cc Road Test	*Cycle*	Mar		1952
22	30.50 Competition Scrambles for 1952	*Cycle*	Mar		1952
24	498cc AJS and Matchless Parallel Twins	*Motor Cycle*	Apr	5	1951
28	Competition Conversion - Matchless Twins	*Cycle*	Mar		1953
32	The 347cc AJS and Matchless	*Motor Cycle*	Aug	27	1953
36	The 348cc Overhead-camshaft 7R AJS	*Motor Cycle*	May	27	1954
41	Big Binders for AJS - Matchless	*Cycle*	Feb		1954
42	1955 Improvements for A.M.C.	*Cycle*	Feb		1955
45	"Sports Twin" Matchless G9 Super Clubman	*Cycle*	Apr		1955
48	The 349cc Three-valve AJS	*Motor Cycle*	Apr	28	1955
52	Bigger Parallel Twins - 592cc AJS and Matchless	*Motor Cycle*	Sept	8	1955
55	A.M.C. Competiton Models	*Motor Cycle*	Nov	3	1955
56	AJS and Matchless Modifications	*Motor Cycle*	Sept	27	1956
59	347cc G3LC Matchess	*Motor Cycle*	Feb	14	1957
60	New AJS & Matchless Scrambler Twins	*Cycle*	Dec		1957
61	Ready for Roughin' - Matchless Super Hurricane	*Cycle*	Jan		1958
64	New Two-fifties from Woolwich	*Motor Cycle*	Mar	13	1958
68	Hookey to the Seaside - 248cc Model 14 AJS	*Motor Cycle*	Feb	26	1959
71	Grown-up "Boy Racer" for Cross-country - 500cc	*Motor Cycling*	Feb	26	1959
72	AJS Typhoon Scrambler Road Test	*Cycle*	Apr		1959
75	347cc Model 16 AJS & Sidecar Road Test	*Motor Cycle*	Sept	3	1959
78	The 646cc Matchless Model G12CSR Road Test	*Motor Cycling*	May	4	1961
83	646cc Matchless Model G12 de Luxe Road Test	*Motor Cycle*	June	8	1961
86	The 347cc o.h.v. AJS Model 16 Road Test	*Motor Cycling*	Aug	31	1961
89	348cc Matchless G5 Matador Road Test	*Motor Cycle*	Sept	28	1961
92	The 248cc Matchless G2S 'Monitor Sports'	*Motor Cycling*	Nov	9	1961
95	498cc AJS Model 18 Statesman Road Test	*Motor Cycle*	Mar	1	1962
98	The 646cc Matchless G12 'Majestic' Road Test	*Motor Cycling*	Jan	25	1962
102	Matchless G-50 Engine Technical Analysis	*Cycle World*	Apr		1962
106	Sports AJS - Sceptre 348cc Road Test	*Motorcycle Mechanics*	May		1962
108	AJS Hurricane	*Motorcycle Mechanics*	June		1963
109	Mercury and Grand Prix - Matchless G3	*Motor Cycle*	July	5	1962
112	248cc Super Sports Matchless Road Test	*Motor Cycle*	Aug	9	1962
115	646cc AJS Hurricane Road Test	*Motor Cycle*	Mar	21	1963
118	Matchless G-80 CS Road Test	*Cycle World*	July		1963
122	248cc AJS CSR Road Test	*Motorcycle Sport*	Apr		1964
126	248cc AJS Sapphire Ninety Road Test	*Motor Cycle*	Feb	25	1965
129	745cc Matchless G15 Road Test	*Motor Cycle*	Oct	29	1964
132	750 G15 Road Test	*Motorcycle Mechanics*	Feb		1965
134	Matchless 250	*Motorcycle Mechanics*	July		1965
136	Mighty Matchless	*Motorcycle Mechanics*	July		1965
138	745cc Matchless CSR Road Test	*Motor Cycle*	July	15	1965
141	Impressions of a Modern Three-Fifty: the AJS 16	*Motorcycle Sport*	Sept		1965
144	Super Sapphire - 14 CSR Road Test	*Motorcycle Mechanics*	Jan		1966
146	AJS and Matchless Two-Fifty Singles	*Motor Cycle*	Feb	24	1966
151	Matchless G85CS Scrambler Road Test	*Cycle World*	Nov		1966
154	646cc AJS 31 CSR & Sidecar Road Test	*Motor Cycle*	May	4	1966
158	AJS and Matchless Big Singles 1956 to 1966	*Motor Cycle*	Aug	11	1966
163	AJS	*Cycle World*	Sept		1968
167	Matchless	*Cycle World*	Apr		1969

Brooklands Books

Acknowledgements

With more than 30 years experience of producing motoring anthologies for enthusiasts, Brooklands Books are now widely recognised as an outstanding reference source for information about the world's most sought after cars. The success of the Gold Portfolio series, now with more than 100 titles, led to motorcycle enthusiasts enquiring if something similar could be published covering their vehicles. Hence this book.

We have asked Jeff Clew, a leading authority on vintage and classic motorcycles, to pen an introduction to this post-war collection of AJS and Matchless road tests and stories. He has drawn on his specialised knowledge of these machines, not only as an historian but also as an active rider now enjoying his 50th consecutive year of motorcycling, both on and off-road.

Like the many other titles in the Brooklands Books series, this book could not have come about without the generous assistance given by the world's leading motorcycle magazine publishers. In this instance we are indebted to the management of *Cycle, Cycle World, Motor Cycle, Motor Cycling, Motorcycle Mechanics* and *Motorcycle Sport* for allowing us to reissue their copyright articles.

Our thanks also go to Ernie Merryweather and his colleagues at the AJS & Matchless Owners Club for supplying the colour photographs that appear on our front and back covers.

R.M. Clarke

AJS and Matchless can both claim early involvement with the British motorcycle industry, originally as separate companies with their own identity. It was not until 1931 that AJS was acquired by Matchless Motor Cycles (Colliers) Ltd. when the recession of the late twenties forced the former into voluntary liquidation. AJS production was transferred from Wolverhampton to the Matchless factory in Plumstead, South East London, the two separate marques retaining their individual characteristics initially. As time progressed, production economies brought about rationalisation of design although the AJS and Matchless models were never identical in every respect, such was the strength of brand loyalty.

When the production of civilian models resumed after World War 2, only 350cc and 500cc single cylinder ohv models were at first listed, but it was not long before each marque had its own 500cc vertical twin. They created quite an impact when they were launched at the 1949 Earls Court Motor Cycle Show, appearing with pivoted fork rear suspension at a time when others used the older, less satisfactory plunger-type units. The singles followed suit, with the exception of the competition models, and it was not long before pivoted fork rear suspension was adopted by most other manufacturers on their standard production models. The same could not be said of the crankshaft centre bearing arrangement used on the twins, however. It made an engine more difficult to dismantle and reassemble, whilst some considered it undesirable to have a crankshaft that was too rigidly mounted. Unswayed by popular opinion, this design feature was retained, even when the twins were increased in capacity in the never ending search for improved performance.

AJS and Matchless also made an outstanding contribution to world-wide competition events, irrespective of whether it was in trials, scrambles or road racing, both before and after the war. Perhaps the most notable development was that of the AJS 7R racer, which made its debut in 1947, to offer the all-conquering Manx Norton of similar capacity serious challenge. Later came the Matchless G50, a 500cc version that was reputed to handle better on short circuits than the contemporary Norton. Only the Matchless G45 twin failed to make any lasting impact, largely due to its limited power band. During March 1958 lightweight 250cc models were introduced, built on a semi unit-construction basis. Somehow, they lacked the aesthetic appeal of their pre-war predecessors, as did the 'cleaned up' versions of their 350cc and 500cc singles. When commonisation of parts became rife following earlier acquisitions of Norton, James and Francis Barnett, Associated Motor Cycles found themselves heading for oblivion and the company was acquired by Manganese Bronze in 1966. It was a sad end to two of Britain's best-known makes who had produced some very memorable motorcycles in their time.

Jeff Clew

Two A.J.S.s Enter Production

Scintillating Quality Machines Based on the Famous Army Mount

TWO A.J.S.s, two gleaming black, gold and chromium models, are about to enter production. Both are overhead-valve singles, one of 347 c.c. and the other of 498 c.c. The first machines will be available some seven or eight weeks hence. Since Associated Motor Cycles are continuing the manufacture of motor cycles for the Army, three-quarters of the factory's production capacity being devoted to the supply of W.D. motor cycles and spares, the machines for the civilian market have to be based on their Army model. They are, however, thrillingly civilian in their finish, and have their acid-dipped crankcases, polished magneto chain covers, scintillating levers and even rubber for footrests and kneegrips, though not at the moment for the foot-change lever. Note, too, that there is a five-hundred as well as the three-fifty, which is the capacity of the W.D. machine. The prices are to be announced shortly.

Perhaps one of the most striking points about the 1945 civilians-for-the-pleasure-of models is that they do not give the slightest impression of being "Army." Even the three-gallon tanks, with their black and gold finish, look deeper.

The machines have the proved-on-every-battlefront duplex cradle frame of the G/3L and the now famous "Teledraulic" front forks, which provide a total possible movement, hydraulically damped, of $7\frac{1}{8}$in., divided roughly into 3in. upward and 4in. on the rebound. Four-speed Burman gear boxes are employed. These give "Civvy Street" ratios, namely, 5.8, 7.5, 10.3 and 15.4 to 1 on the 350 c.c. machine (the ratios of the W.D. three-fifty are 5.8, 7.5, 12.2 and 18.5 to 1) and on the 500 c.c. model 5.25, 6.8, 9.3 and 13.9 to 1.

As usual, the A.J.S. has a shielded magneto in the front of the engine. The compression ratio of the 350 c.c. model is also different from that of the W.D. mount. There is no compression plate, which means that the ratio is 6.3 to 1 instead of 6.0 to 1. The compression ratio of the five-hundred is 5.9 to 1. Lo-ex pistons are used in both cases.

The 498 c.c. engine is a single-port similar to the engine marketed in late 1939, with the latter's heavier flywheels and with a new inlet-valve timing. A $1\frac{1}{8}$in. bore Amal two-lever carburettor is fitted. A cylinder head steady running from the rear near-side rocker-box bolt to the top tube of the frame is fitted in the case of this model. On the 350 c.c. engine there is a 1in. bore carburettor, as against $\frac{7}{8}$in. on the W.D. machine. Like the latter, the 350 c.c. A.J.S. has the exhaust pipe above the off-side footrest; the five-hundred, which has a $1\frac{3}{4}$in. pipe, has it set below.

Other features of the machines are 3.25—19 Dunlop tyres, internally illuminated speedometer heads on top of the cross-member of the "Teledraulic" forks, built-in type rear number plate, the neat A.M.C. prop-stand, spring-up rear stand, tubular front stand-cum-mudguard stay, fabric-covered handlebar grips, a four-pint separate oil tank, with fabric oil filter, a 7in. head lamp and hand adjusters for the front and rear brakes and the clutch.

Black, gold and gleaming chromium! The 500 c.c. single-port A.J.S., a machine with an exhilarating performance

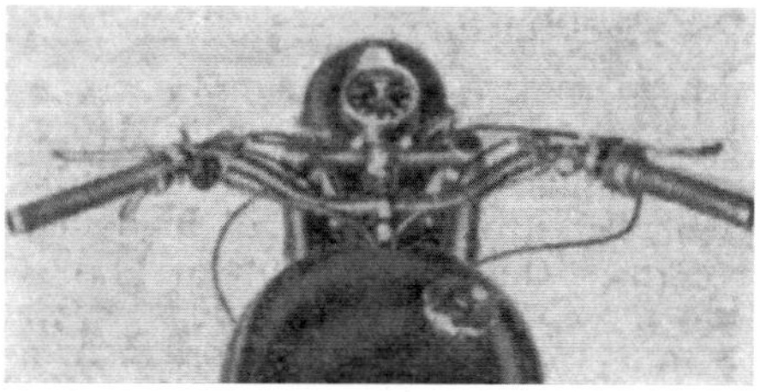

Control layout of the A.J.S. The speedometer is internally illuminated. Fabric-covered handlebar grips on Army lines are employed

A close up of the 500 c.c. engine unit. Note the shielded magneto and the accessible position of the dynamo brush gear

A Flip on a 1945 Model

"Torrens" Gives His Impressions of the New 500 c.c. A.J.S.

"WHAT about taking it away with you?" said Mr. Manufacturer. I was decked out in pin-stripe city suiting—a very wartime, much-worn suiting—and had no goggles, no gauntlets and not even a mackintosh. Everything had pointed to a day in the office. Then there was this bolt from the blue: the A.J.S. programme for next week's issue.

The day was hot and seemed likely to remain fine. I need not ask you what you would have done in my case, for, like me, you would have jumped at the opportunity of riding "something" 1945. What model A.J.S.? The overhead-valve five-hundred. I particularly wanted to ride this machine. As the description on page 455 of this issue states, the two A.J.S. models are based on the A.M.C. factory's G/3L—are necessarily so since three-quarters of the company's production capacity continues to be usurped by the Army. But the latter machine, as you well know, is of 350 c.c., and here was an engine of 500 c.c.—150 c.c. greater capacity. What would the model be like?

"One of the Nicest"

This aspect—the extra power—interested me, for I know the G/3L well, having ridden dozens of Army ones during the course of the war. This, the first of the new models, came as a surprise to me. I was expecting "just a G/3L with a bit more urge," and found one of the nicest, liveliest 500 c.c. singles I have ridden for a long time. The speedometer showed 190 miles, and I had only the opportunity of adding 51 to that total; hence this little article is in no sense a road test, but merely a record of my impressions over this short distance.

First, a dull note. It was possible to kick a number of times without anything happening unless the ignition was fully advanced. At the works it was remarked that it was better to start on the exhaust-valve lifter than to adopt the standard dodge of finding compression, easing the engine over on the exhaust lifter and giving a swinging kick. However, with the magneto fully advanced all was well, so presumably it was largely a case of the particular magneto. There were very few other dull notes and a number of very pleasant surprises.

The engine is flexible. One can trickle along at 18 m.p.h. in the top gear of 5.25 to 1 and then, if one retards the ignition a trifle, accelerate away quite happily. It is also pleasantly quiet mechanically. The exhaust is "beefy," if you gather my meaning—beefy without being in any way harsh and obnoxious.

It was the "urge" that surprised me most. At 55 or 60 m.p.h. one is moving along on quite a small throttle opening without the engine appearing to be "at work." "Umph!" you say to yourself. "The rest of the throttle opening won't mean much," and you are wrong—utterly wrong. Tweak the twist-grip open with the machine doing sixty and things happen. I will not say that the machine wooshes forward, but the acceleration from a mile a minute is very useful indeed. There is a liveliness about the machine that makes it a real joy, and it is something which I do not recall with the previous five-hundreds. What has brought it about?

The engine, I gather, is basically the same as the 1939 five-hundred—the late 1939, which had heavier flywheels than the early ones—but has a different inlet-valve timing. The results are impressive. And what I also liked about the machine was that it feels such a quality job. Riding it one has the sensation of handling a mount that has been built and not merely assembled.

If you know the latest G/3Ls you will be aware of the excellence of the steering and road-holding—I refer to the models that are as they left the factory, and not after someone has put too much oil, or the wrong oil, in the "Teledraulics." Whether the 500 c.c. engine with the fore and aft disposition of the magneto and dynamc respectively makes any odds I cannot say, but it seemed to me that the general navigation was particularly good, and I especially revelled in the way the model heels over on a bend.

Single-finger Light

Both brakes were really good, with the front one, as I proved, single-finger light. The riding position is G/3L, which means better than average. The prop-stand, as you know, is excellent, and the position of the speedometer head—on the top cross-member of the "Teledraulic" front forks—could not be bettered. My mileage was not sufficient to answer that point so many raise in these days, namely, whether the engine can be counted upon to remain free from oil leaks. At the end of the 51 miles there was not even seepage from any of the engine joints, but a little from the gear box.

There is little else to relate except to state that I was very favourably impressed. The machine I rode gave me real pleasure and, if they are all like that, the model is certain to make a name for itself.

Two Matchless Models

Civilian "Club-man" Mounts, One of 347 c.c. and the Other of 498 c.c.

" Sleek " is a word that seems apt with the 347 c.c. civilian-model Matchless, the " Clubman G3/L ". Note the large ground clearance

SINCE the Service contract for Matchless motor cycles—for the popular G3/L—is to continue, the two civilian model Matchlesses that will become available a few weeks hence are necessarily based on the Army machine, one of the Army's favourite motor cycles, incidentally. Both have the frame, forks —the famous "Teledraulic" front forks —and the general features of the G3/L, the lighter special military mount

A split clamp type of fixing is employed for the front-wheel spindle. Hand adjusters are provided for the brake cable, which has a quickly detachable nipple

developed from the G3 of the early war days, but one is of 347 c.c., called the "350 Clubman Model G3/L," and the other of 498 c.c., the "500 Clubman Model G80."

Thanks to chromium plating, the lustrous black enamel, which in the case of the fuel tanks is lined with black and silver, and such items as rubber kneegrips and footrest rubbers, it is difficult to imagine that the new "Clubman" models have any connection with the W.D. machine all know so well. They look what they are, a pair of sleek, modern overhead-valve singles.

A Burman gear box that gives closer bottom and second gears than are provided by the box fitted to the G3/L is employed. Top gear in the case of the three-fifty remains 5.8 to 1, and third gear 7.5 to 1, but second and bottom become 10.3 and 15.4 to 1 instead of 12.2 and 18.5 to 1. The compression ratio of the 347 c.c. "Clubman G3/L" is also different; there is no compression plate, as on the W.D. mount, which means that the compression ratio is 6.3 to 1 in place of 6 to 1. A further feature is the larger-bore Amal carburettor, namely 1in. as against ⅞in.

The lively 500 c.c. engine fitted to the larger-capacity model is similar to the engine listed for 1940. It has heavier fly-wheels than the early 1939 engine and a different inlet-valve timing. The compression ratio is 5.9 to 1. Gear ratio. in the case of this model are 5.25, 6.8, 9.3 and 13.9 to 1. Another feature of this, the larger machine, is a cylinder head steady running from the rocker box to the top tube of the frame.

Both machines are of the single exhaust port type, with the exhaust pipe arranged to pass above the off-side footrest on the three-fifty and below it in the case of the 1⅜in. pipe of the five-hundred. Both, of course, have totally enclosed valve gear, and the wheels and rear chain are the only exposed moving parts—the rear chain guarded as to its top run and with a deep flange running downward to protect the chain from mud from the tyre.

Among the other features are separate Lucas dynamo and magneto, fabric oil filter, internally illuminated speedometer head mounted on the top cross-member of the "Teledraulic" front forks, 7in. head lamp, easily operated prop-stand, fabric-type handlebar grips, 3-gallon petrol tank, with twin taps, ½-gallon oil tank and 3.25-19 Dunlop tyres. Prices are to be announced shortly.

A close-up of the nearside showing the neat, effective prop-stand and the arrangement of the brake pedal and footrest

THE A.J.S. RACING TWIN

(Above) The horizontal disposition of the power unit enables the huge six-gallon fuel tank to be accommodated in a layout notable for its low overall height and centre of gravity. Observe the miniature "Teledraulics" featured in the rear suspension. (Right) Imposing from any aspect the new racer is particularly impressive when viewed from this angle. The widely splayed dual down tubes permit an unobstructed flow of air to the cylinder head. The massive brakes are typical of this designed-for-a-purpose machine.

IN a talk he gave to a party of clubmen some months ago, Mr. G. L. Hack instanced, as one reason why manufacturers should support racing, the stimulative effect which an interest in the firm's success in competitions has on the *esprit de corps* of a factory in general. Evidently Associated Motor Cycles, Ltd., subscribe to Mr. Hack's opinion, for, when "Motor Cycling's" men visited Plumstead last Thursday to inspect the new A.J.S. racer, they were asked to "hang fire" for a few minutes because the machine was on view in the canteen! Around it were gathered the whole works staff, taking in the details of an assembled whole on which many had for months been working, preparing the bits and pieces but with no idea what the finished job would look like.

From the pictures you can judge for yourselves something of what they saw. Jock West's first remark, after we had absorbed an eyeful, was "What do you think of the beast?" And that word "beast" does somehow seem to fit it. There is an almost animal-like ferocity about it—something strongly reminiscent of a bull buffalo with his blood up!

The Power Unit

But you want details, not comparisons. Very well, then, the engine first. It is a parallel, even-firing twin with the cylinders (68 by 68.5 mm., 499 c.c.) lying almost, but not quite, horizontal. They are held down, each by four long bolts that pass up through the one-piece, deeply finned, cylinder head to an enormously massive and rigid, box-like crankcase-cum-gearbox unit. The cylinder barrels themselves are surrounded by lateral finning radiating all round the bores with each fin some 2 ins. deep and 2½ ins. long. Iron liners are shrunk into the light alloy material which is used for the cylinder block, head and crankcase units.

The high-duty alloy connecting rods have split big-ends, with shell bearings which work on a one-piece crankshaft carried in three main bearings. The centre main is a plain journal and the outer ones are of the roller type. The nearside end of the crankshaft is geared direct to the gearbox mainshaft by plain, straight-toothed gear wheels and, as there is no idler wheel, this means that the engine "runs backwards," i.e., the crankshaft rotates in the opposite direction to normal practice.

On the timing side a train of gear wheels, contained in a Y-shaped casing, takes the drive to the two camshafts which lie across the cylinder head. Each pair of valves is set in a hemispherical combustion chamber at approximately 90 degrees. That is to say, there is one inlet and one exhaust to each cylinder.

The camshafts are housed in heavily finned chests that not only enclose the whole of the valve gear but also incorporate intricate oilways leading to metering jets provided to spray lubricant straight on to the working surfaces. In the drawing of the cylinder head a number of hexagon-headed plugs can be seen let into the camshaft casings; they are the outer ends of these jets.

Roller bearings with split cages carry the camshafts and short tappets are interposed between the cams and the valve stem ends. "Interlaced" hairpin springs are used.

At its crankshaft end a vertical extension of the timing case will be observed. This contains the gearing to a shaft which runs right across the engine to drive the main oil pressure pump, which is a most substantial and sturdily ribbed component. Midway along this shaft, gearing is provided for the B.T.H. magneto which is placed forward of the oil pump and supplies a single 14 mm. sparking plug in each cylinder.

Thorough Lubrication

The lubrication system is most comprehensive and thorough. Drawing its supply direct from the 1½-gallon tank, the main pump first passes all oil through a Tecalemit cartridge-type filter carried over the gearbox. Thence a pipe leads straight into the engine and, via internal oilways, to main bearings, big-ends, etc. Additionally, oil is fed under pressure to each camshaft. Surplus from the inlet camshaft chamber drains by gravity through a short pipe to the exhaust chamber where a scavenge pump returns it to the after end of the engine. Here another pump is at work scavenging the sump and returning oil to the tank. A drip feed control, located on the gearbox housing, supplies oil to an annular groove cut on one face of the driving sprocket. From this trough, which is kept filled by centrifugal force, drilled holes lead to the sprocket teeth.

All the clutch-operating mechanism is enclosed, as also is the foot gear change. An interesting feature is an outrigger bearing on the offside of the gearbox mainshaft, while the internals of the box, which of course provides four speeds, are straightforward and laid out to facilitate rapid changes of wheels to provide the desired sets of ratios for varying types of race circuit.

A most ingenious system of mounting the carburetters has been employed. The two mixing chambers, fed by a common central float chamber, are supported on a pair of bonded rubber induction pipes. Thus, the whole carburetter assembly, including the float chamber, is elastically mounted and can, in fact, be quite freely moved about by hand on its resilient supports.

The method of holding the power unit in the frame is simple. At the forward end two split clamps bolt rigidly round the cross tube connecting the lower ends of the duplex front down tubes. At the rear four bolts pass through a platform across the back of the frame into the gearbox casing.

The frame itself has as its backbone a big diameter straight-top tube and from below the steering head duplex members fork outwards and downwards to pass on each side of the cylinder barrels. From the head, to the point where they are joined by the engine-supporting cross tube, these members are elliptical in section, but rearwards, where they curve inwards and upwards and meet another cross member behind the gearbox, they are of round section. They then travel upwards, becoming elliptical again, to meet with a forked rear extension of the top tube, and at this juncture lugs are provided to anchor the upper ends of the two oleo-pneumatic suspension members which are part of the "spring heel." All joints in the frame make-up are gusseted and welded.

Detailed Description of a Double o.h.c. Machine Produced for the "Senior"

The pivot end of the sprung rear wheel fork is carried on plain, self-lubricating phosphor-bronze bearings, whilst Silentbloc bushes are used in the eyes of the telescopic legs.

On to the loop of steel tubing which forms the rear mudguard support is built the remarkably extensive Dunlopillo padded seat.

Teledraulic front forks, very similar in design to those already well known, look after forward suspension and there are 3 ins. of movement for each wheel.

Both brakes are of 8¼-in. diameter and each has double leading shoes operated by coupled cams; the rear anchor is cable controlled. Magnesium alloy hub shells are fitted with exceptionally light gauge, straight-pull spokes passing into Dunlop light alloy rims.

Although it looks enormous, the fuel tank is actually of the pannier type with

(Right) Twin cams expand the shoes in both brakes. The cable-operated brake arms are coupled by adjustable link members. (Below) The rear brake is adjustable from the saddle by means of a knurled nut. Note the ingenious footrest adjustment. The cup of the hanger is a push-fit over a spigot member and is secured in the required position by a transverse bolt. The drilling of clutch, engine plates and other parts is extensive.

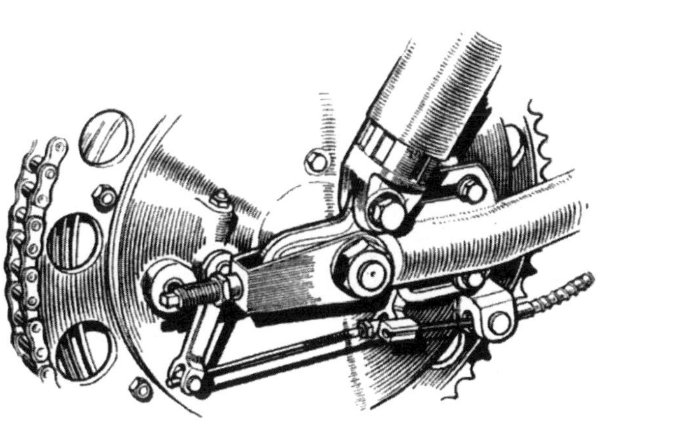

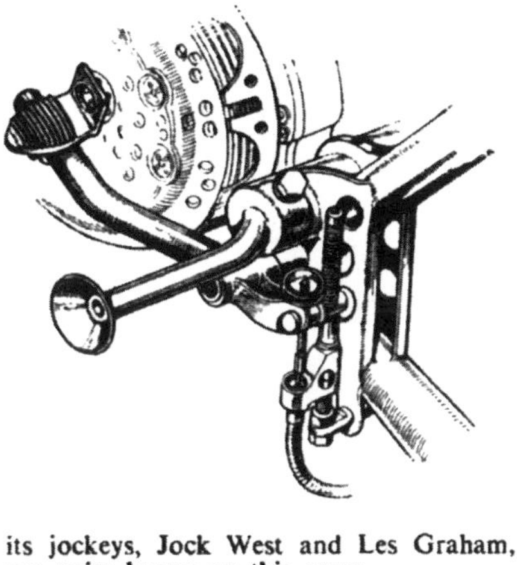

The comprehensive finning of the one-piece cylinder head and the o.h. camshaft boxes. A cross-pipe transfers oil from the upper to the lower box.

a considerable tunnel or gap down the centre. This is used as a chute to direct a current of air straight on to the oil tank which has two 2-in. diameter tunnels passing longitudinally through it—to provide an ingenious combination of oil cooling and tank strengthening. Both fuel and oil tanks are rubber-mounted and a feature of the former is that it has no flat surface; thus it cannot "drum." It holds six gallons. All the controls are quite normal in layout, though the handlebar is remarkably narrow at 22 ins. only.

Weight? So busy have its sponsors been in working against the T.T. deadline that the complete machine had not yet been weighed up to last week. The estimate is 308 lb.

Performance? So far the prototype has done nothing more than a private gallop over a stretch of none-too-good road, but its navigation on this was such that both its jockeys, Jock West and Les Graham, are quite happy on this score.

By the time you read this the first T.T. practice morning will be over and everyone will know a lot more about performance, handling and so forth. Meanwhile, whatever happens, here are congratulations to A.M.C. for their enterprise and especially to the Racing Department which, under the leadership of Freddie Clarke, has striven literally day and night for weeks past to do what was said to be impossible—to get a complete machine to the Island. They have done more. They have got three!

Other T.T. News

THE R.A.C. will again be prepared to cope with the draining of petrol tanks at Liverpool and Douglas. Petrol rationing still being in force, the R.A.C. cannot carry out its pre-war practice of supplying a small amount of free petrol to each rider on arrival at Douglas. However, the petrol removed from tanks at Liverpool will be stored, and, until stocks are exhausted, each returning rider will be issued with sufficient to get him to the nearest filling station.

The R.A.C. patrols equipped with the necessary tank-draining apparatus will be on duty at Liverpool and Douglas from June 2 onwards. The special Registration Forms, which should be completed on the boat and taken to the temporary I.o.M. Highway Board on the pier, will be supplied by the patrols.

* * *

A WEST COUNTRY enthusiast who has built a number of successful "bitzas" in his time will be in Douglas from June 8 to June 14, and is anxious to do anything he can to assist a Clubman's T.T. competitor. He will have a friend with him who is also anxious to help. The name is H. P. Lake, and the address, 18, Coombe Road, Dartmouth, S. Devon, if you want to get in touch.

* * *

VICTOR HORSMAN, ex-racer and nowadays the proprietor of a leading motorcycle agency in Liverpool, has presented the Horsman Cup for the fastest B.M.C.R.C. finisher in the June races.

* * *

THE first entries received for the Peveril Club's Isle of Man Grand National include Harold Rowell, Colin Frye, of the Gloucester and Cotswold Club, Bill McVeigh, of Scunthorpe, and Tom Arter, of Barham, Kent. Entries for this Open scramble, which takes place on Sunday, June 8, should be sent to S. Wardell, Peveril M.C. and C.C., 2, Windsor Road, Douglas, I.o.M. The Island Government are allowing two fuel units per competitor for this event.

* * *

THE Sidcup and D. M.C.C.'s team for the professional T.T. comprises Jock West, R. L. Graham and A. R. Foster. MERCURY.

"What looks right, is right!" After days and nights of work against time smiles of approval are accorded the new model by sales manager Jock West, development engineer Fred Clarke, rider Bob Graham and director D. S. Heather.

ASSIGNMENT AT ASSEN

A "Motor Cycling" Sports Reporter Covers the Dutch Grand Prix Using a Much-Borrowed Staff 1947 Model 18 498c.c. A.J.S.

"Bon Voyage!" Bernal Osborne, the "proprietor," watches Cyril Quantrill depart for Dover to report "The Dutch."

A CONFOUNDED nuisance, at times, is "Addie"—E N Adlington, Esq. Advertisement Manager of Temple Press Ltd.!

It is several years now since he ceased to be on the staff of "Motor Cycling." In his present exalted state he should have acquired a large motorcar complex, but he still is altogether too enthusiastic about motorcycling.

If a new road test machine arrives at the back door, "Addie" descends upon it like a kite-hawk and—before we've had time to get it thoroughly dirty—he's galloping off home for the week-end on the "Douggie," the Triumph or whatever else happens to have been temporarily bequeathed to us by a benevolent manufacturer.

A trip home for "Addie" doesn't mean 10 miles to Thornton Heath Pond and then a "bash" along the Croydon By-pass. It means 120 fast miles in each direction, for the man lives in Gloucestershire. It is a tribute to the durability of modern machines that his correspondence-column averages have yet to result in a burst bicycle.

Some time ago "Addie" borrowed Bernal Osborne's Model 18 A.J.S. He came back with a tale of a most preposterous average . . "48.8," I think he said it was, and he'd timed it by his Grandad's gold hunter, so he was sure it was right!

As it came back. The 498 c.c. Model 18 after its strenuous return trip to the Continent, described in these pages.

The Advertisement Manager borrowed the model the next week-end, and again the next.

Ho-hah! Must be something in all this! So, after promising Bernal faithfully that I'd clean it when I got back, I borrowed the "Ajay" myself.

I used it to gallop down to the race meeting at Bryanston Park in Dorset. Practising was due to start at about the time I usually thump the alarm clock out of action and turn over for "another five minutes." That meant being on the road just before dawn.

It Can be Done!

When I ran out of petrol between Shaftesbury and Blandford the sun was still so low on the horizon that I was forced to believe that bar-parlour averages *can* be maintained on English roads—if the conditions are right, and you've got a willing bicycle.

I came back with "Winco-Mac," of the R.A.C., who had borrowed the "Gent" for the week-end, and the Ajay hung on gamely to the Ariel "thousand," although the plug began to "whisker" towards the end of the day

Several times, now, I have borrowed the "Ajay"—particularly since my own Norton has grown a third wheel—when a fast solo has been called for It has always been a most reliable mount.

Recently there came the problem of getting to and from the Dutch Grand Prix. I could fly—but I wouldn't unless I absolutely had to. (I'm old-fashioned: at least, my tummy is!) I could go by boat and rely upon "train, tram, omnibus or tube" at the other end, or—well, what was wrong with taking a bike?

I took one, Bernal's A.J.S., crossing to Ostend and riding up to Assen—about 300 miles each way if you travel after dusk and have to go south to avoid the ferries over the Maas and Maal, both of which, it seems, close at night.

At Dover on the Wednesday I met Peter Goodman, Vic Miller (his Man Friday), Tommy Wood and Les Dear. As it was after 9 p.m. (European time) when we'd cleared the Customs at Ostend, it seemed a good idea to tag along behind the heavily laden "V8" van instead of rushing off to get lost on my own, especially as Tommy had already announced that he knew the route from Ostend to Antwerp "like the back of his hand."

In the darkness I suppose it was quite understandable that Tommy should mistake a tributary of the Oofenhuglwegl for the canal at Bruges, but I had a *feeling* we were on the wrong road when we got mixed up with the circus equipment down a cul-de-sac in Ghent!

But by then, of course, Tommy was asleep. So were Vic and Les. Wedged uncomfortably between them, Peter drove the Ford, kept awake by the unsynchronized snoring of his companions.

In North Belgium

We'd been having several stops, as something had gone wrong with the "electrics" on the van and Peter was running on the dimmest of head lamps. I hadn't the heart to put my own headlight on, for fear of dazzling him, and so bumped along "blind" over the never-ending pavé of Northern Belgium, in and out of the potholes which could scarcely have been avoided even had I been able to see them.

As the Ford had poked a con-rod through its basement on the way to Dover, we kept down to about "30 per" most of the way, the Ajay plonking on hour after hour, scarcely off the pilot jet until we passed through the Customs

By **CYRIL QUANTRILL**

—the Dutch officials showing a great deal of enthusiasm when they spotted the racing Velocettes in the van—and headed north for Breda.

Early in the morning there was a time when everybody woke up rather sharply. The Dutch roads are generally good, but an unforeseen pothole jolted Tommy into the realization that "things were rattling in the back."

They were indeed! "T.L.'s" box of spares had toppled off the shelf and the precious contents were nicely distributed over the floor of the van—and 2 a.m. is not the best time of day to discover a thing like that!

It seemed that a cup of coffee might help to restore morale, so I forged ahead to look for the local "Bedford Drivers' Club." All shut! For mile after mile the "Ajay" cracked towards the brightening Eastern sky. It was lovely to give the motor its head. It was also rather chilly.

At Gravé, a few miles west of Nijmegen, I spotted a filling station with lights burning in the house alongside. In like a flash!

Had they got coffee? Yes. Food? Yes.

Five coffees, five foods, please. And I waited for the van to come rumbling into sight. There it was. I waved frantically and Peter swung into the yard.

Mr. Barten, mine host of the Elftwegl roadhouse, and his charming wife served five hungry Englishmen with as nice a brace of eggs apiece as ever I've set my eyes on. Then we slept in armchairs for an hour or more.

Anxious to "make" Thursday morning's practice, the Velocette boys pushed on while I enjoyed the luxury of a wash.

Through Nijmegen

I suppose I must have passed them while Tommy was finding the way through the back streets of Nijmegen, for I never spotted the Ford again that morning.

Nijmegen, by the way, now looks very trim and tidy, as does Arnhem, only the areas near the bridges showing signs of the historic battles fought around them not so very long ago.

On the magnificent new road between these two towns I held the A.J.S. flat out. As on the run to Dorset, a few months earlier, it was the plug which cried "Enough!" A change to a slightly harder one improved matters, but, nevertheless, I was becoming distinctly whisker-conscious by the time Assen was reached. If only that little trouble could be cured, I felt sure that the "Ajay" could be walloped all day without protest.

Apart from a certain amount of oil which had "boiled over" from the filler cap, there was nothing in the appearance of the model to show that I had been galloping it at a steady 65-70 m.p.h. for two hours and more. And after that caning it ran just as sweetly and smoothly in top at well below the "30" mark when required to do so.

During the stay at Assen I managed to extract a K.L.G. 718 C. plug from Rex Mundy, who shared with "Dicky" Davis, of Dunlops, the honour of being one of the two "Trade Barons" to cross to Holland for the race.

"This isn't the standard recommendation," he pointed out, "but if you're really going to belt that machine you'd better have one."

And whether I wanted to or not, I had got to "belt" the A.J.S. on the Saturday night. "Alongside 7.30 a.m., said my sailing instructions from the R.A.C. After I'd taken leave of my hosts it was half past nine, and when I'd reached the outskirts of Zwolle, after riding in a traffic stream which would make the Brighton road on August Bank Holiday look deserted, it was getting dark.

The traffic jam in the town itself, where there is a Bailey bridge, which must be one of the longest in Europe, was just nobody's business. I tried to by-pass it by using the cycle track, but a Dutch policeman who could speak a little English—I'd met several who couldn't—pointed out, most politely, that it wasn't the "done" thing.

It was one of those wretched sort of traffic jams which moved forward a couple of yards every time you stopped the engine in the hope of letting it cool. Fortunately, the Ajay started first kick on every occasion—what a relief *that* was!

Thanks to the "Teles."

Eventually I got clear and picked up the road for Apeldoorn. For several miles it is like a trials section—for in Holland they undertake the rebuilding of roads on a grand scale—and I gave A.M.C., Ltd., a mental pat on the back for thinking up their "Teledraulics."

By the time Arnhem was reached I was getting weary and looking forward to the supper I had ordered in advance at Gravé, now only 30 miles farther on.

And—good!—there at last were the lights of the café. After supper I had another little sleep in the armchair and, with the sky once more getting light, set off for the final 200 miles to Ostend.

I don't like cutting things fine—I do it all too often through a disinclination to get up in the morning—so, rather than maintain a steady "50" for the rest of the journey, I decided to hammer along at a considerably faster bat. The result was that I arrived back on the coast far too early for breakfast and had time to have a look around the town. In due course Madame at the Hotel de la Gard started taking down the shutters, and I went in for coffee and yet more eggs.

In the middle of shaving I was assailed by the good lady, who was clearly of the opinion that it would be better for me to sit down to eat with one half of my face shaven and the other bearded than that the eggs she had cooked with her own fair hands should spoil.

I had intended to put my notes into ship-shape order on the boat, but as soon as I sat down in the bar to write I went to sleep. . . .

Finally, there was the run on Sunday evening from Dover to home and the 16 miles into the office early on Monday morning.

Bernal pounced on his machine. "What've you broken?" He looked long and hard.

There was one nut missing from the saddle mounting—the vibration caused by trying to do a mile a minute over pavè having done that—and the rear chain looked a bit dry. The head lamp glass was covered with squashed flies and beetles, and there was some oil on the off side of the tail end.

But the engine still had its excellent tickover, it was still beautifully quiet in traffic and once the "no limit" signs were passed it would still accelerate like a thoroughbred up to and past the figure "70" on the "clock."

It is a very nice machine, the Model 18 A.J.S. I phoned Jock West as soon as he was back from Holland to tell him so, and with most becoming modesty Jock said, yes, he, himself, thought it was quite fair value for the money!

THE NEW A.J.S.

The 1948 version of the 350 c.c. "Cammy Ajay," of beloved memory, bears little resemblance to its famous predecessor save that the o.h. camshaft is driven by chain.

THERE are far too few British racing machines. In fact, were it not for the persistent efforts of a handful of manufacturers—whose enterprise benefits the whole industry —our prestige abroad would be considerably lowered, with the result that overseas sales of British motorcycles might not be as good as they are to-day.

Of the few names which have consistently figured in race programmes, one of the best known is that of A.J.S. There were the early 2¾ h.p. side-valve machines, then the amazing fast "big port" of the early twenties—the only 350 ever to win a Senior T.T. These were followed, in 1927, by the "Cammy Ajay," which, although it was raced in all three capacities (Jimmy Guthrie won the "Lightweight" on one in 1930), became established in popularity as a "350." The air-cooled "four"—which never quite "made the grade"—and its water-cooled successor which might have been a world beater if the recent war and subsequent F.I.C.M. ban on superchargers had not shortened its useful life, have been followed by a twin which is the outstanding post-war design.

Another "Cammy Ajay"

Now the twin, which has attracted so much attention during the past season, will be joined by a "Junior" machine. And, with a fine respect for tradition, its manufacturers, Associated Motor Cycles, Ltd., of Plumstead, S.E.18, have made it another "Cammy Ajay."

"But," says Jock West, A.M.C. sales manager, "the only resemblance to the old one is the chain drive to the overhead camshaft."

A lot of us would like to see the new A.J.S.—or the Model 7R as it is called—appear as a "250," to combat the overwhelming Italian menace, but, having expressed this wish to the manufacturers, I can only sympathize with their reply, which can be summarized as:—

"We have already produced one entirely new racer since the war . . . " (I happen to know that the cost of that little venture can be assessed in figures that only astronomers *really* understand)." . . . We are now producing another. Why doesn't somebody else get cracking? Anyway, we only *sell* '350s' and '500s,' so would it be right for us to *race* a lightweight?"

This new machine is a different proposition from the A.J.S. twin. *That* beautiful device—the twin—is scarcely likely to appear in an A.J.S. catalogue, if only because there is such a limited number of millionaires interested in buying racing ironmongery. But it is intended to have the "350" in production in time to provide a season's racing *this year* for lucky customers, and I understand that it will cost no more than one expects to pay for any other blue-blooded, pedigree racer. At the moment the prototype is being built and I was down at the factory last week to see how it was progressing.

The 7R, when it appears on the grid for the first time, will be an easy machine to distinguish. Its double-loop frame, "Teledraulic" front forks, rear springing and massive brakes are similar to those of the twin. There is the same long one-piece saddle-cum-mudguard pad carried on a U-shaped tubular extension of the frame, and the petrol tank, although not possessing the bulging sides characteristic of the "twin," will be mounted in a similar manner, pannier-wise, on the single top tube.

The 348 c.c. power unit has an impressive, heavily webbed crankcase into which the cylinder is deeply spigoted, the finning, both on barrel and cylinder head, being generous. A distinctive feature of the engine is, of course, the chain-drive for the overhead camshaft. The one-piece rocker-box casting will be partially masked by the sides of the fuel tank. There is an independently mounted four-speed Burman gearbox of entirely new design.

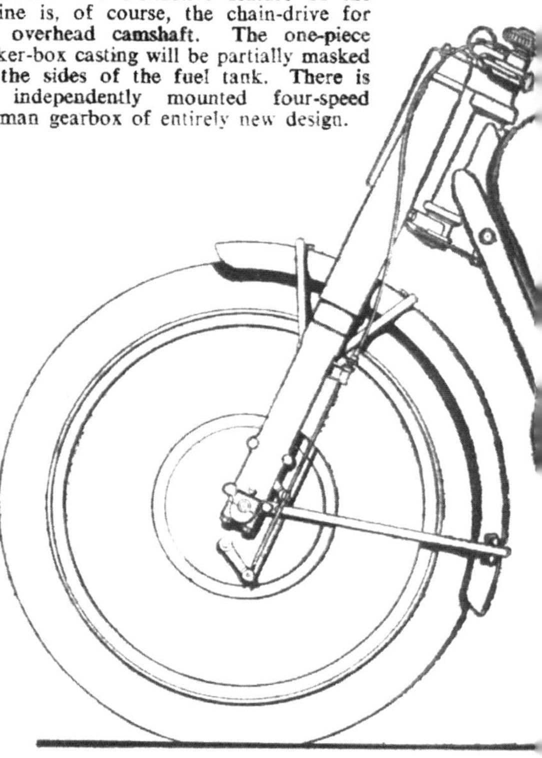

A "Motor Cycling" artist's impression of the new A.J.S. with the petrol tank "broken away" to show the method of bolting the top of the cam box to the main frame tube and the provision for driving the rev. counter from the camshaft.

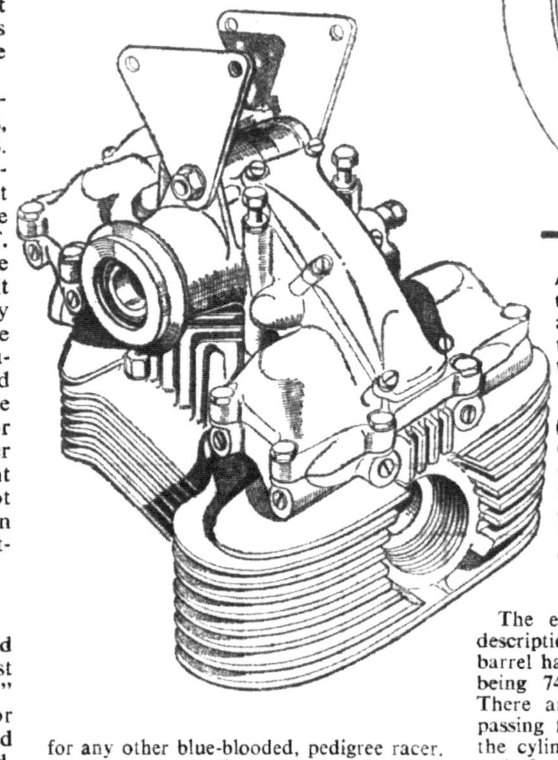

(Left) The aluminium alloy cylinder head has cleverly designed cooling. Note how the fins round the centrally disposed exhaust port are "swollen" to reach maximum depth between the port and the camshaft-drive chaincase.

The engine warrants a fairly detailed description. Its aluminium-alloy cylinder barrel has a shrunk-in iron liner, the bore being 74 mm. and the stroke 81 mm. There are four long holding-down bolts passing from the crankcase right through the cylinder head.

A forged aluminium piston of conventional racing design is employed and the massive I-section steel connecting rod has a bushed little-end and a duralumin-caged, single-row big-end bearings of exceptionally generous proportions. There are circumferential webs to the eyes of both big and small ends to provide vital strengthening.

The solid-steel flywheels are drilled, and

"JUNIOR" RACER

Details of the 350 c.c. Chain-driven o.h.c. Machine Shortly to be Placed on the Market by A.M.C. Ltd.

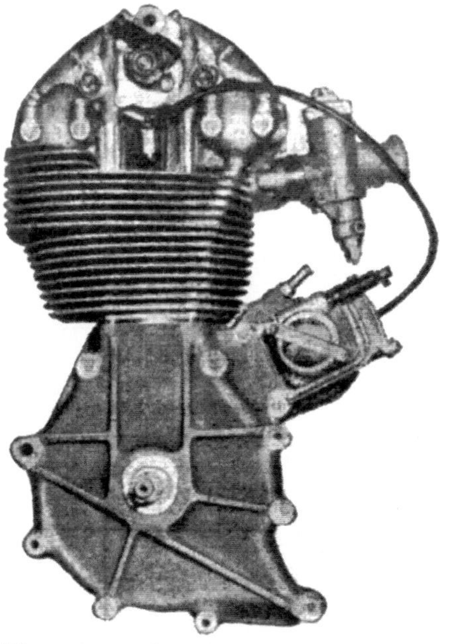

The new engine is very "clean" with massive webbing to give adequate strength to the magnesium alloy crankcase. The valve clearances can be checked without disturbing the valve spring covers.

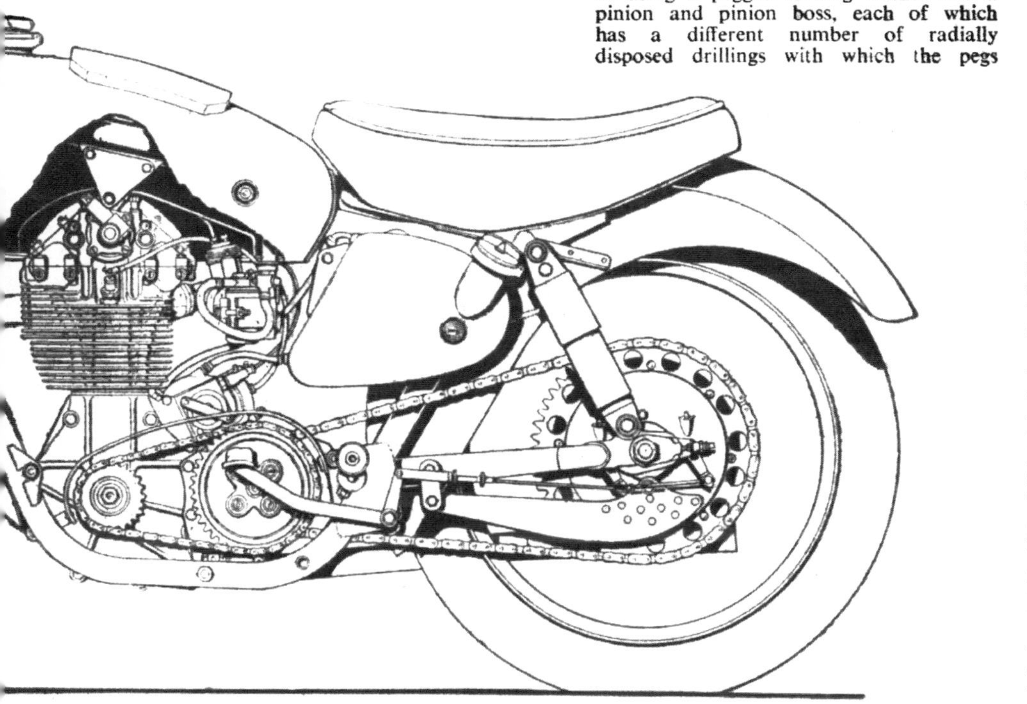

recessed on their inner faces to accommodate the big-end bearing, thus keeping the lateral width of the assembly to a minimum. The crankpin, following normal A.M.C. practice, consists of a toughened alloy-steel pin on which a hardened race is pressed.

The timing side mainshaft has a single, large-diameter journal ball-race while, on the drive side, there is, first, a double-caged roller-bearing and then, on the outside, a further single roller-bearing. At the extremity of the shaft comes a normal cam-type shock absorber.

The magnesium-alloy crankcase has strengthening webs, arranged in the best possible manner to withstand all stresses involved.

The timing cover encloses the gear-type oil pumps, both of which have cylindrical bodies. Each pump takes its drive co-axially from a pinion in the timing gear train, the delivery pump being driven from the magneto-drive idler pinion and the scavenging pump from the main half-time wheel. There are only two external pipes—short ones running between the oil tank and the pumps. The main feed from the delivery pump goes direct to the big-end bearing; another feed goes, by way of a pipe cast integrally with the magnesium-alloy chaincase, straight to the overhead camshaft which is drilled so that oil is led to the lift faces of the cams. The one crankcase breather valve is located in the drilled drive-side mainshaft and exhausts on to the primary chain.

The timing case has a small cover plate, giving access to the magneto-drive pinion. The timing of the strap-mounted racing Lucas magneto has a vernier adjustment, produced by the well-known A.J.S. method of using a pegged locking washer and a pinion and pinion boss, each of which has a different number of radially disposed drillings with which the pegs mate. The same type of adjustment is also used for the camshaft.

A Weller tensioned chain takes the drive to the overhead camshaft, the flat chaincase casting having a central division in its upper half.

The massive but light aluminium-alloy cylinder-head casting has unusual finning. Round the centrally disposed exhaust port the horizontal fins "swell out" until they reach their maximum depth between the port and the o.h.c. chaincase. They squeeze in to accommodate this, then swell out again. Beneath the rocker-box there is vertical finning, running diagonally at an angle of 45 degrees from the drive side to the rear of the chaincase on the timing side. Both rocker-box and camshaft housing are of magnesium alloy.

The method of attaching the rocker-box to the head is interesting. By using 7/16-in. trunnion bolts housed in drillings in the head and an equal number of ¼-in. holding-down bolts, screwed into the trunnions, it becomes possible to remove the rocker-box without taking the engine out of the frame and, at the same time, the nightmare of stripped threads in the alloy head is defeated.

The camshaft runs on twin ball-races, and at its "free" end there is provision for the rev. counter drive. Both cams are of exceptionally generous proportions. The rockers, of toughened alloy-steel with deposited hard-metal bearing pads, are short, to keep reciprocating weight to a minimum, and the rocker spindles, which run in plain bushes, are mounted eccentrically, affording means for setting valve clearances without recourse to screwed adjusters or shims.

There are, in fact, no end caps to the valves, which receive the same treatment as the rockers and have a chemically deposited surface of bearing metal on their ends. Fully enclosed interleaved hairpin springs are used, and these springs are retained by circlip type collets. The valves are set in the cylinder head at an inclusive angle of 80 degrees and the inlet port is downdraught and offset. A T.T. Amal carburetter is employed, with remote float chamber connected by a flexible junction pipe.

In addition to the normal bolt and engine plate attachment between crankcase and frame, the engine has two triangulated stay plates bolted to the rocker-box and to the top frame tube.

There is no saddle tube of the accepted type and the gearbox—extremely compact for a four-speed unit—is carried within the rear engine plates, having a top and bottom mounting, and provision for adjustment of the ½-in. by 5/16-in. primary chain being made in the usual manner but with the drawbolt at the base. The clutch is of normal multi-plate design and the rear drive is by ⅝-in. by ¼-in. chain.

Considerable thought has been given to the positioning of the pivot point of the foot-operating gear-change, so as to bring the pedal into line with the rear-mounted footrests.

The welded frame which houses this power assembly is very similar—on the drawing board—to that of the racing twin. The same oval tubes are used; there is the large section single top tube and the upward curving U-shaped loop, forming an extension to the "backbone" and carrying the one-piece saddle-cum-mudguard pad; the twin loops which, running from steering head to saddle bracket, form the whole of the main frame, follow similar lines; and both "Teledraulic" front forks and the oleo-pneumatic rear springing layout are identical. But, in fact, with the exception of these last-named parts and the saddle loop the frame has had to be completely redesigned to

Continued on page 101

A 350 c.c. Racing A.J.S. FOR THE

Chain-driven Overhead-camshaft Model to

A RACING three-fifty is announced by A.J.S.s! This is a production model, although output will be limited, and intended for the private-entrant racing man. The design is straightforward and businesslike, which is especially attractive to the rider who carries out his own tuning and maintenance.

The frame and fork are basically similar to those used with such success on the racing twin. It will be remembered that the frame is of the duplex cradle type in alloy steel tubes with all joints welded. For the new model the width between the tubes forming the cradle is narrower and the angle of the front down tubes to the vertical is less than with the frame for the twin.

Rear suspension operates on the swinging-arm principle with Teledraulic legs on each side. These legs give an easy movement round the static-load position and a rapid hydraulic damping build-up under depression. A Teledraulic front fork is, of course, used.

Hubs and brakes are also similar to those employed on the twin. The hubs are magnesium-alloy castings and are conical in shape. The brake drums have cast-iron liners with an internal diameter of 8in ; there are cooling fins on the outside of the drums. The brakes are of the two leading-shoe type with two cams and cam levers for each brake. Lining widths are 1¾in front and 1¼in rear.

Rigid Flywheel Assembly

The 348 c.c. engine has a bore and stroke of 74mm x 81mm. It is of robust construction and full use is made of light alloys. The crankcase—a magnesium-alloy casting—is freely webbed, especially on the driving side where three rows of caged roller bearings support the mainshaft; on the timing side there is a single-row journal ball bearing.

In the interests of rigidity, the flywheel assembly is relatively narrow and compact ; the flywheels are solid steel discs and the large-diameter crankpin carries a single row of ⅜ x ¼in rollers in a duralumin cage. The steel connecting-rod has a phosphor-bronze bush at the small-end, and the piston is an aluminium-alloy forging. There are two compression rings and one scraper ring on the piston, and the hollow gudgeon pin is retained by circlips.

In the mouth of the crankcase are four long studs for holding down the cylinder and cylinder head. The cylinder has an iron liner cold-shrunk in the aluminium-alloy finned block.

Special Head Finning

The cylinder head is in light alloy with austenitic-iron valve seatings shrunk in. Finning round the exhaust port is particularly deep ; the finning on the top of the cylinder head runs at an angle from near-side three-quarter front to off-side three-quarter rear with an air outlet through a cutaway in the chaincase of the camshaft drive. It has been found that this arrangement gives extremely good cooling of the crown of the cylinder head where it is shrouded by the cam box.

Ball bearings support this camshaft in the magnesium-alloy cam box. The cams have a vernier device which

PRIVATE OWNER

be Produced in Limited Numbers for the 1948 Racing Season

allows the cam positions relative to each other to be altered. Straight-line rockers have eccentric spindles to give adjustment for rocker clearance. The rocker gear arrangement is extremely neat and it is claimed that the reciprocating weight involved is little, if any, more than with a double camshaft design.

Overlapping double hairpin valve springs are used. These springs fit in wells formed in the cylinder head casting and the covers are part of the cam box. Attachment of the cam box is by four main holding-down bolts and 12 periphery bolts. The nuts for the periphery bolts are cylindrical and fit in bosses in the cylinder head casting; these nuts have screwdriver slots in the faces to aid positioning and may be readily withdrawn when dismantling. Small covers on the cam box gives access to the rockers.

Lubrication

Camshaft drive is by chain with a Weller tensioner. This is the only likeness of the engine to the single-cylinder racing A.J.S.s of yesteryear. Drive to the Lucas racing magneto is by spur gears. The intermediate pinion of this drive operates the supply pump of the lubrication system and the pinion on the mainshaft operates the return pump; both pumps are of the gear type.

Engine lubrication is achieved without external pipes. The supply pump forces oil to the big-end bearing and up to the camshaft through an oilway in the drive chaincase; the oil emerges through a flank of each cam, lubricates the rocker and valve gear and drains down from the valve spring wells through two drillings in the cylinder head and cylinder barrel to the crankcase.

From the small sump at the rear of the crankcase the return pump forces the oil back to the light-alloy one-gallon oil-tank fitted below the racing seat.

The carburettor is an Amal 10 T.T. with downdraught mixing chamber and remote-mounted float chamber. The light-alloy fuel tank has a capacity of 4¼ gallons. It is rubber mounted and retained by horizontal bolts in the manner adopted for the special racing twins.

A completely new Burman gear box is used. This is of the type described in the issue of *The Motor Cycle* dated October 17th, 1946, but which is not yet in quantity production.

(Above) Massive external webs on the driving side of the crankcase ensure rigidity. The finning is "diagonal" across the top of the cylinder head

(Left) Looking down on the engine. The cam box is retained by 16 bolts

A special drawing of the new 350 c.c. racing A.J.S., by Max Millar

Briefly, this new four-speed gear box is specially attractive because it has short, stubby shafts, is relatively light in weight and incorporates an ingeniously simple foot-change mechanism operating on a drum and plunger principle. On the racing A.J.S. a normal multi-plate clutch instead of the torque-assisted clutch will be used. The gear box is mounted in light-alloy plates; it pivots at the top and swings at the bottom for primary chain adjustment.

For Racing Purposes Only

Steel wheel rims are standard, but it is expected that light-alloy rims will be available as an extra. Tyre sizes are 21 x 3.00in front and 20 x 3.25in rear. Other details are light-alloy mudguards, a steering damper with friction discs at the base of the steering column, a sponge rubber chin pad on the tank, and a revolution counter. The wheelbase is 55½in.

If the steering and roadholding characteristics of the racing twin employing similar suspension details may be taken as a guide, the handling of this new three-fifty should be up to the highest standard.

No announcement can yet be made regarding price. As intimated earlier, the number of these machines that will be produced is limited and they will be supplied through normal channels for racing purposes only. Deliveries will start in time for the 1948 racing season.

A more detailed description of this interesting new model will be given in the near future.

348 c.c. RACING SINGLE

Technical Details of the Model 48/7R Overhead-camshaft Racing Engine : A High-performance Power Unit of Sound, Straightforward Design

ANNOUNCED in February, the 348 c.c. (bore 74 mm, stroke 81 mm) single-cylinder overhead-camshaft A.J.S. has already appeared in a few Continental races, and a large number of these models will be in the T.T. Races next month. The machine was conceived especially for the private-owner rider. It is thus a straightforward business-like job on which the owner with an average knowledge of racing practice can readily carry out his own maintenance.

A manufacturer with a wide racing experience has faced the very high cost of development and production to turn out a special machine—a machine representing the best that the lavish resources of a huge factory can do for the enthusiastic racing man.

Robust Construction

A description of the complete machine was given in the issue of *The Motor Cycle* dated February 5th, 1948. It is now possible to add a detailed description of the power unit.

As might be expected, the engine is of robust construction to withstand high revolutions for long periods. In fact, it is designed to run comfortably at 6,800 to 7,000 r.p.m. on 72 octane fuel. The flywheels are discs in 35-ton steel and are machined all over. On the timing side the mainshaft has a diameter of 1in, and oil-hardened alloy steel is the material used; for the $1\frac{3}{8}$in diameter driving-side mainshaft the material is case-hardened mild steel. Both these shafts are a keyed parallel fit in the flywheels and are locked in position by nuts.

A similar method of fitting is used for the $\frac{7}{8}$in diameter crankpin (oil-hardened alloy steel) on which there is an interference fitted sleeve of case-hardened alloy steel to form the journal of the big-end bearing; the outside diameter of this sleeve is $1\frac{1}{2}$in and there are plain steel washers between the sleeve and the flywheel faces.

The big-end bearing is a single row of $\frac{5}{8}$in × $\frac{1}{4}$in rollers in a duralumin cage. Heat-treated alloy steel of 70-tons tensile strength is used for the highly polished H-section connecting rod, which has two circumferential ribs round the big-end and small-end eyes. In the big-end eye there is a bearing ring of case-hardened mild steel and at the small end there is a phosphor-bronze bush.

Taper bored the fully floating gudgeon pin is retained in the forged Specialloid piston by circlips; the outside diameter of the gudgeon pin is $\frac{7}{8}$in. The piston, which has a full skirt, is oval and tapered and is diamond turned; at the crown it is domed with clearance flats for the valves. There are two $\frac{3}{64}$in compression rings and one slotted scraper ring. The standard compression ratio is 8.45 to 1.

Magnesium alloy is employed for the deep, sturdy crankcase, webbed externally to provide support for the driving-side bearing boss. In the boss there is a case-hardened sleeve, and the bearing consists of three rows of $\frac{1}{4} \times \frac{5}{16}$in rollers; the two rows nearer the flywheels are in one duralumin cage, and a separate cage is employed for the remaining row. On the timing side there is a journal ball race of $2\frac{1}{4}$in outside diameter; this race is inserted from the outside and seats on a cup formed in the crankcase casting. A large washer clamped by the back half of the timing case against the machined face of the crankcase holds the bearing in position. The complete flywheel assembly is located by the inner ring of the bearing, being pinched between the shoulder on the timing-side mainshaft and the half-time pinion which is a parallel fit on the shaft; this pinion has two keys and is held in place by the left-hand-thread nut.

Cylinder Sleeve

The deeply finned aluminium-alloy cylinder has a shrunk-in austenitic-iron sleeve. This sleeve is sunk into the deep crankcase mouth and lightly clamped in position by the topmost crankcase bolts. The flange which seats on the crankcase mouth is an integral part of the sleeve.

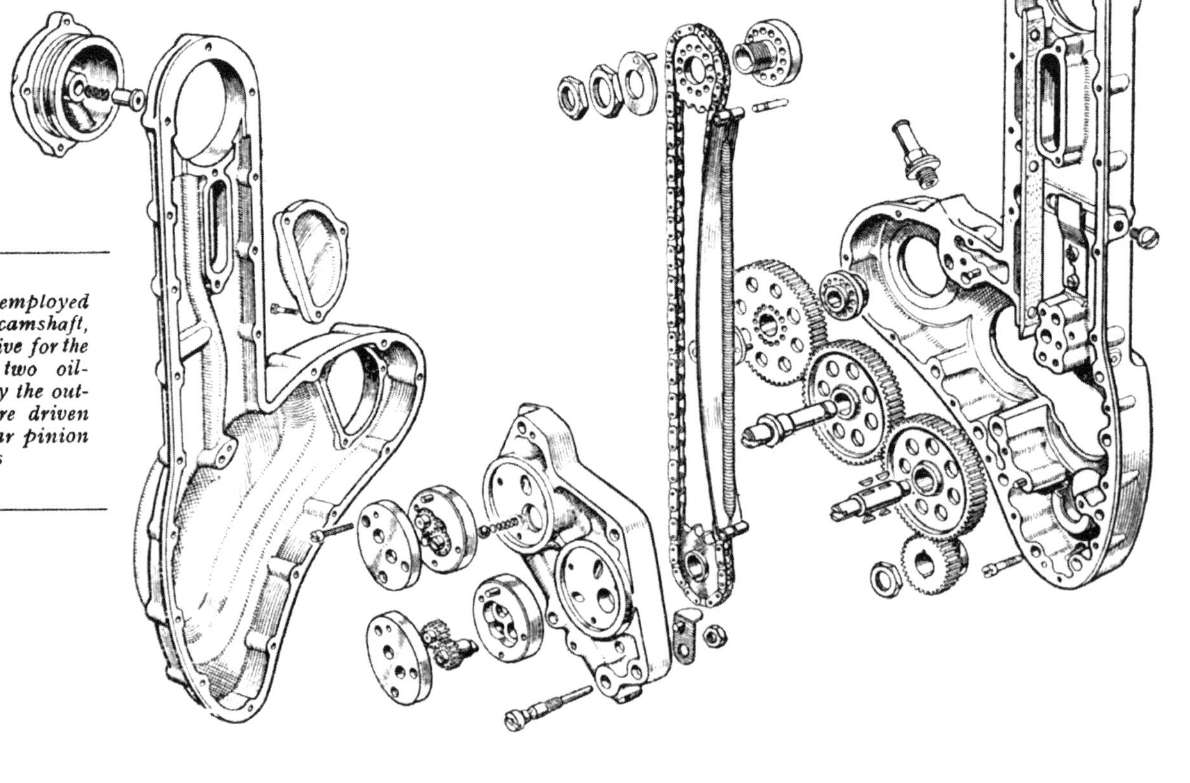

Chain drive is employed for the overhead camshaft, and spur gear drive for the magneto. The two oil-pumps carried by the outrigger casting are driven by the timing-gear pinion spindles

A.J.S.

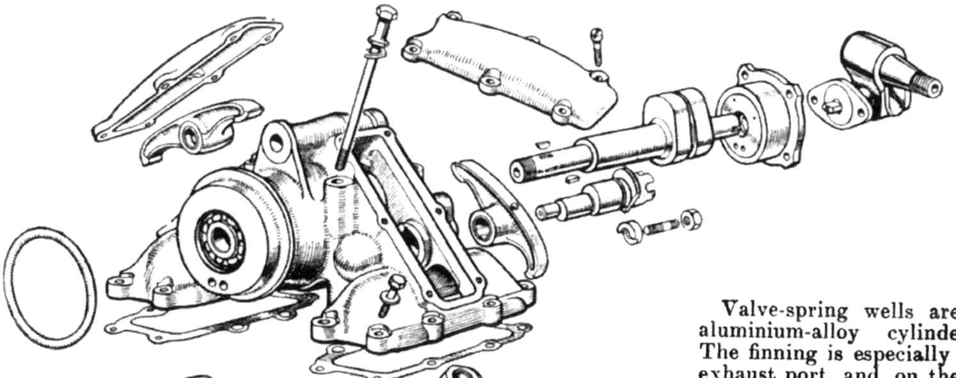

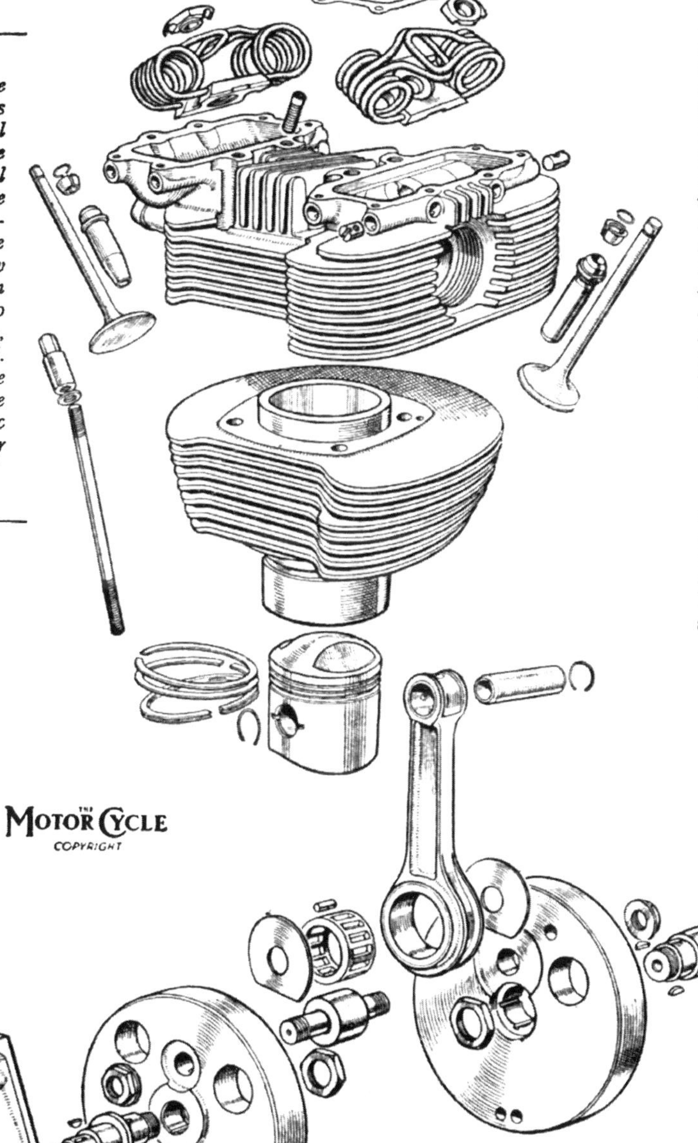

On the drive side the one-piece camshaft is supported by a journal ball race and on the other side by a special aluminium - bronze bush. This bush displaces the ball race fitted to the first few engines. Oil-ways in the bush direct oil on to the faces of the cams, which are copper plated. Rockers are of the straight - line type mounted on eccentric spindles to give rocker clearance adjustment

Valve-spring wells are formed in the aluminium-alloy cylinder-head casting. The finning is especially deep around the exhaust port, and, on the top of the head —under the cam box, that is—the fins run at an angle from three-quarter front to three-quarter rear, nearside to offside. Shrunk-in valve seatings are employed; that for the exhaust valve is aluminium-bronze and for the inlet is austenitic iron. Both valve guides are in phosphor-bronze, and the valves are in K.E.965 steel with the stems Stellite tipped. Valve dimensions are—inlet, $1\frac{9}{16}$in diameter at the throat and $\frac{11}{32}$in at the stem; exhaust, $1\frac{3}{8}$in and $\frac{3}{8}$in respectively. Port diameters are —inlet, $1\frac{1}{8}$in at stub; exhaust, $1\frac{5}{8}$in at outlet. The inlet port is slightly offset and downdraught, and the stub is fitted with a fibre distance piece to reduce heat transference from the head to the carburettor.

Hairpin Valve Springs

Overlapping double hairpin valve-springs are used which exert 100lb pressure with the valves closed and 170lb at full lift. The springs are held in position and prevented from turning by a seating block in each cylinder-head well; the spring holder on each valve stem is re-

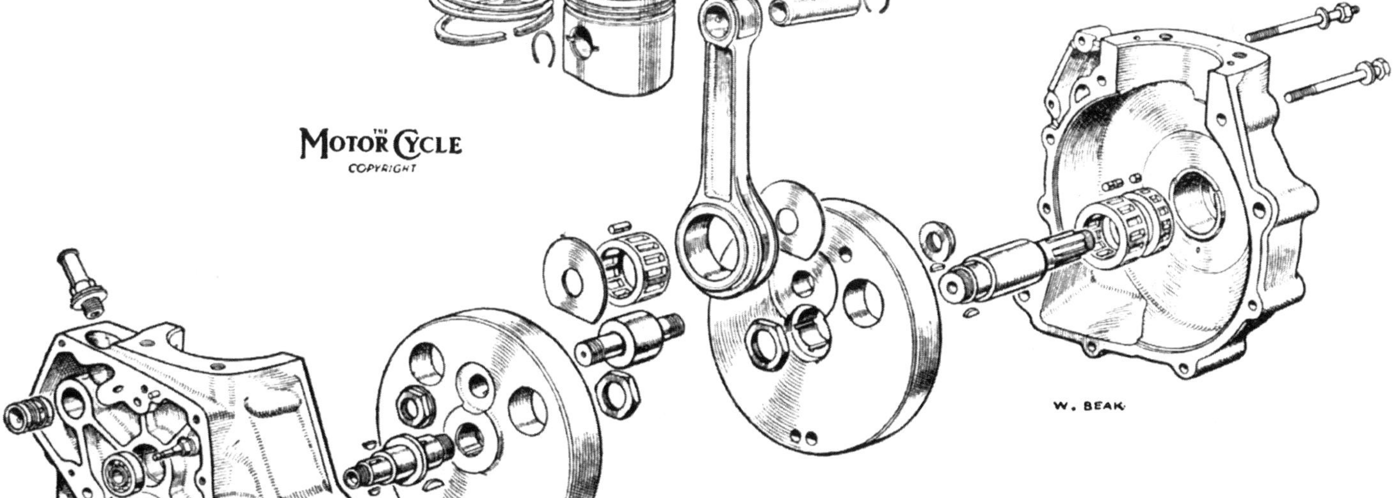

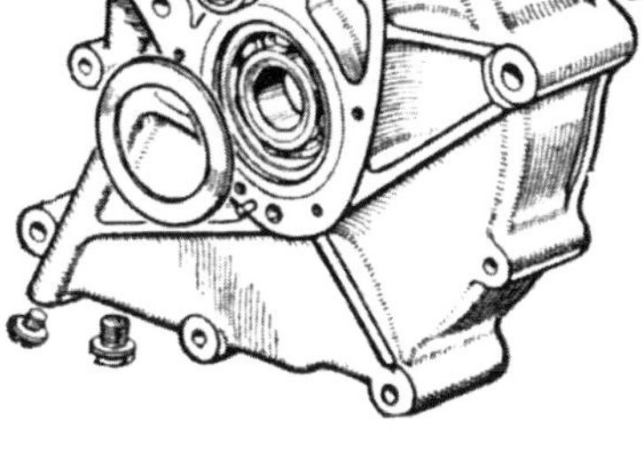

The magnesium-alloy crankcase is a massive casting. Three rows of roller bearings support the driving side mainshaft, and a journal ball race is employed on the timing side

tained by tapered split collets which bear against a spring ring seating in a groove round the valve stem.

Lapped faces make the joint between the cylinder head and the cylinder barrel, which are retained by four long high-tensile steel studs in the crankcase mouth fitted with special deep nuts.

Four inserts in the cylinder head are provided to take the main cam-box holding-down bolts. There are, in addition, six bolts round the periphery of each valve spring well—16 holding-down bolts in all. The periphery bolts screw into special cylindrical nuts which are located in bosses in the cylinder-head casting. These nuts may be turned to the correct position to receive the bolts by means of the screwdriver slots in the outside faces.

Camshaft Bearings

A journal ball race on the drive side and an aluminium-bronze bush on the near side carry the camshaft. The overall diameter of the ball race is $1\frac{9}{16}$in and its width is $\frac{7}{16}$in; the diameter of the camshaft at both bearings is $\frac{5}{8}$in. This camshaft, with its cams, is turned from the solid, and is in case-hardened mild steel; the cams are copper plated on the faces. The nearside end of the camshaft is grooved to take the tongue of the revolution indicator drive, the gear box of which is mounted on the end cover.

The polished straight-line rockers, which are Stellite faced at both ends, have phosphor-bronze bushes and pivot on case-hardened mild-steel spindles eccentrically mounted in the cam box. At the external end of each spindle is a flange and a slotted cup. By means of a small tommy bar in the slots, the spindle can be turned to give the correct rocker clearance (0.015 in inlet and 0.018in exhaust measured between rockers and the cams when the rocker covers are removed), and then locked in position by a circular clamp bearing on the spindle flange.

Valve timing, obtained with no rocker clearance, is—inlet valve opens 68 deg before top dead centre and closes 76 deg after bottom dead centre; exhaust valve opens 72 deg before b.d.c. and closes 50 deg after t.d.c.

At the peak of the cam-box casting is a lug for the engine steady plates, which are attached to the top tube of the frame.

Gear drive is employed for the Lucas racing magneto clamped on the back of the crankcase. Above the 30-tooth half-time pinion on the mainshaft is a 60-tooth idler pinion, the spindle of which also carries the lower sprocket of the camshaft drive. Next there is the distributor pinion, and finally the pinion on the magneto armature spindle. Ball races support the idler pinion spindle—one race in the back half of the timing case and the other in the outrigger casting which carries the oil-pumps. The distributor pinion—so-called because it has that function in the lubrication system, as will be explained later—is similarly supported by phosphor-bronze bushes.

From the sprocket on the idler pinion spindle to the camshaft sprocket is a $\frac{3}{8}$in chain with a long Weller-type spring tensioner on the slack-run side; the tensioner is provided with a spring-loaded damper. On the drive run of the chain is a fibre rubbing strip.

The camshaft sprocket and the magneto pinion are provided with vernier couplings to give a quick and easy method of obtaining timings to an accuracy of about 4 degrees of flywheel rotation.

Lubrication Details

Lubrication is especially thorough, and is achieved without external oil pipes apart from the supply and return pipes between tank and engine.

The two separate oil-pumps are attached by screws to the outrigger casting and driven by tongued couplings from the distributor and idler pinion spindles; the former drives the supply pump and the latter the return pump. Capacity of the supply pump is 26 gallons an hour at 7,000 r.p.m.

The Model 48/7R A.J.S. specification includes light-alloy tanks, a special racing seat, and swinging-arm rear suspension

From the supply union in the back half of the timing case the oil passes through to the pump. The main supply is forced through drillings in the outrigger casting into the hollow timing-side mainshaft, and thence through the timing-side flywheel and crankpin to the big-end bearing.

A second supply passes along the distributor pinion spindle and emerges through two holes of different diameters where the spindle operates in the bush in the back half of the timing case. Similar holes in the bush match up with the spindle holes at every revolution and allow the oil to pass into annular grooves in the outer circumference of the bush.

The smaller-capacity feed passes oil through drillings in the crankcase to an annulus on the cylinder sleeve from which the rear of the piston is lubricated by means of two holes in the sleeve.

Hollow Camshaft

From the larger-capacity feed the oil passes up a conduit cast in the outer half of the timing case. Next the oil passes through the camshaft inspection cover, and by means of a spring-loaded gland which bears on the ground end-face of the camshaft is directed along the hollow shaft. When the oil reaches the aluminium-bronze bush supporting the camshaft—this bush is at the nearside end from which the revolution indicator drive is taken—it passes through two tiny holes drilled at the correct angle to ensure that the oil is directed on the faces of the cams.

Lubricant which passes the annulus in the camshaft bush escapes at the end of the camshaft, drains round the back of the bush and travels along the base of the cam box to the drive side, where it lubricates the chain and also keeps topped up the semi-circular bath into which the half-time pinion on the mainshaft dips.

Flexible Oil Pipes

Surplus oil drains from each valve-spring well, through the cylinder head, cylinder barrel, cylinder sleeve flange and crankcase mouth to the small sump at the rear of the crankcase. From this sump, the return pump draws the oil up drillings in the timing side of the crankcase, in the back half of the timing case, in the outrigger casting and thence the oil is forced through outrigger, the back half of the timing case and the crankcase to the return union situated just behind the cylinder. Flexible hose-type pipes are fitted between the engine and the large oil-tank.

An Amal-type 10TT carburettor with remote float-chamber is fitted. The bore diameter is $1\frac{1}{8}$in with V3U choke. A No. 440 main jet is used and a 0.109 needle jet. Standard gear ratios are 5.14, 5.84, 6.94 and 9.95 to 1, obtained with a 21-tooth engine sprocket and a 54-tooth rear-wheel sprocket. With this top gear, 6,600 r.p.m. represents 100 m.p.h.

GULLEY KING

AJS Competition Single Tested Following Big Bear Victory

By Bob Greene—Editor

HERE IT IS! Hot on the tail of the annual Bear chase, a test on the bike that has been cleaning up in every type gulley scrambles from Kokomo to Cucamonga.

A week ago, I could only have told you that for some strange reason riders of very short experience could climb aboard one of these mechanized mountain goats and "go like the wind," finishing up in the money overnight . . . Why? . . .

Deciding that hearsay and personal observation weren't enough to satisfy either CYCLE'S readers or myself, a call was put through to Frank Cooper, California distributor of this famed single cylinder acrobat. First plan was to commandeer Del Kuhn's very own machine, but was dismissed when we found that it had been plugging away the better part of the year without a coke job. So in all fairness a brand new replica of the winning Ajay Competition Scrambles was checked out to your tester. There's no mistaking the purpose of this special breed. Each part is fitted with one thought in mind: that this machine be capable of getting through the worst imaginable terrain with the highest speed, comfort and ease.

A 21 inch block tread tire fitted up front boosts frame clearance, adds delicate steering. At first it was felt that the narrow front hoop would prove too sensitive in sand or soft ground. This might ordinarily be the case, but some factor in the Ajay's makeup—presumably steering head and fork angle—make the bike feel as though each end were actually suspended by strings in mid-air. This sensation is so pronounced that I, not being accustomed to the Ajay, noticed a slight hesitancy to turn the bars when entering my first corner. To check my reaction, another rider took over and came back with the same unprompted comment. By the end of the day, however, this inherent tendency to remain upright had changed from a strange sensation to one of assurance.

In contrast to the slight appearance of the front forks and wheel are the brutish, strictly business hind quarters. Massive 3⁄16 in. wheel spokes carry the knobbiest tire possible. When this wheel turns, something moves, and it's usually the bike. Near impassable

General Specifications

ENGINE. 500 cc displacement, bore 3.25 inches, stroke 3.65 inches. Compression ratio 7.5 to 1. Aluminum alloy cylinder heads, cast-in cast steel valve inserts. Hair pin valve springs enclosed in oil. Aluminum alloy pistons, aluminum alloy cylinder with a permanent cast-in, cast-iron liner. Extremely large lower end of three rows of 1/4x1/4 in. rollers. Two large ball bearing races on drive side of main shaft. Magneto and generator both operated by chain drive, contributing to longer life of electrical parts.
CARBURETOR. Amal 1 3/32 in. diameter venturi. Air cleaner optional at extra cost. Choke lever on handlebar.
TRANSMISSION. A constant mesh direct gear to gear system eliminates usual spool or sliding gear. Shifting by dogs, similar to synchromesh.
LUBRICATION. Circulating dry sump system; two-way plunger type oil pump actuated by sliding worm gears. Built-in filter in oil tank easily replaceable. Oil system through internal passages. Only outside leads are flexible mounted pipes from oil tank to engine, pressure lubrication to top end.
IGNITION. Lucas magneto. Chain drive and shielded by metal deflector, manual control.
ELECTRICAL. All electrical equipment is Lucas, new long type heavy duty 6-volt generator has voltage regulator, chain driven. High and low beam headlight, taillight and parking light.
CLUTCH. New heavy duty 5-stud with fibre inserts. Multiple disc dry type.
DRIVE. Pre-stretched Mark 10 Reynolds racing chain, shock absorber on engine sprocket. Single row 1/2 in. pitch by 5/16 in. wide roller primary chain. Rear chain is 5/8 in. pitch x 3/8 in. wide roller, same size as American big twins. 18 tooth sprocket instead of standard 21 tooth.
GEAR RATIOS. Top 5.83; 3rd, 7.47; 2nd, 12.20; 1st, 18.44 to 1.
FRAME. Single loop front section with rear suspension swinging arms set in large, wide journal bearings. Huge rear main springs are 6 in. long by 3 in. in diameter. 3 1/2 in. travel on rear swinging arms.
FORKS. Teledraulic, fitted with metered pistons for constant control with oil on both sides of piston to buff counter recoil as well as recoil. This same system applies to rear forks.
MUFFLER. Short, upswept, pipe chrome-plated plus extra straight pipe for competition at no additional cost.
HANDLEBARS. Chrome-plated competition, rigid mounted. Left bar: clutch, compression release, spark control, dimmer switch. Rear bar: front brake, throttle, choke lever and horn. Clutch and throttle have extra control cables at no additional cost.
WHEELS. Mounted on Timken type roller bearings. Safety washers on hubs prevent spokes from tearing out. Rear wheel: 3/16-in. spokes and extra heavy rim.
BRAKES. Internal expanding 7-in. front and rear, shoes 7/8 in. wide.
TIRES. Front: 300x21 Trials Dunlop. Rear: 400x 19 Dunlop Sports knobby.
TANKS. Pressed steel 2.8-gallon competition type with double outlet, single filler cap.
INSTRUMENTS. Smith's 120 mph illuminated speedometer, ammeter and light switch in headlight shell.
SADDLE. Wide Lycette, leatherette covered. Sprung with conventional coil springs.
FINISH. High temperature baked enamel black finish. Very hard and durable.
FENDERS. Lightweight high polished aluminum alloy sports type with braces over outside of fender.
TOOL KIT. Complete set of tools, grease gun, tire irons and ignition tools. Tire pump mounted on left side of frame.
WEIGHT. 375 lbs. with oil.
OVERALL LENGTH. 85 1/4 in.
OVERALL WIDTH. 29 1/2 in.
WHEELBASE. 55 1/4 in.
GROUND CLEARANCE. 7 in.
SADDLE HEIGHT. 32 1/2 in.

sections of boulder and brush-studded open countryside were challenged with deliberate cockiness. Usually the Ajay proved itself just a shade more flippant than its tormentor. At times when it seemed certain that the machine would hang up or stick, the bike and I would catapult over the mess without so much as ticking the frame. On rare occasions when it would dig in after hanging on fallen trees or rock, it was only necessary for me to lean the bike over, while still astride, kick dirt or rock beneath the rear wheel with my heel, right the little job again—and be gone.

Special consideration is given to riders' comfort by the soft, swinging arm suspension system similar to the McCandless type used on the Isle of Man Velocette, Norton and Gilera. Unusual in appearance, this design was determined one of the most practical rear suspensions yet to be tested.

Although the rear of the machine did creep sideways when the pressure was turned on in a fast pavement corner, the blame

The unusually soft, long action rear plungers, like the patented Teledraulic front forks, are oil cushioned at both ends of their travel. Result —no metallic bottoming. Quick-release headlight plug is novel feature

High Scrambles seat position makes it easy to uncoil a leg and stab at the ground when cornering on a slick surface. Rear Dunlop Trials tire held well in snow. Rear shock has compressed almost an inch during turn

was attributed directly to the radical knobby dirt tire, for when heeled well over, the entire weight of rider and machine was concentrated on the single outer row of knobs. Here at last is a rear fork that has a slow and soft full length travel. That this design reflects thoroughness is shown by the fact that each rear shock unit is additionally cushioned with 85 cubic centimeters of hydraulic fluid. I can't recall a metal to metal bottoming once during the test.

On the timed run over Angeles Crest mountain range, speed fell off. This challenge to engine power and handling qualities was further enhanced by intermittent patches of ice. The nasty road condition and previously mentioned radical rear tire were the slowing forces that clutched at the speedometer all the way over the mountain. This cushioned my surprise when the watch was whipped out at the end of the run. The usual 35-minute, 35-mile ride had taken exactly 39 minutes.

Because of their short wheelbase and other obvious handicaps, motorcycles have always posed a problem in springing. AJS approaches the epitome. The long lazy action of the rear member works hand in hand with the front forks, taking the sting out of sharp dips with a minimum of counter-recoil whip. Both front and rear plungers are not only oil dampened, but have the same liquid cushion at each end of travel.

A full appraisal of the Scrambles' handling ability and clearance can be appreciated through one of its many tricks. Riding between the rails of an ordinary railroad track at an approximate 7 mph it was possible to turn sharply and ride right over and out of the rails without hesitating. Even here the frame didn't as much as touch the rails when crossing over.

In over 200 miles of rough and tumble riding, it is easy to assume that a few shifts would be missed. Not on your life! This new gear box, an adaptation of the one used in the famed AJS Boy Racer, can hardly be compared to anything you've ever experienced. Very little more than the weight of your foot on the shifting pedal is needed for each change. Once the pedal has been depressed it seems to suck itself into the next gear. Direction of travel is the same as in the past, up for low, then down through to high. This gear box is a direct take-off on the Boy Racer model except that it is not full ball bearing throughout. Practically speaking, it is impossible to miss a shift with this new arrangement.

While the braking figures were not phenomenal, it must be remembered that the bike is rigged strictly for competition, and that neither the narrow block tread on the front nor the accented knobby on the rear are conducive to braking efficiency over pavement. Both pedal and hand lever action were easy and progressive. With more rational tires fitted, it is believed that this model would show high braking efficiency in view of its springing range.

Everything is up out of the way on the Scrambles including the rider. Although the rigid gull-wing footpegs are nearly a foot off the ground, the relative high location of the saddle keeps the rider's legs at such an up-

Continued on page 101

Running between railroad tracks at 5 mph it was possible to turn sharply and ride over and out of the rails without pause or footing. Contributing factors here were bike's short turning radius and high frame clearance. Note contoured exhaust system is out of the way

Rear wheel is practically geared to road for slogging. Fender braces are mounted outside to help keep inside of fender from loading up in mud

30.50 COMPETITION SCRAMBLES FOR 1952

Since their adaption of the McCandless swinging arm rear suspension in 1949, the most important improvement in the AJS line is the new Burman B52 transmission as developed on the 7R Boy Racer. The new gear box, now featured on their standard machines, has short, rigid gear train and positive, easy working foot-change. Mainshaft is only 4½ inches between end supporting ball bearing races. Notice that all gears are in constant mesh. There are no sliding gears and re-routing of power is done entirely by dogs. Chance of the gears' clashing has been eliminated, since they are never apart. The only other possibility of trouble (that of springing a shifting fork) has been minimized to the nth degree by the replacement of former aluminum-bronze selector forks with steel forks

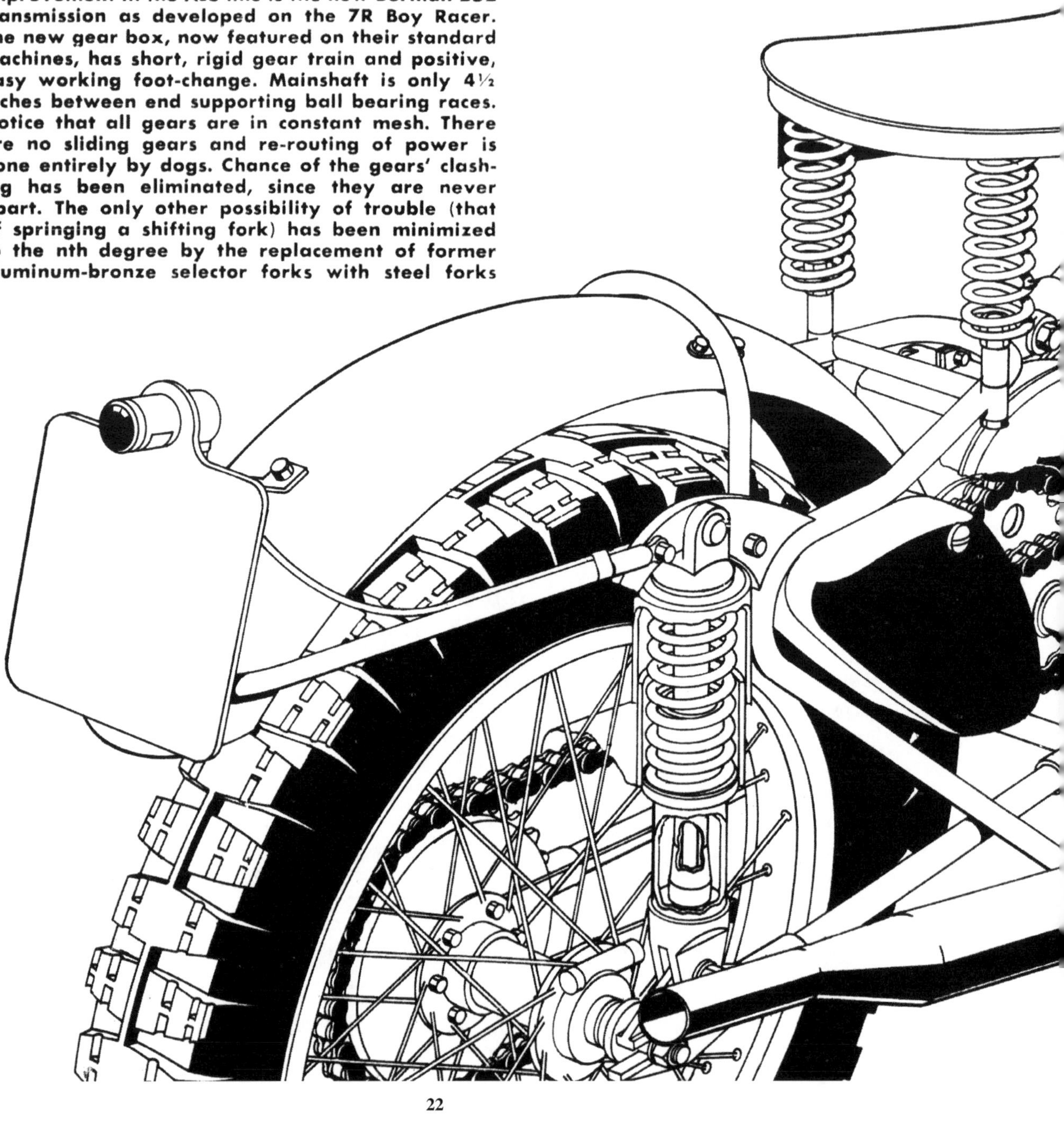

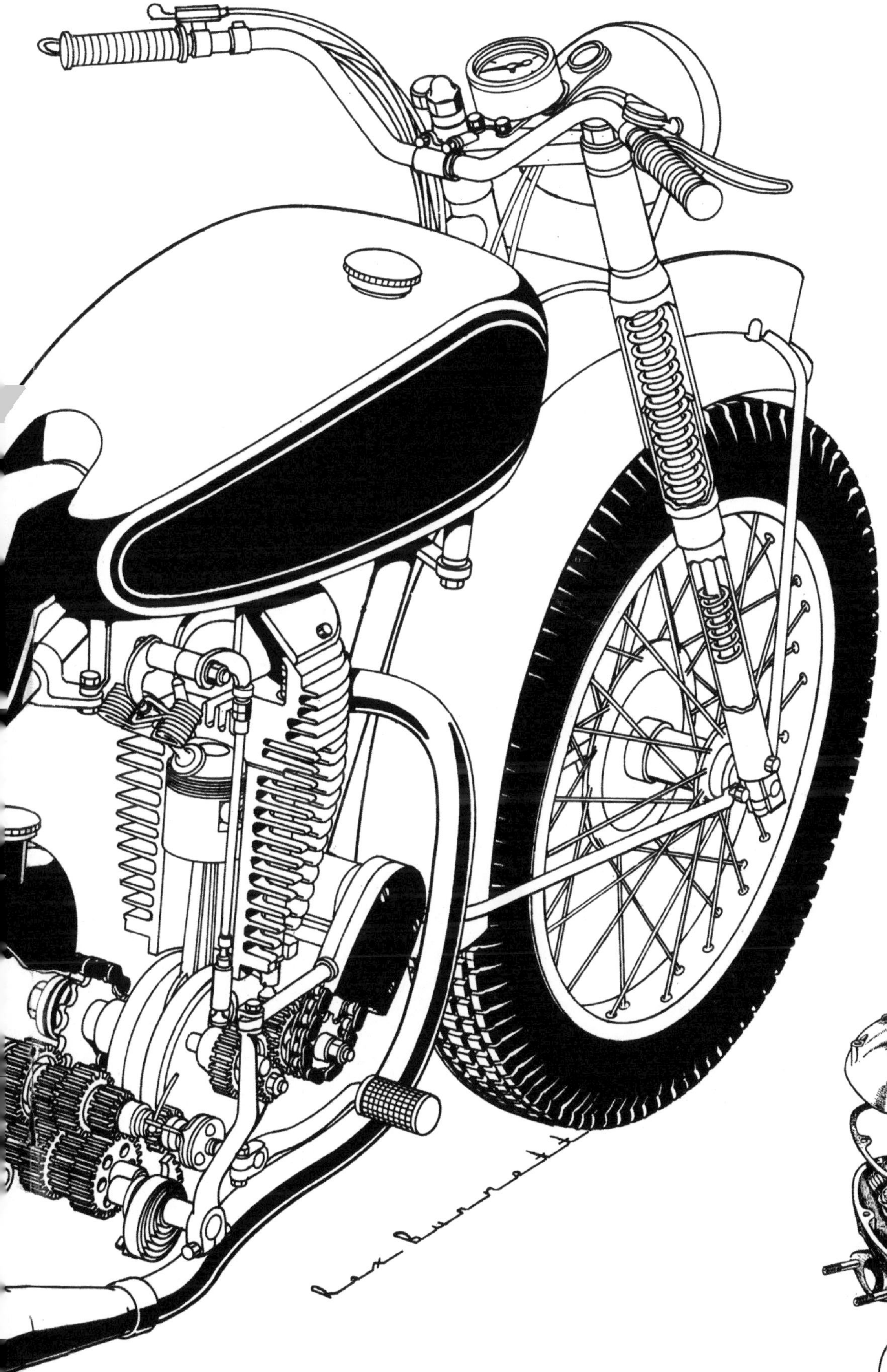

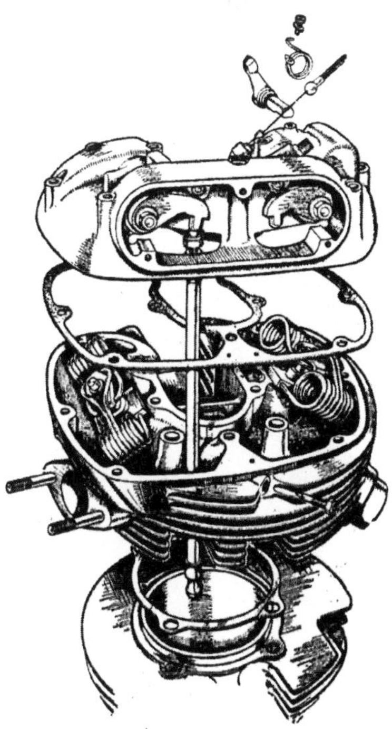

Unusually robust in construction the, A.M.C. twin has a cast-iron crankshaft supported by two roller bearings and a shell bearing in the middle. Connecting-rods are in Hiduminium RR56 light-alloy, with shell bearings at the big-ends. Cylinders are deeply sunk in the crankcase. Cylinder heads are light-alloy castings

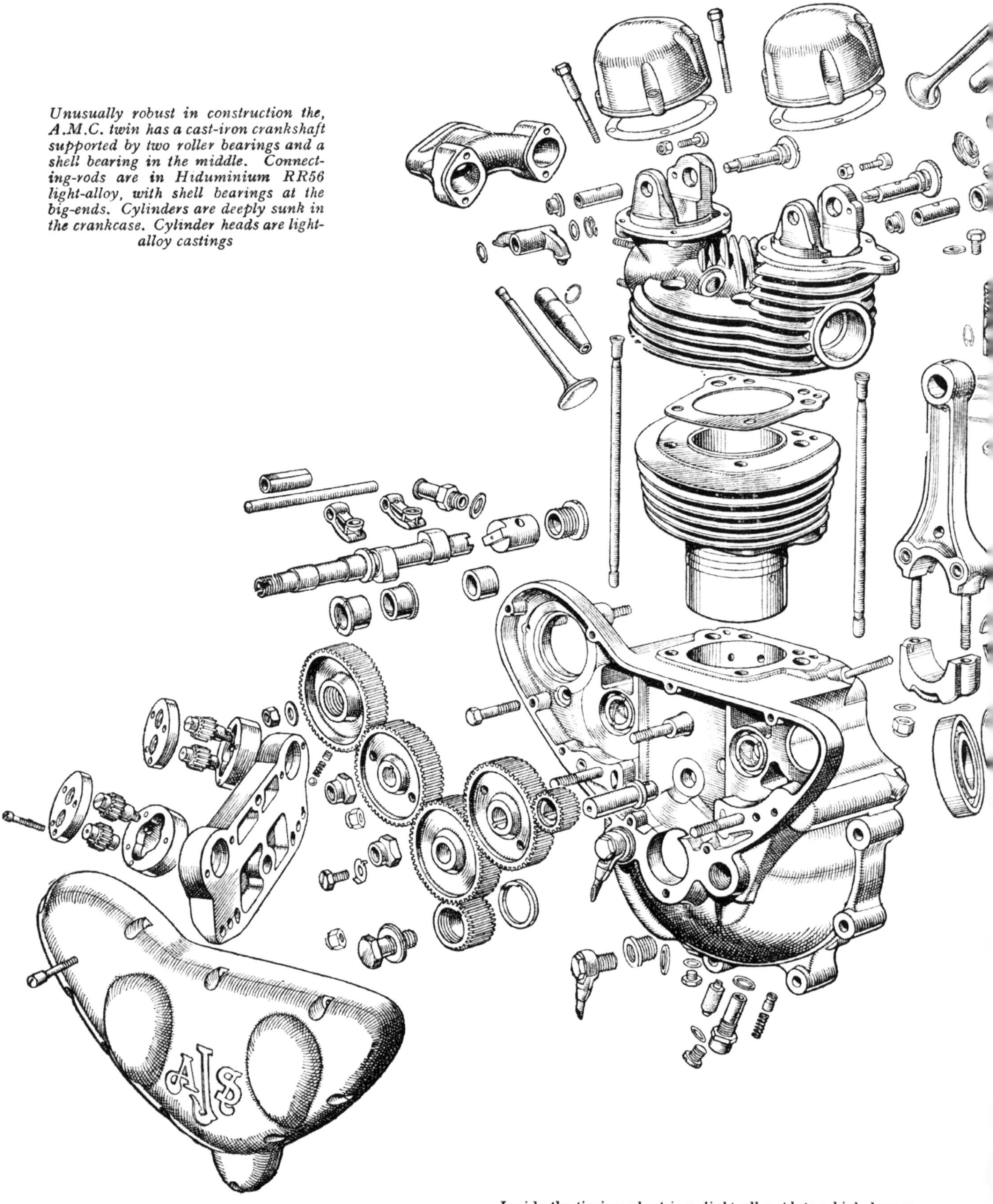

Inside the timing chest is a light-alloy plate which houses the two gear pumps. This plate also provides support for the idler-pinion spindle. The pumps are driven by the camshafts, the exhaust operating the supply pump and the inlet the return

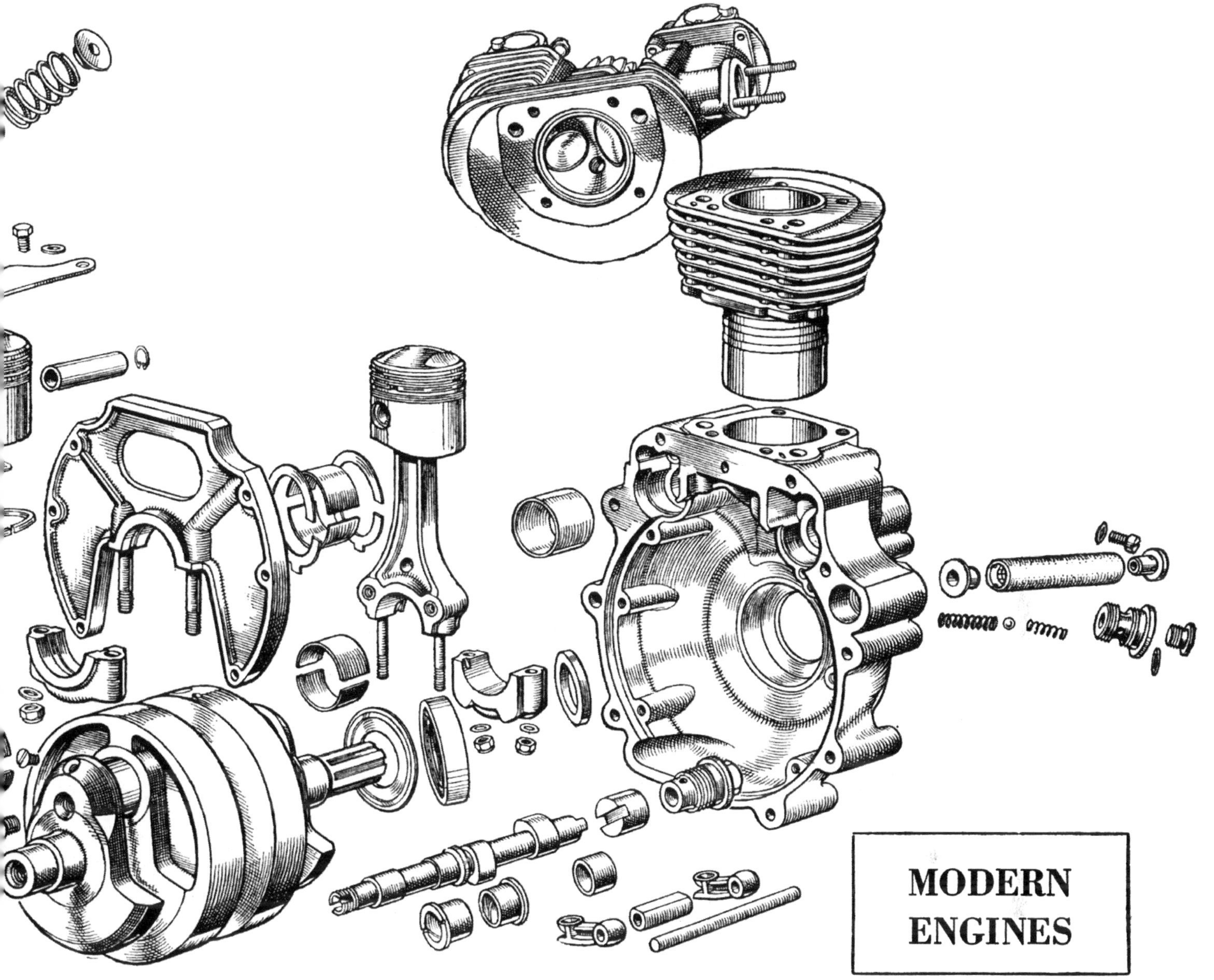

MODERN ENGINES

498 c.c. A.J.S. and Matchless Parallel Twins

A Catechism on the Technical Features of a High-performance Power Unit, with Answers from Joint Managing Director D. S. Heather, B.Sc., and Designer P. A. Walker, A.M.I.Mech.E.

By HARRY LOUIS

AT the end of 1948, Associated Motor Cycles introduced the now well-known A.J.S. and Matchless 500 c.c. parallel twins. More than usual interest was aroused because the power unit, almost identical for both models, was of most advanced design and incorporated features obviously derived from racing experience with parallel-twin-cylinder engines. For me, the opportunity of obtaining answers to my questions had a special attraction because my personal machine since mid-1949, on which over 30,000 miles have been covered, has been a Matchless twin—the Super Clubman, as it is called.

My first question was, "Among present-day parallel-twin-cylinder motor cycle engines, this unit is unique in that it has a three-bearing crankshaft, and is unusual in that the shaft material is cast-iron. May I ask why you adopted a cast-iron crankshaft as opposed to the more popular forged-steel shaft? Aside from the obvious attraction that a fairly complicated design can be cast where it probably could not be forged, what are the pros and cons that induced you to employ cast-iron? "

Answer: "Having decided that we wanted a three-bearing crankshaft with integral flywheels, the selection of a cast shaft was economically inevitable—machining allowances on a steel stamping of the size required would be colossal as compared with those on a cast shaft and, in addition, a great deal of metal would have to be machined off the flywheels to obtain the correct crankshaft balance. In the case of a cast shaft, this material can be cored away in the casting.

"The high-grade alloy-iron used for cast crankshafts provides an excellent bearing surface without heat treatment. Further, the less complicated production processes and easier machining of the cast shaft are economic virtues, which far outweigh any possible advantages that could be obtained from the higher tensile values of a steel shaft in this application."

Question: "The middle bearing, one assumes, ensures almost complete rigidity of the crankshaft assembly. Does experience show that the middle bearing, in fact, takes any appreciable load? If the answer is no, then is the middle bearing an advantage, since a slightly longer shaft is necessary? "

Prevents Whirling

Answer: "Owing to the inherent rigidity of the crankshaft design, the centre main bearing is not heavily loaded, but it does, of course, take its fair share of the load. In view of the necessity for ample space between the cylinder barrels to ensure adequate cooling, the use of a centre main bearing does not, in fact, increase the overall length of the crankshaft. A most important advantage of the use of a centre main bearing is the opportunity which it provides to feed oil to a point centrally between the two big-end throws and thus ensure equal oil supply to each. In addition, it provides a perfect means of end location of the crankshaft."

Question: "One line of thought is that a crankshaft should be as inherently rigid as possible, but that any whip that does take place should be allowed to express itself rather than to be constrained by a bearing. What are your views on this theory? "

Answer: "We hate to think of a crankshaft behaving like a skipping-rope. There is, of course, a safe rotational speed for any shaft beyond which whirling will prove fatal. The centre main bearing prevents whirling and so permits higher safe r.p.m. with no danger of crankshaft failure and with less loading on the outer crankshaft bearings than will be encountered with a two-bearing shaft."

Question: "The connecting-rods are of unusual design in that the big-end cap studs are anchored in steel trunnions. This, I believe, arises from your experience with racing engines and has the advantage of eliminating the difference in connecting-rod section usual with the orthodox form of retaining bolt, and thereby obviates a possible source of weakness. What other advantages does this form of stud-fixing give? "

Answer: "As you have suggested, the design was evolved for fatigue-strength considerations only. We had in mind throughout the design and development of this engine, that its performance in standard production form would be sufficiently good to encourage keen people to try to make it go still faster, and we endeavoured to incorporate in the design every feature which would give a high safety factor if the engine was run at unusually high r.p.m."

Question: "Obviously each connecting-rod assembly is immensely strong. As a point of academic interest, what is the approximate tensile strength of a rod when assembled in an engine? That is to say, assuming the engine revolutions were allowed to mount to a point where the rods would be overstressed, then at what point would fracture take place and at approximately what loading? "

Answer: "It is quite impossible to contemplate one of these connecting-rods failing under tensile load when assembled in an engine. The ultimate strength of the material is 29 tons per square inch and, allowing for eccentricity, etc., the approximate calculation shows that there is no less than 11 times the amount of new (unfatigued) metal required to meet the load at 7,500 r.p.m.

Full Revs—50 Years!

"With light-alloy connecting-rods, it is not only the maximum tensile strength which is important, but also the number of times the maximum stresses are applied. This is true, of course, for all material, but the fatigue life of high-duty aluminium alloys is particularly affected by the ratio of stress applied to ultimate strength, and this is one of the reasons for the bulky nature of such connecting-rods.

"As now in production, the connecting-rod will, we believe, tend ultimately to fail at the gudgeon-pin eye through fatigue fracture, but the engine would have to be run at full revolutions for about 50 years before that fatigue condition would be reached."

Question: "The connecting-rod material is Hiduminium RR.56. This material does, I believe, make an extremely good bearing surface when operating on forged steel; you do, for instance, allow it to bear directly on the gudgeon-pin—there is no small-end bush. Would it be equally good when bearing direct on cast-iron, assuming that there was a comparatively soft metal in the big-end cap to pick up and absorb grit or other foreign matter that may conceivably get between the bearing surfaces? "

Answer: "There is a lot to be said for the use of RR.56 operating direct on the crankpin from the heat flow and weight points of view, assuming the oil is clean. Such an arrangement, however, means that if wear does occur, a new connecting-rod is needed. This would be a very expensive replacement when compared with the price of the readily replaceable Vandervell shell bearing."

Question: "My personal experience with the Matchless twin-cylinder engine over a mileage exceeding 30,000 is that bore wear is astonishingly small. Presumably that is why you do not fit a chromium-plated top ring to each piston as on your single-cylinder 500 c.c. engines? "

Answer: "Our experience is the same as yours with regard to cylinder bore wear on this engine, and consequently we do not fit chromium-plated piston rings."

Question: "You employ separate castings for the cylinders. It is appreciated that the separate castings facilitate cooling between the cylinder bores and also eliminate any chance of distortion of the type that can occur with a one-piece casting embracing both bores. However, the attraction of rigidity given by the one-piece casting is lost. Is it found that with the cylinders deeply sunk in the crankcase and with the tie plate between the cylinder heads full rigidity is obtained? "

Answer: "Separate cylinders are used to improve cooling and to eliminate distortion of the bores and distortion at the cylinder-head joint. The necessary rigidity is provided by the crankcase design. It will be noted that both halves of the crankcase completely encircle their respective cylinders and that the bridge between the cylinders is sufficiently deep to provide adequate rigidity.

"If no engine steadies were fitted, the tie plate linking the cylinder heads would be unnecessary, and considerable uneventful mileage has been covered under those conditions. The tie plate is necessary because the head steadies are not parallel to one another in plan."

Over-cooling?

Question: "Cylinder heads are in light alloy, which seems a modern trend on production motor cycle engines. The higher conductivity of the light alloy no doubt permits higher compression ratios to be used for a given fuel, but I have frequently suspected, after much experience with engines so fitted, that the light alloy tends to result in over-cooling other than when an engine is being driven very hard. In other words, I am suggesting that for all normal purposes, such as the average rider's requirements, cast-iron has its advantages. What are your views? "

Answer: "It is hard to know where to draw the line between the cheaper and quieter, but heavier, cast-iron cylinder head, and the light-alloy head, but the use of light alloy for racing purposes should demonstrate that iron is a barrier to development of high-speed efficiency, and we were resolved to eliminate from our original design any handicap to progress that could be economically avoided."

Question: "Rocker-clearance adjustment is by means of an eccentric spindle. As with other components on this engine, one can see the lessons of racing emerging. Would you please outline the advantages of this method of clearance adjustment as opposed to the more common method of a ball-end screwing into a rocker arm and fitted with a lock-nut? "

Answer: "The eccentric-spindle method of valve-clearance adjustment reduces the reciprocating weight of the valve-gear parts. This is another example of our endeavour to avoid, as far as possible, limitations on r.p.m. A further advantage of this method is that it is not one that could be made inaccurate by slack threads that would permit a lock-nut to move a set-screw endwise in its socket."

Question: "The rocker arms which bear on the valve stem appear to be extremely short and of such curvature that they apply a fair amount of side thrust to the stems. I have not studied the geometry of the rocker movement and perhaps I am under a misapprehension, but have you anything to say on this topic? "

Answer: "A short rocker arm has less reciprocating weight than a longer one and is therefore an advantage in itself. The question of side thrust is certainly important and the geometry of this job is such that the last $\frac{1}{32}$in of valve movement on to its seat occurs at that period in the rocker movement when the point of contact of the rocker-pad radius and the valve end travels in line parallel to the valve stem; the slight amount of angular roll of the rocker pad balances the foreshortening effect of the arm swing.

Chain or Gears?

"Side-thrust during the valve opening has some advantage near the valve-float r.p.m., but in any case is not serious with the geometry and radius of rocker pads which have been very carefully selected."

Question: "I see that the valve-guides are located by spring circlips. What is the attraction of this method of location against a collar forming part of the guide? "

Answer: "The circlip method of valve-guide location is lighter and cheaper than the collar method."

Question: "The timing-gear pinions of $\frac{9}{16}$in width are obviously of extremely robust dimensions for an engine of this size. One imagines, also, that they are a comparatively expensive item. Do you hold any strong views for or against chain drive for the timing gear, which could obviously be provided at less cost? "

Answer: "We do not think any advantage is to be obtained by the use of chain drive instead of spur gears for the timing gear, except perhaps a very slight reduction in cost. We have thought it preferable to provide a timing gear which may be expected, under normal conditions, to outlast the life of the engine without replacement becoming necessary."

Question: "Each camshaft, I note, is supported by three phosphor-bronze bushes and the cams actuate short followers, in the cups of which the push-rods seat. It is, of course, practicable to operate the four overhead-valves from a single camshaft placed either at the front or rear of the cylinders and thus obviate one camshaft and its drive. What were the considerations that made you decide in favour of two separate camshafts? "

Answer: "It was our endeavour to allow free passage of air between the two cylinder barrels and the two cylinder heads and also to have the shortest possible unsupported portion of the camshaft subjected to cam load. We think that we have achieved both our objectives in our design."

Question: "Further, why did you have a follower for each cam as opposed to a tappet which would serve the same function of dealing with the side thrust or 'wipe' from the cam? "

Answer: "This is a further instance of a design intended to keep down the reciprocating weight of the valve gear to the minimum possible. It may be somewhat academic to point out that the power absorption by tappet friction is equal to the side thrust multiplied by the friction co-efficient, whereas in the case of the cam follower it is geared down, the same loading being applied at the radius of the pivot pin and reacted into the engine at the radius of the rocker arm, a reduction of 80 per cent."

Question: "I see you employ separate gear pumps for supply and return of oil, that these are housed in a light-alloy outrigger plate in the timing chest, and that the pumps are driven by the camshafts. The pumps appear to be of large capacity. How much oil is passed per hour at peak r.p.m.? "

Answer: "Approximately 24 gallons."

Question: "There is, I note, a positive feed of oil to the rocker gear from the supply side of the pump. The amount of oil reaching the overhead rockers is considerable. Is trouble never experienced with oil leaks or with oil finding its way down the valve guides? "

Answer: "No trouble is experienced with oil leaks or with oil finding its way down the inlet-valve guides, as the supply to the inlet rocker housing is carefully metered and the valves are a close fit in the honed guides. Adequate drainage facilities are provided. There is, of course, no danger of oil going down the exhaust valve guides which are pressurized during the exhaust-valve opening."

Question: "It is often stated that a gear pump to be thoroughly efficient should operate submerged; this is because its powers of suction lift are below that usual with, for example, a plunger pump. In this engine the return pump has to suck oil through an internal pipe from the crankcase sump and then return it to the tank. Is any difficulty likely to be experienced due to the low suction lift of the gear pump? "

Answer: "The scavenge pump is directly primed by a metered bleed from the delivery pump and is thus prevented from drying out and thereby failing. A wet pump is, in effect, a submerged pump."

Better Starting

Question: "Magneto ignition is employed, but you retain a manual advance-and-retard control. What are your reasons for not adopting automatic control? "

Answer: "You, with your experience, can probably answer this question just as well as we can. We have not discovered an automatic ignition advance control which can cope effectively with variations of fuel rating and temperature to the extent which is desirable in high-efficiency engines. For this reason we retained manual control, which can be set for any slow-speed condition.

"We found that even the cold-starting advantage, usually claimed for automatic-advance systems, was a myth. In very cold weather, there was certainly no difficulty in pushing the engine over fast enough to get a good, fat spark, but the ignition was then so much retarded that the initial impulse given by the first explosion was not sufficient to keep the engine spinning and so one had to kick about six times before the engine would continue running. With manual control, one soon gets to know how to set the ignition to give the right combination of spark intensity and ignition advance to ensure first- or second-kick starting in any weather."

Question: "Will you please give me the peak r.p.m. figure and the b.h.p. figure for this engine? "

Answer: "As you know, we do not like quoting b.h.p. figures because the horses employed on test benches at different factories vary from Shetland ponies to Clydesdales. Our test-bench horses have thick legs! "

INFORMATION BOX

Technical Data

CAPACITY 498 *c.c.*
BORE 66 *mm.*
STROKE : 72.8 *mm.*
COMPRESSION RATIO : 7 *to* 1.
STANDARD PISTON-RING END GAP : 0.006*in* ; *side clearance*, 0.002*in.*
PISTON CLEARANCES : *top land*, 0.016 *to* 0.018*in* ; *top of skirt*, 0.0006*in* ; *bottom of skirt*, 0.0004*in.*
VALVE-SPRING LENGTH : *outer*, $1\frac{3}{4}$*in* ; *inner*, $1\frac{19}{32}$*in* ; *renew when* 3/16 *to* $\frac{1}{4}$*in less than standard free length.*
ROCKER CLEARANCE : 0.006*in both inlets and exhausts when engine cold.*
VALVE TIMING : (*with* 0.012 *rocker clearance*) *inlet valve opens* 35 *degrees before top dead centre and closes* 65 *degrees after bottom dead centre* ; *exhaust valve opens* 65 *degrees before b.d.c. and closes* 35 *degrees after t.d.c.*
IGNITION TIMING : $\frac{3}{8}$*in* (39 *degrees*) *advance with standard contact-breaker gap of* 0.012*in* ; *Lucas magneto.*
ENGINE DIMENSIONS : *crankshaft roller bearings* $1\frac{3}{8}$*in internal diameter*, 3*in external diameter*, $\frac{11}{16}$*in wide* ; *crankshaft middle bearing and crank-pin journals*, 1.62625*in* (*max.*) *to* 1.62575*in* (*min.*) *diameter* ; *crankcase middle-bearing housing*, 1.7715*in* (*max.*) *to* 1.7710*in* (*min.*) *diameter* ; *connecting-rod big-end bearing diameter*, 1.7715*in* (*max.*) *to* 1.7710*in* (*min.*) ; *small-end*, $\frac{3}{4}$*in* (+ 0.0005*in*) *diameter* ; *big-end to small-end centres*, 5.75*in* ; *gudgeon pin*, 2.232*in* (*max.*) *to* 2.228*in* (*min.*) *long by* 0.7499*in* (*max.*) *to* 0.7497*in* (*min.*) *diameter* : *inlet valve head*, 1.391*in diameter* ; *exhaust*, 1.297*in diameter* ; *seat angle*, 45 *degrees.*
CARBURETTOR : *Amal* 76 *AG/1AU with* 1*in choke*, 6/3 *throttle valve and* 180 *main jet.*

COMPETITION CONVERSION

Matchless Twin, National Hare and Hound replica, rated by

Dick Day—ASSOCIATE EDITOR

Photos by Felix Zelenka, Bob D'Olivo

"TAKE IT OUT and give 'er hell," was Sales Manager Guy Louis's exact farewell, as photographer Bob D'Olivo and I climbed aboard one of the hottest combinations to ever come out of California Distributor Frank Cooper's Ajay and Matchless shop here in Los Angeles . . . a much-modified 33-cubic-inch Matchless "Scrambles" vertical twin.

With almost every distributor or dealer waiting for their new '53 models, we took advantage of this chance to deviate from our pattern of "strictly stockers," to test a piece of machinery that has been modified for cross-country racing. This same Matchless twin has literally been showing the boys the way, when it came to Hare and Hounds and Scrambles. I say this only because I am one of the boys that viewed this swinging-arm-suspended billy goat from over my steering dampener.

Just for kicks, let's take a gander at the records that these twin barrelled bombs were able to establish in '52. Bud Ekins was able to load his living room mantel with five first place Hare and Hound trophies, three first place Scrambles cups, and several second and third place awards. Recently, we learned that, plus all these glittering goodies, he was named the 1952 High Point Hare and Hounder plus receiving the High Point Scramble Perpetual Award for AMA district 37. One of the biggest victories brought home with this machine was the Rams National Billy Goat Hare and Hound which Johnny McLaughlin won, with Ekins bringing up a fast third.

Now it was our turn to see why this desert jackrabbit could be tooled through bottomless sand at 65 miles per hour with such ease of handling. Once, out in the toolies, we took a short 20-mile jaunt on an off-breed road to let the newly fitted rings and valves seat in a little. Even with the 6.25:1 gearing the Matchless was pulling, it seemed to be versatile enough for pavement riding. There was a very noticeable vibration point at approximately 60 to 65 miles per hour, a point that would vary, of course, with stock gearing. The riding position was most comfortable. There was no strain on the rider's back or arms, a very important factor in long runs. Several fellow club members were given the guinea-pig act as to the riding position and saddle height. Their comments were always—"Great." Upon entering the hills it was necessary to cross several gullies and choppy ground. Our "shutter bug" for this month had never been cow-trailing before and when asked what he thought of the riding comfort for the passenger, he said, looking down at the massive rear suspension, "Nice springs . . ." That isn't half of it, for here's a rear fork that has a slow and soft full-length travel. This swinging-arm rear system reflects thoroughness by the fact that each rear shock unit is additionally cushioned with two ounces of 20-weight hydraulic fluid. The test model had 40-weight oil in its rear forks, which tends to make the rear more rigid for cross-country racing by a stiffer dampening action of the rear section. The front forks were filled with 20-weight—strictly a matter of personal preference.

Chromed twin exhaust pipes are built by Cooper Motors for the "Flying M," cost approximately $20.00 exchange. Straight pipes are made to contour of the machine, but need protective leg shield. Note clearance under the small scrambles gas tank that allows repairs

Sod was literally torn out by the roots when churned over with Avon Gripster rear tire. 21 inch front wheel adds ground clearance to lower front frame which eliminates hanging up in shallow holes or on steep knolls as this

Transporting the photographer to the top of a nearby hill for a vantage point for photos, I was about to learn my first lesson on the vertical twin. Without the taps fully opened, the Matchless lugged down in power and, with a spit back from the carburetor, came to a sudden halt in low cog. Returning to the bottom of the hill, we started up once again with the engine singing, this time maintaining sufficient rpm's that sent us bounding over the top. With its present modifications, the twin was not a lugger in the true sense of the word but was fully up to the job when kept buzzing. Making a few runs up a long steep hill gave forth an enlightening experience. This machine's handling qualities and the lightness of the front end left the front wheel practically at rest for the complete

Mixing it up with some of the boys around a short 1/5-mile track proved that the "Stick" could hold its own, and the transmission offered very versatile gear ratios for any type of dirt riding

climb, only dabbing here and there. This is a good example of the traction afforded by the 4.00 x 19 Avon Gripster tire mounted on the rear. The weight distribution of this machine is such that after riding for any length of time, anyone could become an up-on-the-rear-wheel jockey very easily. High speed cornering on the "Matchstick" was great, once you became acquainted with the squirrelly feeling of the Gripster's outer edge nobbies.

Once, choosing a little more hill than the alloy acrobat could conquer, I was forced to eat crow. The trip down was completed with comparative ease. The brakes, combined with the front and rear suspension, allowed a choice of speed upon descending and not once did the machine feel as though it wanted to sneak out of control. Loading up the camera gear and the photographer, we set out to find some deep sand and soon came across a twisting wash that offered every conceivable type of gulley, water hole, and soft sand. Fortunately, entrance could be made by crossing a set of abandoned railroad tracks, a vicious gimmick for testing suspension. Here the Twin performed like a master. Several trips were made over the tracks without the faintest tick of the crash plate. One could hardly ask for better ground clearance.

Once down in the deep sand, I started off full-bore and found the old vet climbing to the surface in seconds. Dropping into second gear gave the feeling that someone had given you a big shove from the rear. Popping it into third was like an overdrive. Letting the bike have its head at 55 mph proved that this was her shining hour. Shifting down into second gear for a sharp, blind bend found us headed for trouble, for, as fast as you can read this line, the screaming sand hog had bounded over a deep, wide hole. The result was a fast sharp impact but the sting was almost completely absorbed by that marvelous rear suspension. Gripping the handlebars slightly loose, we toured through this half-mile stretch around 65, just gliding on the surface. Ahead, I spied the reflection of some water. After the tidal wave had subsided, it was pleasing to note that the Matchless, although completely doused, hadn't dropped a beat.

Shifting back down into second gear, we made a quick mud-slinging jaunt over 100 yards of soft muddy creek bed. Pulling to a halt, I found myself a believer. The terrain just covered would put a Detroit proving ground to shame but only served as a compliment to the ability of the bike in question.

Going back to the weight distribution, the engine is a squatty unit with no feeling of being top-heavy. The ultimate in traction is obtained by the combination of the severe rear tire and ideal gearing. Rear suspension and the light front end let the rider work easily from a standing position with a feeling of ease and security that I've never experienced before. One very noticeable point was brought to light. When shutting the Twin completely down, as upon entering a corner, there is no coasting period, as with a single, for it comes to a fairly sudden stop. You either have it on or off. This necessitated slipping the clutch slightly to keep the engine revolutions up while negotiating some of the soft, sandy bends.

Checking for demerits at this time we found that the bolt holding the float chamber to the mixing body of the carburetor had become loose, letting the gas seep out like mad. Breaking out the old "King Dick" adjustable spanner, this was remedied in short order. Following a few fast pictures, we loaded up the gear and headed back for Frank Cooper's shop. Taking a short cut through Baldwin Hills, we chanced across a few sand-lot racers practicing on a fifth-mile track that they had carved out of the country-side. The temptation was too much, for after observing for a couple of laps, the Matchless mill was soon spinning about 6,000 rpm's, mixing it up with the boys and doing a good job of holding its own.

Pulling into the driveway back at the shop, looking much as though we had just finished the Jack Pine, little belches of blue smoke began puffing from under the cylinder head . . . a blown right gasket. This could not be determined as a direct demerit for the bike, for following the ring job prior to the test, the head bolts had apparently not been pulled up quite tight. Replacing the gasket was only a matter of thirty-minutes work. The test model had a small scrambles tank which left ample space for the engine to be disassembled with only the disconnection of the gas line. One big feature of this versatile little twin is the fact that each cylinder is a separate unit. When faced with a situation such as this, it only necessitates removing one cylinder head plus the intake manifold, carburetor and the overhead valve cover. This type of trouble happens very little but being able to work on one cylinder at a time without disturbing both barrels is definitely a time saver and a clever bit of design on the manufacturer's part. After replacing the head gasket and buttoning up all the components, I was about to see one of the most unique valve adjustments ever witnessed. Loosening a small lock nut that clamps the rocker spindle, it is only necessary to insert a screwdriver at the slotted end of rocker arm and turn slightly. The rocker arm is actuated by a cam which raises and lowers this unit until .006 clearance is obtained for adjustment. This is such a simple operation that any new owner, with only a small degree of mechanical aptitude could perform the task. Replacing and tightening down the overhead valve cover are all that is necessary

In deep sand the "Matchstick's" low-slung weight showed up noticeably in superior handling. Third gear would net a good 60 mph through deep stuff, purring like a kitten

to button up this thirty-minute job.

While we're still on this technical kick, let's go inside of the bored out 33 cubic inch powerplant and pick up some of the fine engineering points, starting first with the crankshaft that is cast as one unit. One look at the massive crankshaft, with dual flywheels and bobweights, and it's evident that there is brute strength in its large bearing surfaces. The shaft has one main babbitt-type bearing where it passes through a center partition in the crankcases. Heavy duty roller bearings support the crankshaft on the timing and primary side of the crankcases. The connecting rods of the Matchless reflect a finish on a par with your own bathroom mirror, factory-polished with the same gloss that many of us put on our first VL or 101 Scout innards. The crankcases are cast in almost a complete sphere, the ultimate contour for strength. On the right side crankcase of this particular machine there

Stock handlebars were great for leverage and position. Brakes worked smooth and lent to ease of selecting groove while descending. Mag is controlled manually by spark lever on bars

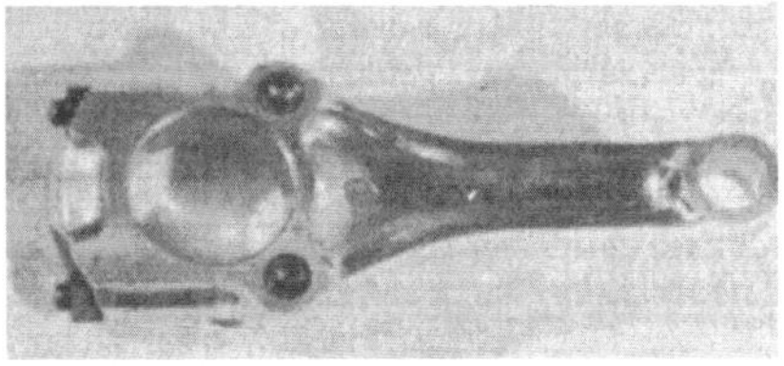

Connecting rods are polished to lessen oil friction. Note upper end of lower section of rod has been drilled and fitted with steel bosses that anchor cap studs for added strength

is an ugly dented scar, a battle wound from one of last year's Hare and Hounds. The dent is concaved approximately $\frac{1}{8}$ of an inch and almost one inch in diameter. There was a slight oil seepage for a few days then it stopped, although there is undoubtedly a crack inside. The point is merely mentioned to indicate the strength of design. Oil lead-in lines to the crankcase are somewhat of a bad feature on the twin, in that they're as vulnerable as Durante's nose, where they fit on the timing case cover. It's easy to see how they could be mashed while traveling through a rock-studded dry wash. On the test machine the oil line had been covered with a fairly heavy rubber aircraft hosing, acting as a cushion for rocks that the lower section might come in contact with. Guy Louis informed me that they are in the process of building a small protective crash shield for this area of the bike which would entirely eliminate fear of mishap. The oil pressure that this engine carries is almost fantastic. Gage-tested, the pressure registered 180 pounds at running speeds and 80 pounds with engine idling. Each cylinder is a separate unit and houses all internal oil line passages, there being no outside tubing. The barrels are centrifugally cast, eliminating blow holes in the cast iron metal. The push rods, also being located in the cylinder, eliminate several outside parts such as: covers, seals, etc. Also, the cylinders being separate components give a greater circulation of air in between each unit, doing away with the familiar air pocket that is found on various other makes. Each separate cylinder is held down by four full-length head bolts extending through the cylinder to the case proving very satisfactory when packing in a few speed goodies, especially more compression. The cylinder head is cast of aluminum alloy metal. Huge light gray fins give the impression that there ought to be more cylinders under all that area. The pistons fitted to these special models are made exclusively for, and sold by, Cooper Motors for the scrambles twins. The size of the slugs enables the use of stock 21 cubic inch hard chromium plated rings. Pistons are manufactured for the following sizes: when the barrels are bored, the stock scrambles pistons are fitted, then there are two more sets of slugs that are .020 and .040-inch over the original scramble bore. Thus, the owner is not left hanging when in need of a re-bore. The scrambles pistons are manufactured of light aluminum and have $\frac{3}{16}$-inch thickness at the top with a compression ratio of 8.5 to 1. When this engine is completely assembled and sitting in the frame, it looks about half again as wide as other twins and twice as condensed from top to bottom. Comparing measurements with others disclosed that variation was usually only $\frac{1}{2}$-inch wider and somewhat shorter in height. This height of the engine differs mainly because of the deep spigotted cylinders that fit well into the crankcases when assembled. Sitting in the frame as low as it does drops the center of gravity of the overall machine down low, which is another main item for ease of handling.

Nine a.m. Monday morning found the weather fair and the road test crew gathered at their favorite spot around the first electric timing eye, sipping black coffee in anticipation of the regular speed run. With everything all set, we warmed the low-geared Matchless up and took off up the strip for the inaugural run. Reaching the end of the approach area, I turned and set sail for the timing traps. In fourth gear, the little bomb was really screaming when, upon entering the trap, it developed a bad sputtering. On the way back to the timer, I let the machine peak out in third gear, for here again the miss was dominant, although not quite so severe. Little did we realize that the first time of 85.10 miles per hour was going to be our fastest time of the day. Checking the plugs, they showed to be burning beautifully and definitely not the trouble. We pulled the carburetor bowl off and checked all the components, but with no indication of trouble. On pavement, the bike felt a little sluggish and slow getting under way, hinting of a flat spot in carburetion. This could be attributed to the $1\frac{3}{32}$-inch carburetor that had replaced the stock 1-inch venturi. With all Hare and Hounds in this area being usually laid out in the Mojave desert, and passing through the surrounding mountains, we are constantly running at altitudes ranging from 3,000 to 5,000 feet. The test machine has been set up with a larger carburetor to compensate for the thin air at those altitudes. When run at almost sea level, it has a tendency to suck in a little too much air compared to the gas comsumption and give the flat, over-carburetted feeling. Do a switch, though, and take the bike up to a higher altitude and it should be perfect. Generally speaking, the stock 1-inch pot will be

Standard to scrambles modifications: dual saddle replaced with Terry solo and pillion, stock fenders replaced with aluminum alloy guards, 4-gallon tank exchanged for 3-gallon scrambles tank, twin cross-over headers substituted for stock exhaust system, 3.00 x 21 front wheel and tire preferred to standard 3.50 x 19 size. All electrical equipment, tool boxes, center stand and chain guard are removed. Throttle and clutch have dual cables for quick change-over in case of one breaking, and large sprocket has been fitted to rear wheel, which relieves a large amount of stress on the clutch

the best when running at lower altitudes. Considering the gearing of 6.25:1, 85 miles per hour seems quite respectable.

The Matchless has one of the most massive heavy-duty rim and spoke set-ups of any bike on the U.S. market today, and it would take some very harsh usage to tweak this rim. There was no sign of clutch slippage throughout the entire test and the transmission was absolutely tops. At first, I missed a few shifts because of being used to a shorter moving shifting lever. After getting acquainted with the distance of travel, missing a shift was practically impossible. Light toe pressure and the easy location of neutral were most noticeable. This gearbox is a direct take-off on the "Boy Racer" model except that it is not a full ballbearing throughout.

Utilizing an S-bend at the end of our paved speed strip afforded a proving ground for determining whether or not the rear swinging-arm suspension would have a tendency to walk on a hard surface while cornering. The answer was negative. The brake figures were not outstanding but it must be remembered that the bike is rigged strictly for competition and that neither the narrow front tire nor the accented nobby Gripster on the rear are the best for braking efficiency over pavement. Both brake levers, front and rear, worked easily and progressively. With a more conservative tire tread fitted to the machine, it is believed that the model would show up a much lower braking figure in view of the springing range.

Putting Editor Bob Greene on the pillion pad, we took off down the road, leaving the test crew securing the timing equipment, in search of some near-vertical hills. We found that the choicest area was a knoll offering two good 60 degree grades of about 25 feet in height on each side of the road. I made a run at the steep hill and with a minimum of work found myself on top of the grassy crest. Back at the bottom, Bob, still not quite believing his eyes, climbed aboard and bounded over the hill in the same fashion and, upon returning, his only comments were . . . Huh, no strain!

With the conclusion of this obstacle, the test on the modified Matchless drew to an end. It had mounted every conceivable challenge that would either make or break it when it came to competition events. Tabulation at the bottom of the test sheet tallied out as such: Performance, good; comfort, traction and handling qualities, the ultimate. For owners who are now riding a stock 30.50 Matchless Twin, or enthusiasts contemplating buying this machine, the complete modification set up, identical to the test machine, can be acquired for the nominal fee of $85 including labor, and the exchange of such parts as exhaust pipes, fenders and the front wheel. For the conversion on just the engine alone such as bore, pistons, and rings, $21.50 will give you those extra suds when grabbing a handful of throttle. And . . . for those of you who are already riding and competing on these Scrambles twins, "Please, gang, don't tear the course up so much. It's rough on us guys back here . . ."

Price of new stock Matchless 30.50 cubic inch Vertical Twin, $875 plus tax and license fees at Los Angeles.

PERFORMANCE SUMMARY

Acceleration

¼-mile drag	15.83 secs.
57 miles per hour average	
1/10-mile drag	8.25 secs.
43 miles per hour average	

Slow Running

High gear without snatch	11 mph

Turning Circle

Minimum diameter	11 ft. 6 ins.

Mileage

Under competition circumstances	34 mpg

Speed

Maximum in low	39.30 mph
Maximum in second	50.14 mph
Maximum in third	72.58 mph
Maximum in high	85.10 mph

Braking

From 25 to stopped, rear brake only	41 ft. 9 ins.
From 25 to stopped, front brake only	38 ft. 5 ins.
From 25 to stopped, both brakes	23 ft. 6 ins.

GENERAL SPECIFICATIONS

ENGINE. Vertical Twin (stock model) 30.50 cubic inch displacement; bore $2\frac{19}{32}$ inches; stroke $2\frac{7}{8}$ inches; compression ratio, 7 to 1. (Test model) 33 cubic inch displacement; bore, $2\frac{23}{32}$; stroke, $2\frac{7}{8}$; compression ratio, 8.5 to 1. Two separate deep finned aluminum alloy cylinder heads, stellite tipped valves and cast-in cast steel valve inserts seats. Individual cylinders deeply spigoted into the die-cast spherical crankcases. Stock aluminum alloy pistons replaced with special cam-ground aluminum pistons when bored out; forged light alloy highly polished connecting rods, twin flywheel crankshaft with two bob weights on each side of shaft, cast in one unit and of exceptional strength. Crankshaft is supported with Vandervell center main and big end bearings (babbitt type). Crankshaft on timing case side and on primary side has heavy roller main bearings.

CARBURETOR. Amal one-inch diameter venturi, air cleaner choke lever on handlebar (has been removed) (test model has Amal $1\frac{3}{32}$-inch diameter venturi).

TRANSMISSION. Oil lubricated, heavy weight, four speed with enclosed positive stop foot gear change and kick starter. A constant mesh, direct gear to gear system eliminates usual spool or sliding gear. Shifting by dogs similar to synchromesh.

LUBRICATION. Circulating dry sump system, twin gear type oil pumps running directly off end of camshafts. Built-in filter in oil tank easily replaceable. Oil system through internal passages. Only outside leads are flexible mounted pipes from oil tank to engine. Pressure lubrication to top end.

IGNITION. Gear driven flange mounted Lucas magneto. Manually operated at handlebar.

ELECTRICAL EQUIPMENT. New long type heavy duty six-volt generator is gear driven, voltage regulator. High and low beam headlight, taillight and parking light (test model has all this equipment removed).

CLUTCH. New heavy duty five stud with fiber inserts. Multiple disc dry type.

DRIVE. Prestretched "mark ten" Renolds racing chain. Shock absorber on engine sprocket. Single row ½-inch pitch x 5/16-inch wide roller primary chain. Rear chain is ⅝-inch pitch x ⅜-inch wide roller, same size as American big twins.

FRAME. Single loop front section with rear suspension swinging arm set in large wide journal bearings. Huge rear main springs are 6 inches long x 3 inches in diameter. 3½ inches of travel on rear swinging arms.

FORKS. Teledraulic, fitted with metered pistons for constant control with oil on both sides of piston to buff counter-recoil as well as recoil. This same system applies to rear forks.

HANDLEBARS. Chrome plated, competition type rigid mounted. Left bar; clutch, compression release (removed on test model), spark control, dimmer switch (dimmer has been removed). Right bar; front brake, throttle, choke lever, and horn (choke and horn have been removed).

WHEELS. Mounted on Timken-type tapered roller bearings, safety washers on hubs prevent spokes from tearing out. Rear wheel, 3/16-inch spokes and heavy duty rim.

BRAKES. Internal expanding 7-inch front and rear; shoes ⅞-inch wide. Brakes can be quickly adjusted by hand, screw type lock nuts are adapted to both front and rear brake controls.

TIRES. 19 x 3.25 front; 19 x 4.00 rear; Dunlop triple stud tread. (Test model has Avon gripster 4.00 x 19 rear; 3.25 x 21 Dunlop triple stud tread front.)

TANK. Pressed steel, four U.S. gallon capacity. (Test model mounted with 3 U.S. gallon scrambles tank.) Single filler cap.

INSTRUMENTS. Smith's 120-mph illuminated speedometer, ammeter and lightswitch in headlight shell. (Test model has ammeter and lightswitch removed.)

SADDLE. Terry, with chromium plated coil springs, separate sponge rubber pillion seat.

FINISH. High temperature black baked enamel finish, very hard and durable.

TOOL KIT. Dual tool boxes, one on each side of rear fender. Complete set of tools, grease gun, tire irons, and ignition tools. Tire pump mounted on left side of frame. (Test model has all this equipment removed.)

OVERALL LENGTH. 85¼ inches.

OVERALL WIDTH. 29½ inches.

WHEELBASE. 55¼ inches.

GROUND CLEARANCE. 7 inches.

SADDLE HEIGHT. 32½ inches.

WEIGHT (dry). Stock twin weighs 400 lbs., after modification twin weighs approximately 360 lbs.

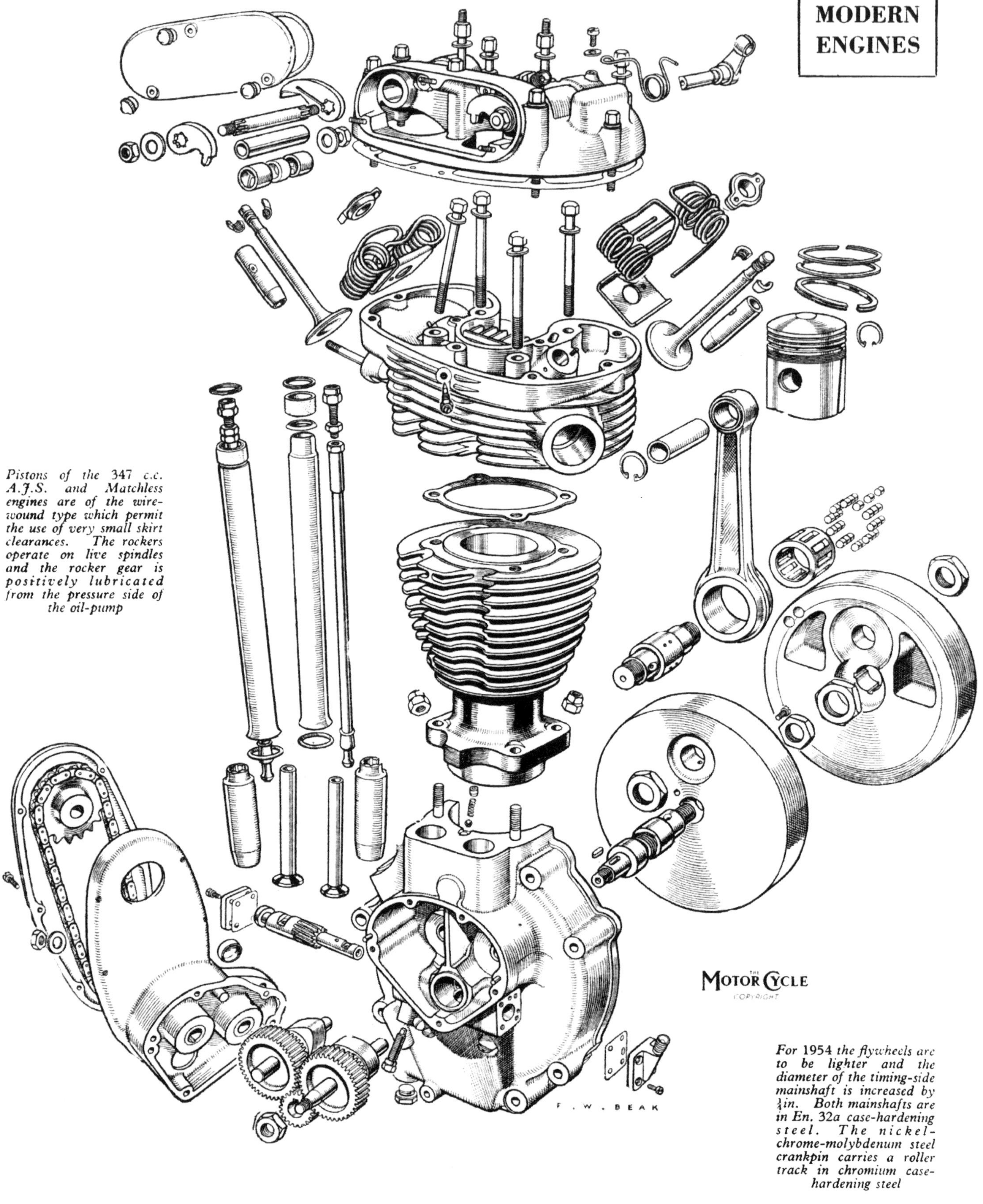

Pistons of the 347 c.c. A.J.S. and Matchless engines are of the wire-wound type which permit the use of very small skirt clearances. The rockers operate on live spindles and the rocker gear is positively lubricated from the pressure side of the oil-pump

For 1954 *the flywheels are to be lighter and the diameter of the timing-side mainshaft is increased by ¼in. Both mainshafts are in En. 32a case-hardening steel. The nickel-chrome-molybdenum steel crankpin carries a roller track in chromium case-hardening steel*

The 347 c.c. A.J.S. and Matchless

The Technical Editor, ALAN BAKER, Discusses a Popular O.h.v. Engine with Designer P. A. WALKER, A.M.I.Mech.E.

A QUESTION that may well be asked when the 347 c.c. A.J.S. and Matchless engines are under discussion is whether they are, in fact, "modern engines." The basic design goes back to before the last war. The Matchless version of the engine, used in vast numbers by the Armed Forces, found favour with Don Rs practically the world over. Power output was high for a W.D. engine. The standard of mechanical quietness was the criterion by which all other engines were judged, and the degree of engine reliability was second to none.

Immediately after the war the engine went into production in exactly the same form as that in which it had been manufactured for the Forces. Since then it has been modified year by year. In 1951 it was equipped with a light-alloy cylinder head and enclosed hairpin valve springs. Competition versions for trials and scrambles were introduced—engines which have been responsible for a vast number of successes both at home and abroad, in the U.S., Australia and Europe.

I called recently at the A.M.C. factory to discuss the engine with Philip Walker, the Associated Motor Cycles designer. Since I had heard rumours of certain alterations to the bottom half, I began: "For 1954 the flywheels are to be lighter by 3-4 lb a pair. Many people, myself included, believe that numerous modern machines are under-flywheeled, with the result that low-speed smoothness is not all it might be. What advantages do you gain by this apparently retrograde step?

Answer: It is true that some loss in low-speed torque is an inevitable result of lightening the flywheels. But the modern single is already extremely docile. My feeling is that, bearing in mind the rapid changes in speed of modern traffic, the need is for increased responsiveness. Engines have to be lively if they are to give the utmost rider-satisfaction. And this characteristic, I feel, is more important than low-speed slogging. It is no secret, of course, that really heavy flywheels are desirable for trials engines where low-compression ratios, retarded ignition and ultra-low speeds are *de rigueur*.

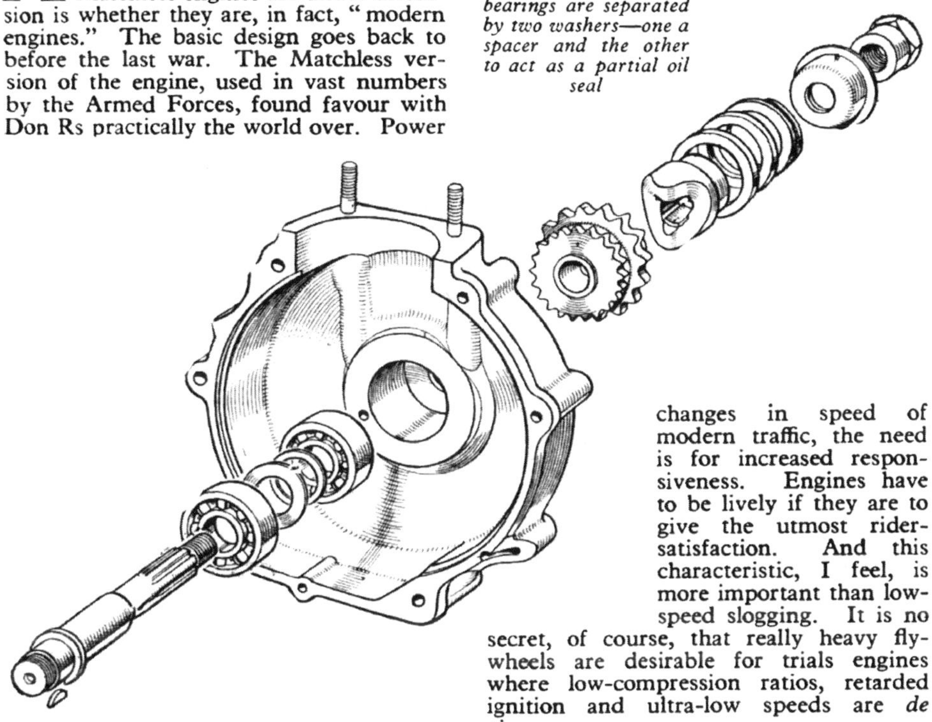

The drive-side main bearing comprises two ball races. The bearings are separated by two washers—one a spacer and the other to act as a partial oil seal

Question: The flywheels, I believe, are of high-grade cast iron with integral balance weights. What proportion of the reciprocating weight do you balance and how are the wheels balanced?

Answer: For these relatively long-stroke engines we balance 62 per cent of the reciprocating weight, and the wheels are checked singly on an Avery balancing machine which indicates where and how much weight has to be drilled out to bring the wheel up to correct balance. The drilling is done on an adjacent vertical drilling machine and the wheel returned to the Avery for recheck.

Increased Diameter

Question: Both mainshafts are of En. 32a case-hardening steel; the drive-side shaft is a parallel fit in its flywheel and it is located by two keys at 90 degrees to each other. The timing-side shaft, however, was formerly a taper fit in the wheel but, for 1954, it is to be a parallel fit with a single key, and its diameter is to be increased by $\frac{1}{4}$in. What are your reasons for the change in size and the revised method of fixing?

Answer: The change in the method of fixing the timing-side shaft has been made partly for production convenience and to take advantage of an increased shaft diameter which permits a shoulder abutment against the flywheel; this abutment provides additional support for the shaft and thus increased rigidity of the flywheel assembly results. The greater diameter provides additional bearing area to meet the increased power output.

Question: The crankpin is of nickel-chrome-molybdenum steel and has a sleeve of chromium case-hardening steel on which the rollers run. Is the sleeve a force fit on the pin and how is the oil fed to its outer surface? And what is the reason for not running the bearing directly on the crankpin?

Answer: The sleeve is a force fit on the pin and has internal grooves to carry oil to the carefully positioned holes feeding the roller bearing. The oil is initially fed into the timing-side shaft, then through the flywheel and into the crankpin, whence it passes to the grooved sleeve. As to the second part of your question, the nickel-chrome-molybdenum steel we use for the pin has a very high tensile value with good impact resistance, whereas the metal used for the hard bearing track requires relatively brittle properties.

Question: I understand that you have a system of selective fits for big-end components which ensures accuracy of the assembly within 0.0001in. Presumably the selection covers the outside diameter of the pin, the bore of the connecting-rod eye and the roller diameter. How many groups of sizes are necessary to ensure such accuracy?

Answer: We grade crankpins and connecting-rods in eight steps of one ten-thousandth part of an inch diameter, and these parts are paired to within one step up or down from the theoretically correct fit.

Longer Life

Question: The big end is unusual in that it employs three rows of $\frac{1}{4}$in $\times$ $\frac{1}{4}$in rollers, in a Duralumin cage, instead of the two rows which are more common. Why did you decide on three rows of rollers? And is your bearing appreciably wider than normal as a result?

Answer: The position is that the rate of wear is more or less related to the width of the roller track used. A wide bearing offers tremendous advantages—a fact which, indeed, has been proved to us by our experience with the racing engines. The crankpin bearing is somewhat wider than that normally used, and we have selected this arrangement, and the close control of the fit, in the interests of longevity.

Question: On the timing side, the crankshaft is supported in a phosphor-bronze bush cut away at the bottom to admit the oil-pump shaft. Why do you employ a plain bearing rather than the ball or roller pattern?

Answer: The use of a plain bearing is dictated by the method of oil transfer to the crankshaft assembly. The bush is

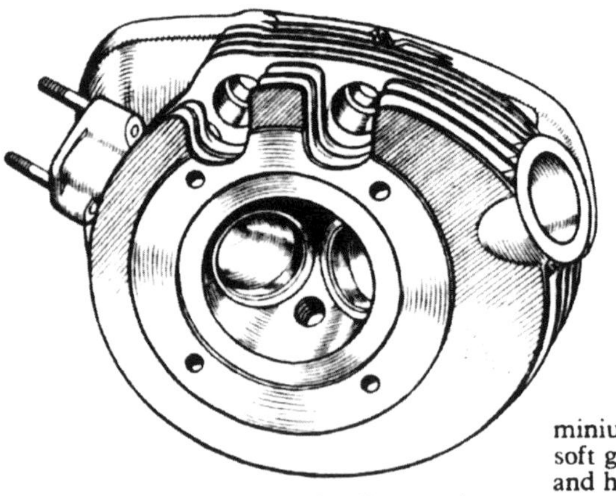

The cylinder head is a light-alloy gravity die-casting in DTD.424, with alloy-iron valve-seat inserts

heavily shouldered at the flywheel end to permit the use of a tight fit at this point, since a tight fit is impossible where the bush is cut away.

Question: The timing gears have live spindles which run in bushes in the crankcase and timing-case cover. How do you ensure the accurate location of the gears which is so essential for silent operation ?

Answer : Gear-tooth pitch-line diameter is accurately controlled during manufacture, and the crankcases are fine-bored in a special machine installed to ensure an accuracy of gear centres within 0.0001in, which we have found essential. Even then, selective assembly of gearing is resorted to whenever a machine fails to achieve the standard degree of quietness.

Question: Cast iron is used for the tappet guides, which have a series of axial grooves to pass the surplus oil from the valve gear; is there any provision for positive rotation of the tappets, to spread the wear ?

Answer: Yes, the tappets are automatically rotated by virtue of an offset between the cam path and the centre line of the tappet.

Mechanical Quietness

Question: The DTD.424 light-alloy cylinder head is a very handsome gravity die-casting with cast-in alloy-iron valve-seat inserts. Why was this material chosen for the inserts, and how do you ensure that they do not work loose ?

Answer: For use in a touring engine, where long life of engine components is of the greatest importance, an alloy-iron for valve seats cannot be bettered. It is true that aluminium-bronze seats, as used in racing engines, provide better heat conductivity but they have a very short life—a factor which is unimportant in an engine built for sheer power output (and reliability over a short period).

Question: One or two firms have found mechanical noise troublesome with light-alloy cylinder heads, yet your engines are famous for their quietness. What is the secret of your success ?

Answer: The quietness results from good workmanship, careful design of cam gear and correct selection of materials for the different components. There is no secret about this ! The cam profile ensures quiet mechanical operation of the valve. The cam pinions are wide to ensure long life. The timing pinion is manufactured from oil-hardened steel. Because of its toughness, an oil-hardened pinion possesses certain shock-absorbent qualities and, in addition, it has high distortion-resisting properties. The cam wheels are case-hardened.

Question: The rockers are not carried in the head but in the aluminium rocker box, and a fairly thick and soft gasket is fitted between the rocker box and head. In view of the zero tappet clearance, I should have thought that a metal-to-metal joint, sealed by jointing compound, would eliminate the danger of the exhaust valve being raised off its seat by bedding down of the gasket. Would you care to comment on this point?

Answer: This is a case where theory and practice are not compatible. All I can say is that the desired results are obtained without causing additional maintenance work. It must be pointed out that there is a very large surface area of gasket to take the loading applied and, with the correct material for the gasket, very little bedding down takes place.

Question: The rockers comprise separate arms splined on a spindle and clamped by nuts to a spacer sleeve which forms the bearing surface. The rocker shafts are thus live and run in bronze bushes in the rocker box. Why do you prefer this construction to the more common arrangement of a bushed rocker with integral arms running on a fixed spindle ?

Rocker Gear Design

Answer: The selection of the whole rocker gear design is dictated by economic and functional requirements. The system provides a substantial bearing area; it is convenient from a production viewpoint and production costs are not excessively high. The advantage of mounting the rocker spindles in the rocker-box covers is that during dismantling the rocker box can be removed complete with the rockers.

Question : The shape of the two rocker arms is almost identical, though one has a machined ball end and the other a thickened-up end to bear on the valve stem. Do you use the same stamping for each or are they separate components?

Answer : The same blank is used for each arm; the pad on the arm at the valve end of the spindle is "upset" in a press tool to provide the necessary width of pad.

Question : The valve guides, I note, have no flanges for location purposes. How do you ensure that they are inserted to the correct depth and have they any tendency to shift owing to their lower rate of expansion than that of the cylinder head?

Answer : Stepped assembly tools are provided for positioning the guides, which are of a type we have used for many years. I agree that a positive location would ensure immobility of the guide, but trouble with our design over the years has been negligible and a flanged guide would add to the cost.

Flame Hardened

Question : There are no caps on the valve-stem ends, but the exhaust valve-stem tip is flame hardened to resist hammering. What is the increase in hardness obtained by the flame-hardening? And why do you prefer hardening the stem to using a hardened cap?

Answer : The hardness is increased from 250 to 720 Vickers Pyramid Numeral, which is approximately the hardness obtained on some alloy case-hardened steels. The reason for not using hardened end-caps is that should the least trace of valve-sticking occur, or the valves float (in the event of the rider missing a gear), the cap jumps off the valve stem. The inevitable result of such an occurrence is too obvious to require discussion!

Question : I note that you use multi-groove valve cotters. What advantages have they over the more conventional pattern with a single, deeper groove?

TECHNICAL DATA

CAPACITY : 347 c.c.

BORE : 69 mm.

STROKE : 93 mm.

COMPRESSION RATIO : Standard, 6.53 to 1. Competition, 7.5 or 9.42 to 1.

PISTON RING END-GAP : Normal, 0.006in ; maximum, 0,030in.

PISTON RING SIDE-CLEARANCE : 0.002in.

PISTON CLEARANCES : Top land, 0.032-0.034in ; at wire winding, 0.001in ; bottom of skirt, 0.0005in.

TAPPET CLEARANCE : Nil, with engine warm (not hot).

VALVE TIMING : Standard engine (with 0.014in tappet clearance) : inlet valve opens 36 degress before top dead centre and closes 51 degrees after bottom dead centre ; exhaust valve opens 50 degrees before b.d.c. and closes 30 degrees after t.d.c. Scrambles engine (with 0.001in clearance) : inlet valve opens 59 degrees before t.d.c. and closes 69 degrees after b.d.c. ; exhaust valve (with 0.005in clearance) opens 74 degrees before b.d.c. and closes 48 degrees after t.d.c.

IGNITION TIMING : Standard and trials engines : On full advance, contact-breaker points begin to separate 37 degrees of crankshaft rotation before t.d.c. Scrambles engine : 39 degrees before t.d.c.

ENGINE DIMENSIONS : Crankshaft drive-side ball bearings, 1in bore x 2.25in outside diameter x 0.625in wide. Timing-side bush, 1.125in bore (+0.000-0.0005in) x 1.547in long. Crankpin, 1.20375-1.2035in diameter x 0.946-0.944in long. Small-end bush, 7/8in bore (±0.00025in) x 1 1/16in long. Connecting rod length, big-end to small-end centres, 6 7/8in. Valve head diameters : inlet, 1.594in ; exhaust, 1.50in. Valve stem diameters : inlet, 0.373-0.372in ; exhaust, 0.3715-0.3705in. Valve seat angle, 45 degrees. Valve lift, 0.358in (standard cams).

CARBURETTOR : Amal Type 76AV/1.E.D. with 3 degrees of downdraught ; 150 main jet, 6/4 throttle slide, throttle needle clip in middle notch (if air cleaner fitted, 130 main jet and needle one notch lower).

Answer: The multi-groove arrangement is one we have used for a number of years. It is particularly suited to the softer valve steels, of which Silchrome is one, by virtue of the increased shoulder area which it provides.

Question: Overlapping hairpin valve springs were introduced concurrently with the light-alloy cylinder head. In what way are such springs superior to coil springs for an engine of this type, and how do you stop them from rotating?

Answer: As compared with coil springs, hairpin valve springs are less likely to deteriorate in tension, as they are remote from the valve guides which, particularly in the case of the exhaust, can become hot enough to blue helical valve springs and cause them to lose length. The seating for the hairpin springs is pegged to stop it turning.

Question: Earlier engines had the ex-

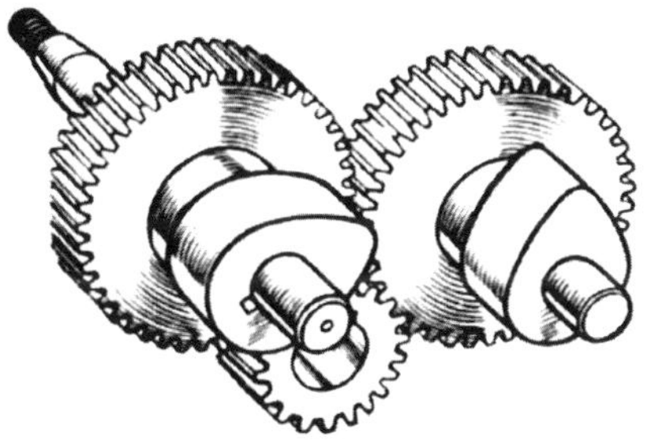

The new cams provide greater lift, increased overlap and less dwell. The crankshaft pinion is manufactured in oil-hardening steel, a material which has high distortion-resisting properties

haust-valve lifter mounted in the timing case, where it operated on the exhaust tappet. With the introduction of the alloy head, the valve lifter was moved to the normal position in the rocker box. Can you tell me the reason for this change?

Answer: There were two reasons for the alteration in design. First, the appearance of the engine was improved by the removal of the operating cable to a less conspicuous position. Secondly, we were able to simplify the tappet and guide assembly.

Question: The standard engine has a cast-iron cylinder barrel held down by short studs passing through the base flange only. In the competition engines, a linered light-alloy barrel is employed, and it is secured to the crankcase by long studs passing through the head. What is the reason for the different construction on the two engines?

Answer: The long studs are used on the competition models as the light-alloy linered barrel is unsuitable as a tension member. Apart from this, there is little space for cylinder-base stud nuts, since the cylinder material at the base is thicker than it is in the case of the iron cylinder.

Question: The wire-wound pistons you employ are unusual. What is the purpose behind this particular feature?

Answer: The wire is a normal medium-carbon steel and five turns are used; it is wound into a spiral groove under a carefully regulated tension and ground overall when the piston diameter is finished. The winding enables the thermal expansion of the piston to be carefully controlled, thus permitting the use of very small clearances.

Question: Selective assembly of barrels and pistons ensures a high accuracy of fit. How many groups are there and what are the intervals between the groups?

Answer: Eleven size groups are available in steps of 0.00025in.

Question: The top piston ring is chromium-plated. I understand that this simple modification has greatly reduced the rate of bore wear. A difficulty with earlier chrome-plated rings was that they bedded down very slowly, owing to the hardness of the chromium. Has your ring any feature to eliminate this bother?

Answer: Yes, the outside diameter of the ring is slightly tapered to ensure rapid bedding down.

Question: Main and big-end bearings are fed directly from the delivery side of the rotary-plunger pump. There is also a feed to the cylinder bore at two points via a drillway from the pump housing, a spring-loaded ball valve and a peripheral groove on the lower face of the cylinder base flange. Am I right in assuming that the loading on the ball valve is such as to pass a fair supply when the oil is cold and very little when it is hot (in which latter circumstances the splash from the big end is greater)?

Timing Gear Lubrication

Answer: That is correct. There is some resistance to flow from the big end with cold oil, owing to the outlet holes pointing towards the crankshaft centre; this resistance is sufficient to overload the ball-valve spring at low engine revs and to allow a quantity of oil to reach the cylinder walls.

Question: The rocker gear is lubricated from the pressure side of the oil pump and not from the scavenge side, as is more commonly the case. I note also that the timing gears and cams are lubricated by surplus oil which drains down the push-rod cover tubes from the rocker gear. Is there any connection between these two features?

Answer: Yes, there is a close connection. Since the timing gear is lubricated by oil draining from the rocker gear, it is better to use cool oil from the pressure side of the pump to feed to the rocker gear. A further reason for using fresh oil is to ensure controllability of the feed by the various jets and restrictions used throughout the rocker gear lubrication channels.

Question: Is there any form of positive lubrication for the exhaust valve guide or do you avoid this because of the risk of carbon formation on the stem and consequent sticking?

Answer: The exhaust valve guide receives a slight wetting, as oil is caught in a suitable

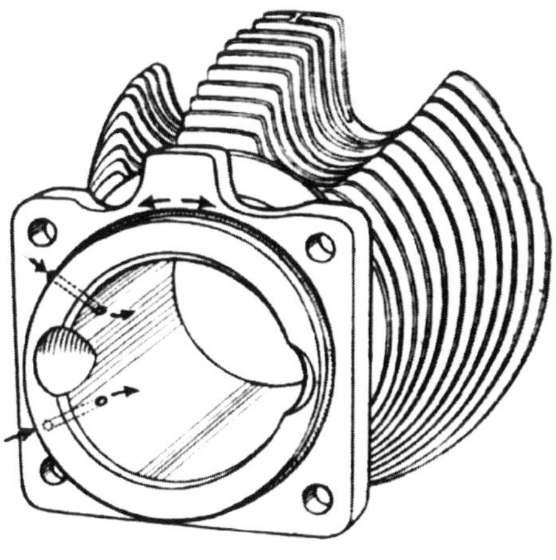

Cylinder bore lubrication is by oil forced under pressure from the oil pump into a peripheral groove in the base of the cylinder. From here it is fed to the thrust face through oilways drilled in the cylinder walls

well in the top face of the cylinder head and drains past a hole in the valve guide. Some oil can reach the valve stem but there is never a head of oil to over-lubricate the valve.

Question: Are the competition engines tuned in any way, such as by polishing flywheels, head and ports?

Answer: No special finish is applied to the components because a reasonable degree of attention is paid to the inlet and exhaust ports on all models; only the few really top-line riders would derive any notable advantage from a high degree of polishing—a very costly process.

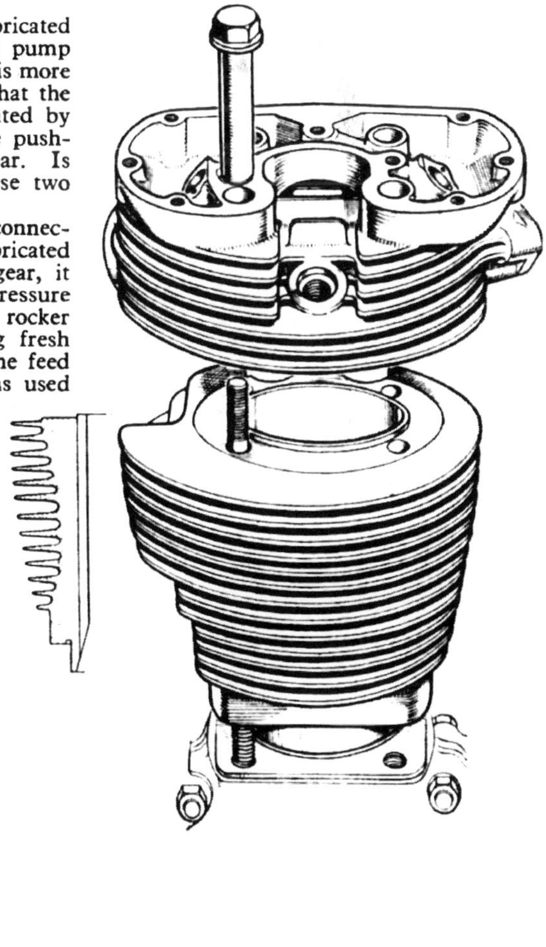

The 347 c.c. light-alloy competition engine was introduced in 1949 for production in 1950. Both cylinder and head are retained by four high-tensile steel studs, protruding from the crankcase cylinder-joint face and passing through the cylinder-barrel casting

MODERN ENGINES

The 348 c.c. Overhead-camshaft 7R A.J.S.

Analysis of a Successful Production Racing Power Unit Revealed by Discussion Between P. A. WALKER, A.M.I.Mech.E., Associated Motor Cycles Designer, and the Technical Editor, ALAN BAKER

THERE can be few machines in the history of motor-cycle racing which established themselves in the esteem of riders so quickly as did the 7R A.J.S. More than this: the "Boy's Racer," as the 7R is affectionately called, has more than maintained its early popularity, as a glance through the entry list of almost any three-fifty race will confirm. A strong reason is the quality of the 7R engine.

Compromise is inevitable in engine design, and the necessity for compromise is at least as prominent where racing power units are involved as it is with touring engines; the difference is one of kind rather than of degree. Happy is the designer who, in Shakespeare's words, can make a virtue of necessity, even the necessity for compromise. That the problems were successfully solved in the case of the 7R is emphasized by the facts that the engine performed remarkably well almost "off the drawing board," and that only minor modifications have been required in the subsequent six years, during which development in the racing sphere has been most intensive.

Not only in Britain has the 7R acquired its fine reputation, and not only in the hands of British riders. On the Continent, in the Americas and throughout the Commonwealth—in fact, wherever racing flourishes—is found this straightforward yet potent three-fifty, of which over 300 have been built since 1948. Among its victories are numbered first place in the Junior Manx Grands Prix of 1950, 1951 and 1952 (in the 1952 event 7Rs annexed the first three positions), and Bob McIntyre's almost fantastic second place in the 1952 Senior M.G.P. This list does not include the many stout efforts of the factory teams, nor the host of lesser successes of private owners, but it does indicate a good combination of engine power and stamina.

Philip Walker, who has many years of service with the Plumstead concern to his credit, was responsible for the design of the 7R engine, and subsequent development work has been in the hands of H. Hatch. The main modifications introduced have been stiffening-up of the crankshaft and main-bearing assemblies, alteration to the cylinder-head geometry, and the substitution of roller cam followers for the slipper type. Each point will be dealt with later in this article.

Before a designer can set to work on an engine, he must have a definite purpose in view and well-defined ideas on how certain requirements are to be achieved. This thought prompted my first question to Mr. Walker.

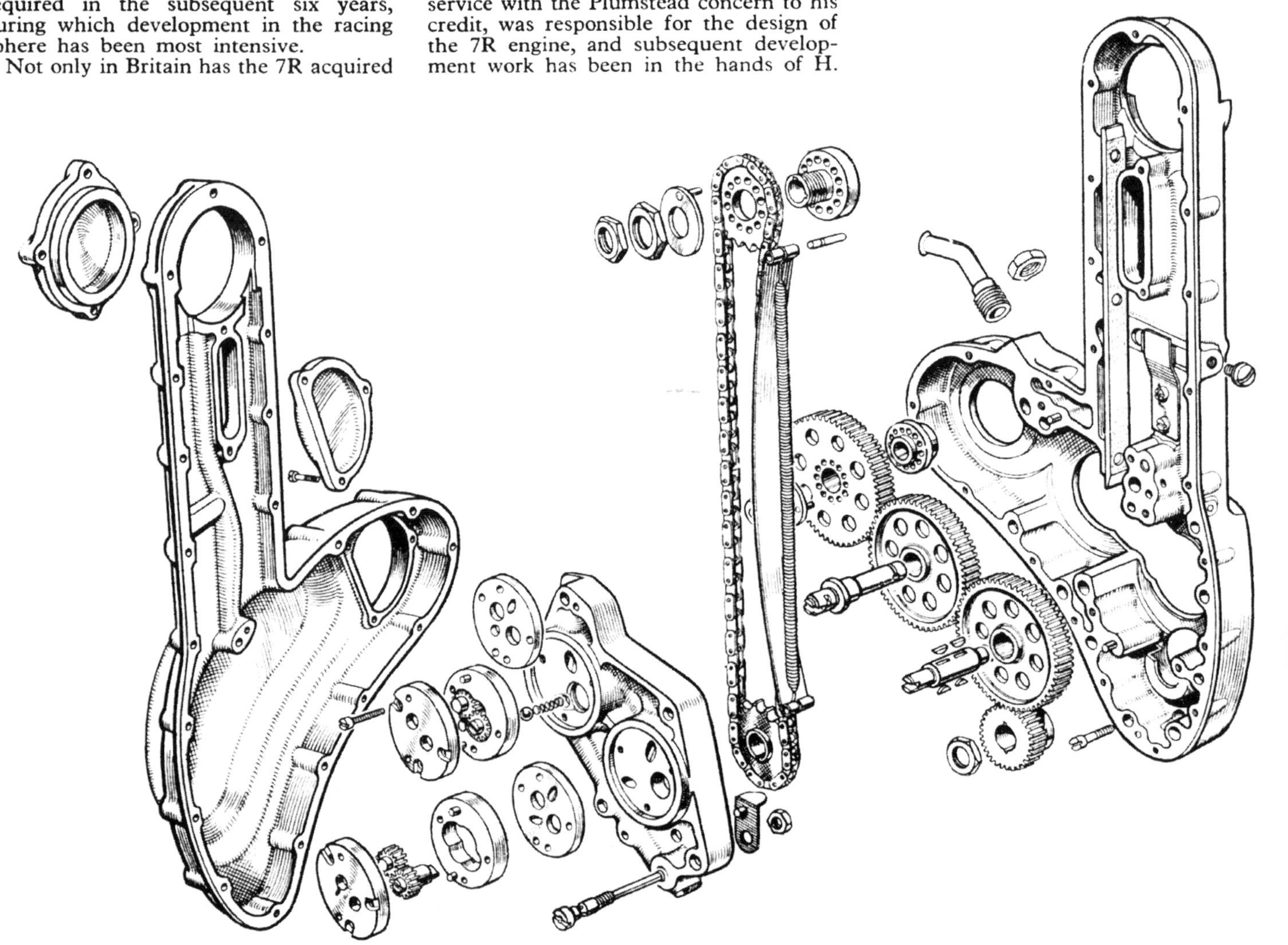

Question: "What were your basic aims when you set out to design the 7R engine?"

Answer: "Although we naturally wanted a high power output, we were equally concerned that the engine should be reliable, durable and as free from temperament as is possible with a piece of machinery of this type. We also sought a reasonable degree of mechanical and constructional simplicity, because the engines would be going to private owners whose workshop facilities or technical skill might be limited. In brief, the 7R was to be a rider's rather than a tuner's engine."

Question: "To what extent did this desire for simplicity govern your decision to adopt the single- rather than the double-o.h.c. layout?"

Answer: "For an engine of this type it is our opinion that the theoretical advantages of twin camshafts make comparatively little difference in practice. The single-camshaft design is certainly simpler and it is also cheaper to produce—a point which we had to consider since we were involved in a commercial proposition. Also—by no means unimportant from the owner's point of view—a single-o.h.c. engine such as ours provides simplicity of valve-clearance adjustment and good oil-retaining qualities.

"A further point which must be appreciated is that a twin-camshaft head is bound to be appreciably heavier than a single-camshaft head. Though the difference in terms of all-up weight would be small, it could adversely affect acceleration unless balanced by a proportionate increase in torque."

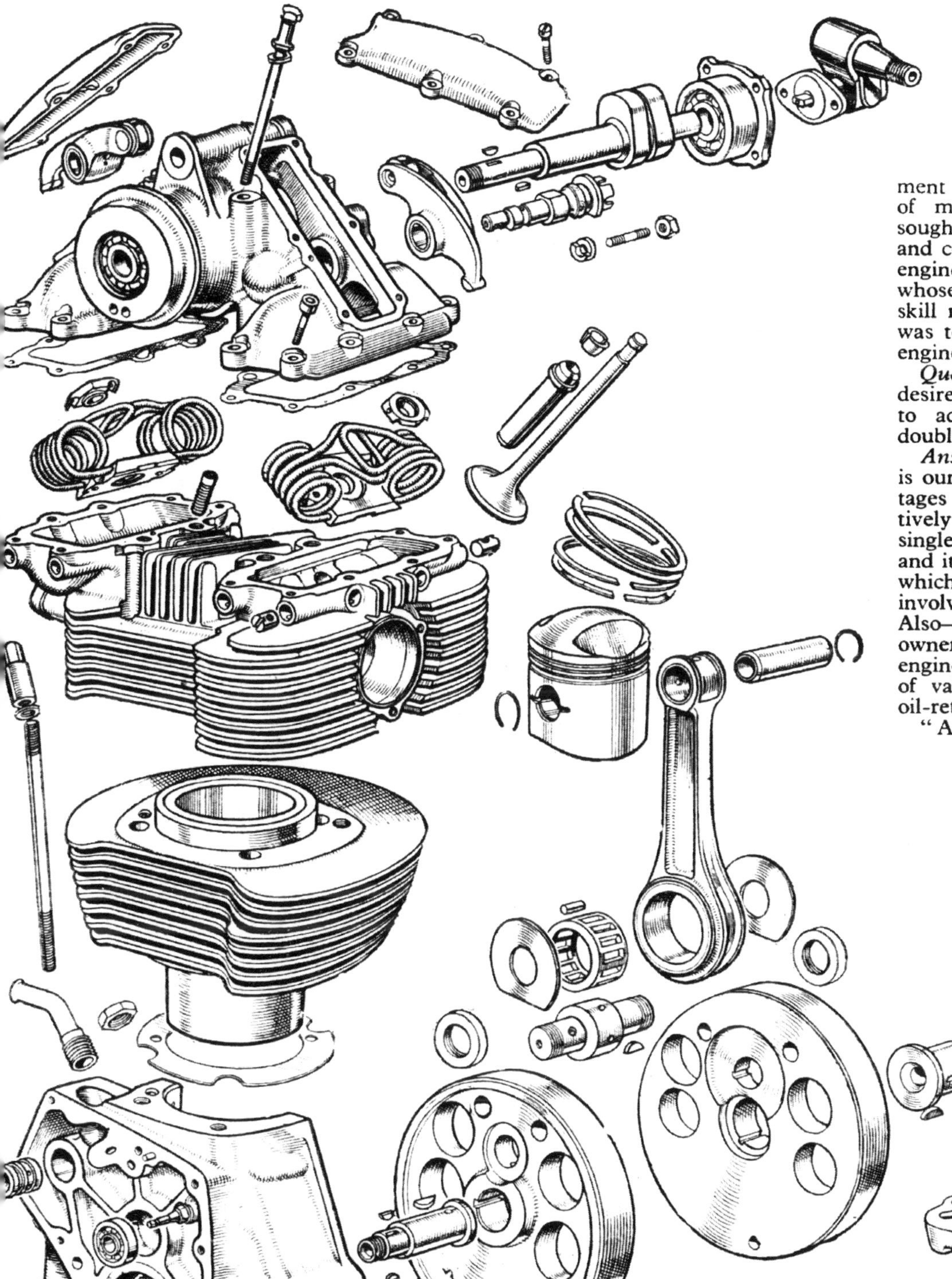

Though essentially straightforward in conception, the 7R has no lack of interesting detail points. Examples are the separate oil pumps housed in the timing case, the rubbing strip and friction-damped Weller tensioner on the chain drive to the camshaft, the roller cam followers and the unusually deep spigot of barrel into head. The crankpin and drive-side mainshaft are of composite construction; the crankpin has been turned through a right-angle in the drawing, to show the two oil-outlet holes. A single row of long rollers is employed for the big-end bearing; the outer race of the drive-side main bearing is positively located in the crankcase

In a racing engine, robustness of the crankshaft assembly is of the utmost importance because of the very high centrifugal and inertia loadings involved in full-throttle running. High power resulting from clever design of the combustion chamber, cams or ports, is of no consequence if the crankshaft cannot transmit

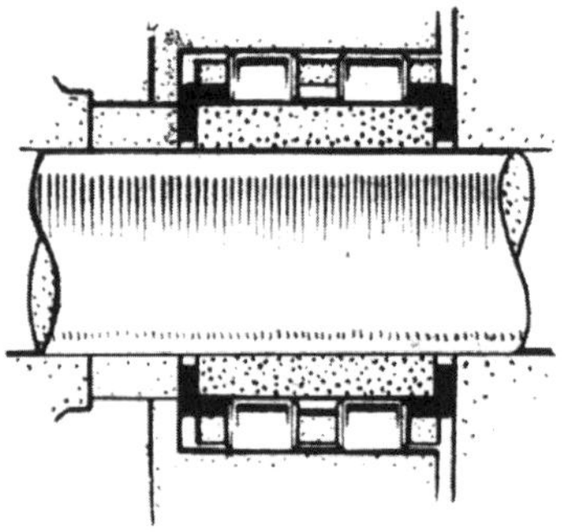

Thrust washers of the drive-side main bearing are flanged to avoid wear of the roller cage

it—and do so for long periods. For that reason the 7R flywheels are forged from KE.805, a high-grade carbon steel, and the mainshafts are of nickel-chrome-molybdenum case-hardening steel.

Question: "The shafts on current engines are flanged at the inboard end and are an interference fit in the flywheels, with key location; on earlier engines (before 1953) the drive-side shaft was inserted from the outside and was secured by a castellated nut. What was the reason for the alteration in construction?"

Answer: "There were three reasons. First, it was difficult to pull the shaft up really tight by means of the castellated nut because the C-spanner tended to yield under heavy pressure. Secondly, for adequate strength, the thickness of the nut had to be greater than that of a flange. Hence, by going over to a flanged shaft we were able to reduce the depth of the counterbore in the flywheel and so have a greater length of shaft supported in the wheel, with a consequent increase in rigidity. Thirdly, whereas the old one-piece shaft had two changes of diameter—always a potential source of weakness—we now have a shaft with only one change of diameter and a sleeve pressed on to form the inner race of the main bearing. The arrangement is more robust and permits sufficient case depth on the sleeve to resist wear."

Reduced Bearing Width

Question: "Another 'bottom-end' modification concerns the drive-side main bearing. Up to 1953, the outer race was a simple sleeve pressed into the crankcase, and there were three rows of rollers—two together in one Duralumin cage and the third in a separate cage. The present engines employ two rows only of $\frac{5}{16} \times \frac{5}{16}$in rollers in a common cage, and the outer race is positively secured in a flanged sleeve which is pressed into the case and held by screws. What lay behind the reduced bearing width and why do you not use a proprietary bearing unit?"

Answer: "To deal with the second point first, we could not obtain a proprietary bearing of the required dimensions and load capacity. As to bearing width, with the stiffened-up crankshaft assembly we decided that we could dispense with the outer, single row of rollers and could shorten the shaft accordingly. By shortening the shaft we could bring the primary-chain line closer to the drive-side flywheel, and so reduce the tendency of the mainshaft to deflect under chain pull. The resultant improvement in chain life has been marked.

"A point of interest on the bearing assembly is that the outer race, which is held into its sleeve by a coverplate and screws, is finish-ground after being fitted, to ensure accurate alignment."

The only engine modification for 1954 was to the cage and thrust washers of the drive-side main bearing. It was found that there was a tendency for the cage to wear against the plain thrust washers fitted last year. The latest washers are shouldered and the cage is recessed to take the shoulders, so that any axial movement results in the roller ends, and not the end of the cage, abutting on the washer concerned.

Forming the inner race of the big-end bearing is a sleeve which is pressed on to the crankpin; as on the drive-side mainshaft, this composite construction gives the desired combination of wear-resistance and strength. The pin is a parallel interference fit in the flywheels and is keyed into the drive-side wheel.

Question: "Since they are devoid of hexagons or castellations, how are the crankpin-securing nuts tightened, and why is the key in the drive end rather than the timing end of the pin?"

Answer: "Prior to 1953 we employed on the crankpin castellated nuts similar to that on the old drive-side mainshaft, and the same difficulty was met over tightening them adequately. We now fit nuts which have a hexagon on an extended shank; after the nut has been tightened, the hexagon portion is sawn off. This modification has given complete security of the crankpin, and the only disadvantage is that removal necessitates mutilation of the nuts.

"We key the drive end of the pin because the other end contains the big-end oil hole and we did not wish to weaken it any further."

Question: "The big-end bearing comprises a single row of $\frac{1}{4} \times \frac{1}{2}$in rollers in a Duralumin cage; prior to 1953, $\frac{1}{4} \times \frac{5}{8}$in rollers were employed. A single row of long rollers has the advantage of greater effective length over two or three rows of shorter rollers (of the same overall length), since there are only two end chamfers instead of four or six. Is it not a fact, however, that longer rollers have a greater tendency to skew, so that a higher degree of accuracy is necessary as regards parallelism of the crankpin? Also, why did you reduce the roller length?"

Shorter Rollers

Answer: "You are quite right regarding the advantages of the single-row bearing and the manufacturing accuracy. We set a tolerance of 0.0001in on the parallelism of our crankpin to ensure that the rollers have no tendency to run off the straight.

"A further advantage of the single-row bearing is that, since the oil outlet is midway along the pin, oil is fed to the middle of the rollers. In a two-row bearing, oil emerges between the rows, unless the feed is duplicated. On touring engines, the oil-feed aspect is not critical, but it can be so on a racing power unit where the inertia loadings at high r.p.m. are extremely heavy.

"We reduced the roller and crankpin length in order to give a shorter and stiffer crankshaft assembly. This modification was introduced concurrently with the redesigned mainshaft and main bearing. In spite of the higher unit loading resulting from the use of shorter rollers in the big end, the life of the bearing remains entirely satisfactory."

The connecting rod is forged from KE.805 steel and is subsequently heat-treated for maximum strength; a sleeve of nickel-chrome, case-hardening steel is pressed in to form the outer race of the big-end bearing. The big and small ends are each stiffened by two peripheral ribs which provide greater resistance to bell-mouthing of the eyes than would single ribs. Hence the rod is lighter than would be a single-rib design of equivalent rigidity at the eyes.

In the top of the small end is a hole through which oil mist from the crankcase passes to lubricate the gudgeon pin. Several radial holes evenly spaced round the middle of the small-end bush serve as oil reservoirs to lubricate the pin immediately after the engine is started.

Since it combines light weight with reasonable strength, Elektron (a magnesium alloy) is employed for the crankcase halves and also for the combined timing-case and camshaft-drive housing.

Separate delivery and scavenge oil pumps are carried in a cast-Elektron housing bolted to the inner half of the timing case. The scavenge pump is driven

TECHNICAL DATA

CAPACITY: 348 c.c.
BORE: 74 mm.
STROKE: 81 mm.
COMPRESSION RATIO: 10 to 1 approximately.
PISTON CLEARANCES: Top land, 0.021in; bottom of skirt, 0.008in.
PISTON RING END GAPS: Compression, 0.013 to 0.008in; scraper, 0.025 to 0.020in.
PISTON RING SIDE CLEARANCE: Compression, 0.003in; scraper, 0.0015in.
VALVE CLEARANCES: Engine cold—inlet, 0.008in; exhaust, 0.012in.
VALVE TIMING: With 0.008in valve clearance—inlet valve opens 50 degrees before top dead centre and closes 74 degrees after bottom dead centre; exhaust valve opens 70 degrees before bottom dead centre and closes 49 degrees after top dead centre.
IGNITION TIMING: Contact-breaker points begin to open 40 degrees before top dead centre on full advance.
ENGINE DIMENSIONS: Drive-side double-row roller bearing (24 rollers, $\frac{5}{16}$in x $\frac{5}{16}$in), 2.0629/2.0626in bore x 2.8168/2.8160in outside diameter x $1\frac{1}{8}$in wide; timing-side roller bearing, 1in bore x $2\frac{1}{2}$in outside diameter x $\frac{3}{4}$in wide; crankpin sleeve, 1.5156/1.5154in outside diameter x 0.696/0.694in wide; big-end bearing, single row of 14 rollers, $\frac{1}{4}$in x $\frac{1}{2}$in; big-end eye of connecting rod, 2.01600/2.01575in diameter x 0.685/0.683in wide; small-end bush, 0.875 (+0.00050−0.00025)in bore x 0.934/0.933in wide; con-rod length, big-end to small-end centres, 6.375 (±0.0025)in. Inlet-valve diameters: head, $1\frac{3}{4}$in; throat, 1.640in; stem, 0.3120/0.3115in. Exhaust-valve diameters: head, $1\frac{17}{32}$in; throat, 1.420in; stem, 0.4342/0.4332in. Valve-seat angle, 45 degrees; inlet-valve lift, 0.449in; exhaust-valve lift, 0.408in.
CARBURETTOR: Amal, type 10 G.P., $1\frac{5}{32}$in choke diameter, 12 degrees down-draught. No. 5 throttle slide; needle No. GP6; 0.109 needle jet; needle position No.2; 230 main jet.

by the shaft which carries the half-speed gear of the timing train and the camshaft-drive lower sprocket. The shaft runs in ball bearings, the inner of which is located in the crankcase and the outer in the pump housing. Drive to the delivery pump is by the intermediate-gear shaft, which is mounted in bronze bushes.

Question: "Why do you employ separate oil pumps rather than a two-in-one layout, and why are they mounted in a separate housing instead of on the inner half of the timing cover?"

Answer: "Were we to employ back-to-back pumps with direct drive, the timing case would need to be deeper and therefore heavier, and the oilways would be less easy to arrange. Worm drive, though permitting a satisfactory installation, would be rather less efficient than direct drive.

"By having the pumps in a separate housing we can provide outboard support for the timing-gear spindles without the necessity of highly accurate location of the timing cover. Also, the cover is not stressed and hence is easier to make oil-tight."

Cam-box Oiling

From the delivery pump the oil follows two routes. An oilway in the pump housing leads to a restrictor in the quill which feeds the big-end bearing through the timing-side mainshaft. The restrictor builds up the pressure to force the remainder of the oil through the hollow pump spindle and thence up a passage in the cylinder barrel and head to the cam box. There oil is fed through the hollow rocker spindles to the bearings, and emerges as splash to lubricate the cams, camshaft bearings, followers and valve ends.

Oilways in head and barrel return surplus lubricant from the valve-spring pockets to the crankcase. Oil emerging from the right-hand camshaft bearing lubricates the driving chain and timing gears.

The crankshaft pinion of the timing train runs in a small trough which is filled with oil draining from the cam gear. Oil splashed from the trough augments lubrication of the timing-side main bearing.

At the lower rear of the crankcase halves are scrapers set close to the flywheel rims. The scrapers deflect oil from the flywheels into the sump and thus prevent excessive oil consumption by reducing the amount of oil in the crankcase.

Question: "I note that there are two oil holes in the crankpin and that they emerge at about 45 degrees on each side of the line joining crankpin and crankshaft axes. It has been argued that it is undesirable to have the oil outlets nearer the axis of rotation than the axial oilway in the pin, because the oil has to overcome centrifugal force in order to flow inward along the radial feed holes. Do you consider the criticism is justified?"

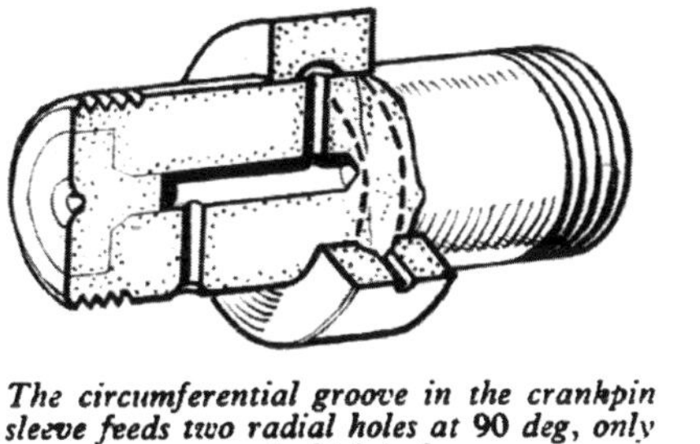

The circumferential groove in the crankpin sleeve feeds two radial holes at 90 deg, only one of which is shown

Answer: "It is, of course, perfectly true that the farther the oil outlet is from the axis of rotation, the greater is the centrifugal force which helps to expel the oil. The effective radius at which the force acts, however, is that from the crankshaft axis to the outlets; what happens inside the pin is really of no importance provided that the oilways offer no restriction to the flow.

"We have found from experience that the areas of lowest loading in a racing big end are about 45 degrees on each side of the centre line. Not only is this the obvious place to feed the oil (since there is the minimum external resistance to its flow) but, as soon as it gets clear of the holes, the oil is centrifuged through most of the circumference of the bearing. The inner 90-degree arc is lubricated by the oil carried round from the "trailing" hole by the rollers and cage."

Question: "The camshaft is driven by a ⅜in-pitch chain on 17-tooth sprockets. A leaf-spring tensioner is employed and there is a rubbing strip of impregnated fabric close to the loaded run of the chain. What reasons, other than tradition, decided your adoption of chain drive rather than gears or shaft? Also, what is the purpose of the rubbing strip and of the friction damper which bears on the chain tensioner?"

Chain Advantages

Answer: "Owing to differential expansion, spur-gear drive is inevitably noisy when hot, and requires high accuracy of machining of the case to ensure correct meshing. Meshing of bevel gears, with shaft drive, also requires some care. Spur gears have the further disadvantage of making it difficult to adjust the compression ratio, since it is normally necessary to fit a different piston. Chain drive, apart from the wider tolerance on centres when a tensioner is used, is lighter and cheaper than the other systems and, when well lubricated, has appreciable shock-absorbing qualities.

"The rubbing strip ensures that, should chain-flutter occur, it would be insufficient to cause overloading of the chain or variations in valve timing. We fit the damper to the tensioner because a camshaft drive is not strictly a one-way drive. Valve-spring pressure acting on the steep closing curve of the cam tends to push the camshaft ahead of its drive, thus causing the normally slack side of the chain to tighten. This reversal of load, with an undamped Weller tensioner, could lead to vibration of the spring blade or even cause the chain to leave the sprockets. For that reason we fit the friction damper, which moves in a slot part way up the drive case and bears on the back of the tensioner spring blade."

Question: "The camshaft-drive upper sprocket is coupled by a 12/13-hole vernier to the camshaft, and a similar vernier is employed on the magneto-drive pinion. This arrangement allows adjustment to within about 4.6 degrees, which

seems rather coarse for a racing engine, particularly as regards ignition timing. Is it possible to obtain a closer setting?"

Answer: "The lower sprocket of the camshaft drive has 17 teeth and two keyways at 180 degrees, so that by turning the sprocket through 180 degrees on the keys a half-tooth variation is possible at the top, equivalent to about 21 degrees of crankshaft rotation. This alteration and five holes on the vernier in the other direction give a two-degree variation of timing. Further, one degree is obtainable by opposing movements of one tooth and nine holes on the vernier, so there is no difficulty in achieving the desired timing to very close limits.

"On the magneto drive, the gear has 60 teeth, so a 12-degree basic variation is possible; and since five holes on the vernier represent 23 degrees, the timing can be set to within one degree by altering the gear meshing by two teeth. One tooth and three holes will give just under two degrees' variation."

The cam box is an Elektron casting and is secured by socket-headed bolts which screw into cylindrical, steel trunnion blocks housed in the cylinder head. Since frequent lifting of the head may be required, the seemingly rather expensive and complicated method of attachment is really very sensible. Were the holding-down bolts to screw directly into the head, a stripped thread would be no impossibility. One end of each trunnion is slotted to permit the use of a screwdriver to align the trunnion hole with the bolt.

Roller Cam Followers

En.32 carbon case-hardening steel is the material of the camshaft, which has integral cams and is carried in a ball bearing at each end. The rockers are stampings of KE.805 steel and their valve end-pads are Stellite-faced.

Question: "Whereas pre-1953 engines had a cam-follower pad of ¾in radius, the rocker is now forked and carries a ¾in-diameter roller which bears on the cam. Since the new rocker must be heavier than the old, what was the reason for the change?"

Answer: "The pad follower was satisfactory if lubrication was 100 per cent efficient; otherwise, fairly rapid wear occurred. We made the change in order to render lubrication less critical; though rather heavier, the roller follower has proved completely reliable.

"By reducing the radius of the

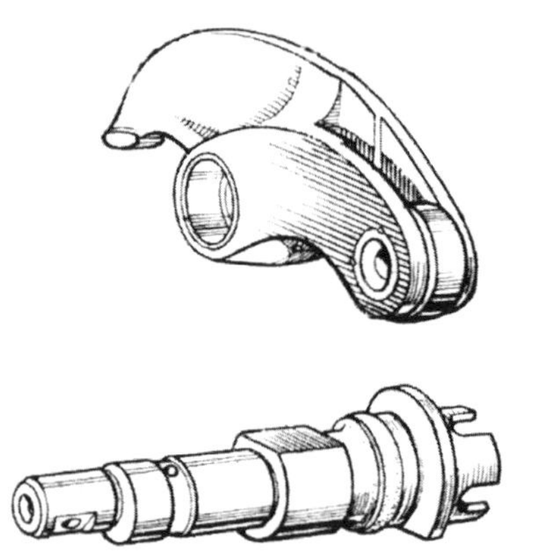

Details of the rocker, which has a roller cam follower and, for clearance adjustment, an eccentrically mounted spindle

follower we altered the acceleration and deceleration rates of the valves, but the effect was not sufficient to warrant re-designing the cam form. A roller of the original radius would, of course, have been too big and heavy to be practicable."

Question: "The rockers are bronze-bushed and run on eccentric spindles whereby valve clearances are adjusted. What is the benefit of eccentric adjustment compared with the more common method of placing shims under the end cap of the valve?"

Answer: "The adjustment is taken out of the reciprocating mass, which must be kept as low as possible to avoid valve float, and is simpler to effect than by shimming. We arrange the eccentricity so that the movement is substantially vertical to the rocker plane when the valve is at one-third lift. In that way we avoid any increase in side thrust on the valve or any variation in valve timing when adjustment is made.

"In practice, a small timing variation is rarely critical, but we have found that incorrect rocker geometry can result in very rapid valve-guide wear and, of course, increased friction losses in the valve gear. It is to the point to say at this juncture that, in our opinion, the only real advantage of twin camshafts for a production racing engine lies in the straight thrust on the valves, with consequent minimum friction and wear."

The cylinder head is cast in RR 53b light alloy and has an unusually deep female spigot to take the correspondingly deep male spigot on the barrel. The arrangement permits additional finning on the head, which is more effective for cooling than would be an extra fin on the barrel.

Valve seats are shrunk into the head. Aluminium bronze, which has excellent heat-conducting properties, is used for the exhaust-valve seat. On the inlet side, however, conductivity is less important than resistance to wear, since wet mixture and the inevitable grit form an abrasive combination. Austenitic iron was therefore selected on account of its wearing qualities. Both materials have a sufficiently high coefficient of expansion for there to be no tendency for the valve seats to loosen in the head at operating temperatures.

Question: "The horizontal fins on the head are wide and nearly square; I assume that this shape is designed to bring the fins as far out into the air-stream as is practicable and to present the maximum length of fin tip to the airstream. However, the vertical fins between the valve-spring pockets run diagonally rearward across the head from left to right. Why is this finning used rather than a longitudinal layout?"

Answer: "You are correct regarding the horizontal fin shape. With enclosed valve springs, the exhaust spring pocket effectively masks the top fins of the head from the direct airstream; hence, longitudinal finning would be of very little benefit.

"We induce air flow over the fins by placing them diagonally and having a tunnel through the upper portion of the camshaft-drive casing. The rush of air past the outer end of the tunnel acts in the same way as air flowing over a carburettor jet: the partial vacuum created in the tunnel causes air to travel over the diagonal fins."

Increased Efficiency

Question: "Originally the valves were disposed at 39½ degrees on each side of the vertical, and the combustion chamber was nearly hemispherical. For 1953, the layout was brought into line with that employed on the 1951 factory engines: the combustion chamber was reduced in depth and the piston-crown shape was altered from a plain curvature to a taper-flanked form; inlet-valve inclination was reduced to just under 35 degrees, as a result of which the downdraught angle of the inlet port was increased to 12 degrees. What was the reason for these various modifications?"

Answer: "The flattening of the combustion chamber reduced the surface area exposed to heat and so increased the thermal efficiency of the engine. By increasing the downdraught angle, we straightened out the inlet port a little and also reduced the angle through which the mixture has to turn to reach the cylinder. Both alterations gave some improvement in volumetric efficiency.

"Offset of the inlet port to the longitudinal centre line has remained the same as before, at 20 degrees; we found by experiment that this angle gives the correct amount of swirl of the incoming mixture to ensure the most efficient combustion."

Question: "Most racing enthusiasts are aware of the important effect of inlet and exhaust system characteristics on power output, and of the importance of matching the two tracts. Can you give me any information on this aspect of the 7R engine?"

Answer: "Considerable care has been devoted to obtaining the best combination of induction and exhaust systems. In our experience the length of the induction tract from carburettor to valve has more bearing on the power output than the overall length including the bell-mouth inlet. The latter affects blow-back, and wastage of fuel occurs if it is too short.

"The best power output at high r.p.m. is given by the present induction-tract length, which measures 6½in from the middle of the carburettor to the centre of the valve opening. If the race circuit is such that maximum r.p.m. are rarely attained, benefit would be gained from lengthening the tract, as that would give better filling at lower r.p.m., with some sacrifice at the very top.

"We consider that the best exhaust pipe length varies comparatively little with cylinder capacity for a given peak speed, but a faster engine usually requires a shorter pipe if full advantage is to be taken of the suction wave. Pipe diameter, on the other hand, is related to cylinder capacity—in the case of the 7R the diameter is 1⅜in.

"Originally the megaphone was merely a means of providing the shorter pipe necessary for maximum output while complying with racing regulations, but it was soon found that the length and cone angle of the megaphone had quite an effect on power output and the range of usable r.p.m.

"We started off with an exhaust pipe length of 31in and a 12½in megaphone with a cone angle of 18 degrees. For use with the modified head we have reduced the pipe length to between 29 and 30in (it is difficult to measure exactly on account of the complicated shape), and the megaphone is now 11½in long with the same angle as before. This layout gives a mild spell of megaphonitis at 5,000 to 5,500 r.p.m.

"You may be interested to know that bench tests revealed that the megaphone system gave an increase in power of between 2 and 2½ b.h.p. at peak r.p.m. over the output obtained with a straight pipe."

In the hope of providing material for comparison, I finished by inquiring whether Mr. Walker could give me power output and torque figures for the 1954 engines. However, it is not A.M.C. policy to divulge such figures; each owner is notified of the peak power and r.p.m. of his engine, to enable him to use it to the best advantage. Peak revolutions are in the region of 7,400, while the torque peak occurs at about 6,500 r.p.m.

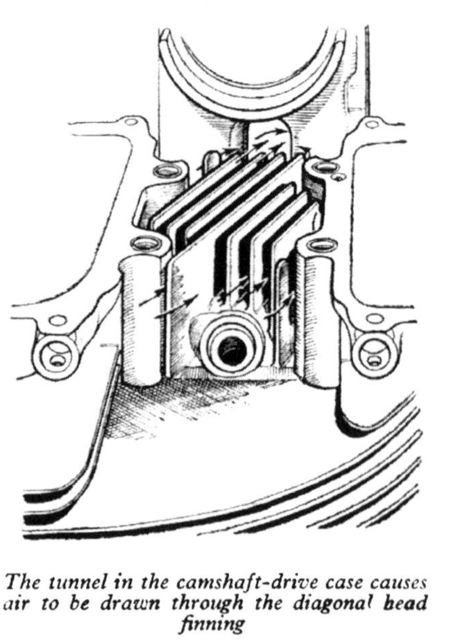

The tunnel in the camshaft-drive case causes air to be drawn through the diagonal head finning

Big Binders For AJS — Matchless

brakes head list of refinements on 1954 models

Above, mammoth front brake is major change in models of both AMC lines. Crankshafts of single cylinder models have been strengthened.

Matchless Super Clubman, a 500 cc. twin, is exact duplicate of AJS model 20. They are the only 3 main bearing verticals available here.

MANY REFINEMENTS aimed at increased power, simpler operation and greater safety have been incorporated in the 1954 AJS and Matchless models produced by Associated Motorcycles, Ltd. of London.

Most apparent at first inspection is the light alloy full-width front hub. The fully finned 7 inch diameter hub contains brake shoes of exceptional width which act on a shrunk-in cast iron drum. Increased stopping power can be expected. Straight spokes are used with the new hub and, since they are inherently stronger than the angled type, a slight reduction of gauge has been made possible. A slight overall reduction of weight has was realized through the use of this front wheel.

Engine modifications on the singles includes a beefed-up crank assembly, lightened flywheels (except on the competition engines), a rotating magnet magneto, new cams, a larger inlet port and bigger carburetor choke size. To obtain the full benefit from the higher lift cams, 1/16 inch has been added to the carburetor choke and the valve timing has been altered a little.

Only one modification to the vertical twin engines has been found desirable. The oil bores in the big end journals have been repositioned to eliminate any build-up of sludge in the oilways. This improvement eliminates the formerly recommended procedure of de-sludging the shaft at 20,000 mile intervals.

The 500 cc. singles feature an automatic ignition advance due to the inability of many owners to use a manual advance to greatest advantage. This change has necessitated a new magneto chain cover with an appropriate bulge to take the centrifugal advance mechanism. To the casual onlooker it will simplify the identification of the engine capacity as a non-bulge case is used on the 21 cubic inch models.

A newly designed fuel tank has been standardized for all 500 cc. singles and twins with the exception of the competition models. The new tank has stitch welded seams and holds 3¾ gallons.

Appearancewise, the dual, tear-drop shaped, parking lights flanking the headlamp will appeal to many. Along these same lines, the voltage regulator will no longer be visible. It has to be removed to a place of concealment beneath the saddle.

Provision is made to give access to the clutch without removing the chaincase outer half. A domed cover, which embraces the entire clutch is held on by eight screws, replaces the small inspection cover formerly employed. Clutches are now lined with a new friction material which is said to grip better when oily than dry, yet does not manifest that aversion to freeing which is evident with cork lined plates running in oil.

These and many other improvements in the line are seemingly minor but are the ones aimed at reducing those many bothersome little annoyances that occur in the life of a motorcycle. They are the things that really count.

Model 18/S AJS is 500 cc. single with automatic advance, rotating magnet, Lucas magneto. 21 inch models will retain manual spark control.

Competition single is one of the most popular models with the hare and hound riders; has fabulous record of wins in Southern California.

Matchless G3 LCS OHV. 350 cc. single cylinder competition model with 7.5:1 compression ratio . . . develops 24 hp. at 6000 rpm.

1955 Improvements

MAJOR improvements as an outcome of their racing experience and many detail modifications are to be found on the new A.J.S. and Matchless motorcycles, manufactured by the A.M.C. Ltd. (Associated Motor Cycles). There are nine overhead singles and a twin in each range. (All are push rod models, except the A.J.S. 7 R which is an overhead camshaft model). The A.J.S. singles are listed as 16 M (350 sports), 16 MC (350 competition) 7R (350), ohc, racing), 18 (500 sports), 18C (500, competition).

All models are available with swinging fork rear suspension as an optional extra which fact is indicated in the catalogue by the suffix S, excepting the 7R which has this feature in standard form. Competition models, whether equipped with rigid frames or sprung rear wheels, can be had in either scrambles or trials form.

The Matchless range comprises the G3L (350 sports), G 80 (500 cc), G3LC and G80C, (350 and 500 cc competition models for trials or scrambles respectively). For the competition enthusiasts the G 45 twin (500 cc, 10:1 compression ratio, 48 hp at 7200 rpm) and the 7R A.J.S. ohc, (350 cc 9.75 :1 compression ratio) will continue to be produced in limited quantities.

On the single cylinder engines a more substantial inner flange on the bronze bush has been provided for the timing side main bearing, and on the drive side the ball bearing adjacent to

Matchless G3 L, 350 cc. OHV. single with rigid frame . . . develops 18 hp. at 5750 rpm.

A.J.S. model 20—500 cc. twin with 7:1 compression ratio . . . develops 29 hp. at 6000 rpm.

For A.M.C.

the flywheel has been increased in diameter by ¼ in.

All models use the new Amal "Monoblock" carburetor. Automatic ignition advance control, previously only to be found on the 500 cc models has now been extended to the 350 cc models.

Racing experience has brought about a revised lubrication system for the ohv rockers, which is now as it was used on the G 45 twin competition machine last year. Oil is forced through the rocker spindles and out through passage to the tops of the rockers.

A.J.S. 18 S OHV single with a compression raito of 6.3:1 . . . develops 24 hp. at 5500 rpm.

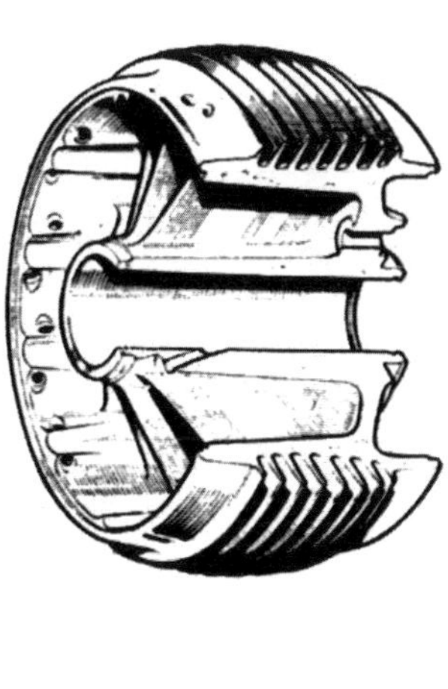

The new full width rear hub with cast in radial ribs for rigidity. Outwardly it resembles the front hub.

An alternative compression ratio (one ratio higher), is available for all except the competition models in both ranges.

Probably the most important im-

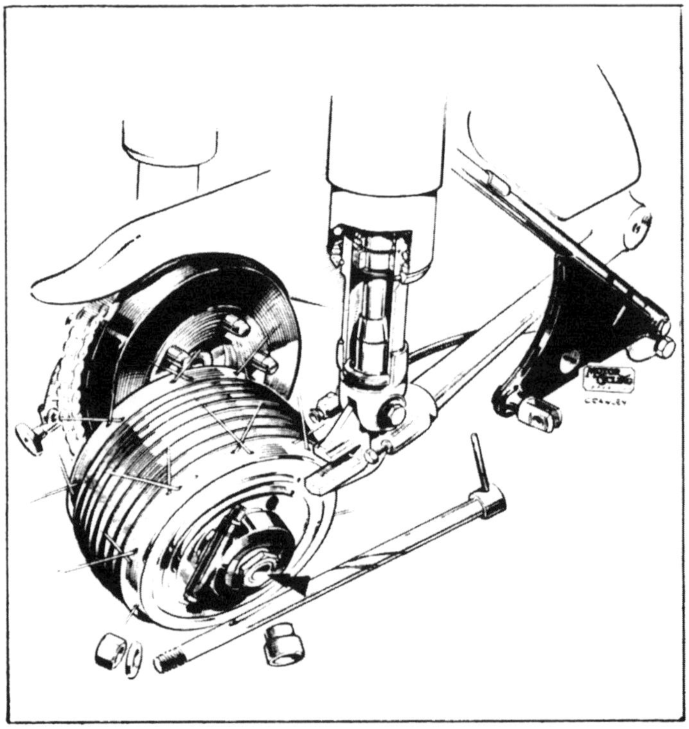

Detail of the quickly detachable wheel and improved chain adjusters. Note side car attachment bracket and lug. Rear hub has full width barrel contour.

provement on the rear sprung models is the completely new design of quickly demountable rear wheel spindle. It has been designed in such a way that the spindle can be pulled out and the wheel removed with the bearing in place without disturbing the drive and brake assembly. The rear wheel brake drum which is integral with the chain sprocket carries 5 shouldered pins which take the rear wheel driving torque. these pins are 7/16 inch on their largest diameter. The accsompanying detail drawing clearly shows the construction, and the new type fork end adjusting bolts and lock nuts, (instead of last year's cam type adjusters).

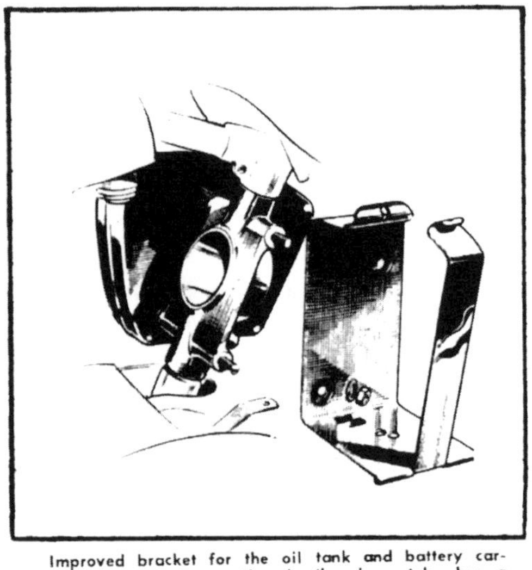

Improved bracket for the oil tank and battery carrier. The bracket casting in the down tube has a hole in the center for the aircleaner tube to pass through.

Rear suspension models are equipped with integral side car fitting lugs for left and right side cars. (See illustration)

Rear suspension units of A.M.C. design and construction are improved in detail, the most important being the internal hydraulic damping tube. Damping is now controlled through four calibrated orifices in the lower flanges which renders the unit more progressive in the limit ranges, which safeguards the suspension members against bottoming. Previously a transverse orifice was used. Also new is the cover retaining ring holding the cover tube in position b ya spring clip which fits into a groove in the lower outer tube.

In the front suspension department, the "Teledraulics" also have been improved. The top part is tapered and carries the bracket for the streamlined headlight and sidelamps. The central main tube has been increased in diameter from 1⅛ in. to 1¼ in. The steering crown, drilled for the speedometer drive cable to pass through, is a very sturdy forging which contains the steering limit stops. It connects the front legs with clamps fitted with Allen screws.

The method of holding the oil tank and battery carrier has been improved by interposing a malleable casting in the frame down tube as can be seen in the accompanying illustration. Optional is an air cleaner which fits in between the frame tube and the rear fender and is connected to the carburetor through the hole in the casting.

The photos clearly show the handsome new design muffler which gives better sound damping.

The front fender now overhangs the wheel freely without supporting stays, a central bridge piece giving adequate support.

The fins of the full width front drum are slightly decreasing in diameter from the center outward, giving the drum a barrel contour. The cast iron brake drum liner is an interference fit in the alloy in the interests of good thermal heat conductivity. The drum is finish bored after the wheel is spoked to the proper tension to nullify the effect of any possible stress deformation.

The rear drum, as the illustration shows, outwardly resembles the front drum, although brake lining area is much less. Radial ribs are cast in for strength.

An interesting new detail refinement is the manner in which the rear brake pedal is secured endwise. In place of the nut previously used, a set screw in the pedal boss fits in a groove in the pivot shaft. This, and the fact that the pedal return spring is now concealed in the central lug gives the assembly a decidedly cleaner appearance.

The new chain guard has been designed with an eye to better chain protection, the competition models also having a bottom guard channel.

Fuel tanks of the regular road models (both 350 and 500) have a capacity of approximately 5.4 gallons. Competition models have 2.9 gallon tanks.

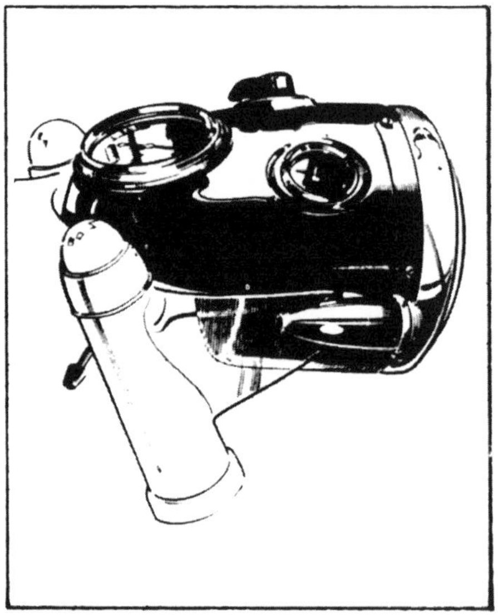

New streamlined head lamp with side lights, exclusively designed and manufactured for A.M.C. Ltd. It contains the speedometer, light switch and ammeter.

A neater attachment has also been provided for the exhaust pipe. In place of the holding strip previously used, an engine bolt has been extended to meet up with the lug on the exhaust pipe directly.

A striking feature is the head lamp assembly with streamlined side lights exclusively designed for A.M.C. Ltd. The headlight body contains the Smiths' speedometer, ammeter and light control switch.

With this impressive array of improvements future owners should be well satisfied with their 1955 A.M.C. products.

Presently AJS and Matchless motorcycles are being distributed in this country by Frank H. Cooper Motors, 5832 So. Broadway, Los Angeles, California.

"Sport Twin"

The Matchless "Sport Twin" proves to be an easy handler in the dirt.

By DON BROWN and HANK ELFRINK

Before giving our impressions of the actual road test of the interesting Matchless "Sport Twin", let's give this machine the technical once-over and see what makes it tick.

The Matchless G 9 "Super Clubman" model, is a vertical twin. The version we tested was a slightly modified G 9, especially prepared for export to the U.S. This model features a slightly larger bore (2 23/32 in.), which gives it a cubic capacity of 545 cc. The G 9 power unit is a development of a new twin introduced in 1949, and the design features a sturdy three-bearing lower end with light alloy connecting rods.

The crankshaft is supported in roller bearings on each side, a grooved plain white metal bearing being used for the center. The center bearing locates the crankshaft endwise. The roller bearings are the free-floating pure-radial type, which take only the radial load.

The main oil supply is directed to the crankshaft center bearing, and from here oil flows through passages in the crankshaft to the connecting rod bearings. The drilled crankshaft oil passages incorporate plugged siudge traps. Lubrication is by a conventional dry-sump system that uses two spur-gear pumps, each independently driven at half engine speed from the ends of the two camshafts. The scavenge pump has much wider gears which give it the extra capacity necessary for efficient scavenging, to keep the crankcase clear of excess oil.

Don Brown and John discuss some of the features of the Matchless Sports Twin.

Oil Circulation

Oil from the oil tank flows by gravity through a coarse-mesh metal filter via an external pipe to the inlet side of a delivery pump, then to an easily detachable felt filter in the crankcase. A pressure relief valve (set at approximately 140 lb./sq. in.) is incorporated as a safeguard when the oil is cold and

John McLaughlin operates the front stoppers in the brake test.

thick. When the engine has warmed up, the oil pressure at idling speed is between 20 and 40 lbs. and rises somewhat when engine speed goes up.

Oil to the overhead rocker assembly is supplied through a metered by-pass, after which it drains into the camshaft tunnels, and surplus lubricant overflows into the timing gear case. For 1955, the lubrication of the overhead gear has been improved. A pre-determined oil level builds up in the timing case and the camshaft tunnels for adequate lubrication of gears, cams and cam levers. Surplus lubricant overflows into the crankcase and sump where the sca-

The rear section of the "Sport Twin" clearly shows the generously finned brake hub, rear shock position.

venge pump returns the oil via another external pipe and through a large felt filter in the oil tank.

To prevent oil from seeping into the crankcase when the engine is stationary, a spring-loaded, non-return valve is used. This valve incorporates a union for attachment of a pressure gauge to check the oil pressure.

Wire-Wound Pistons

An interesting feature of the Matchless power unit is its so-called wire-wound pistons. This type of piston was introduced after exhaustive research and tests in the latter part of 1948.

Five turns of high-tensile steel wire fit in grooves of the piston circumference to control expansion and thus keep distortion within bounds. This result has been achieved to such a degree that the pistons are fitted into the bores with the very small skirt clearance of .001" - .0015", which insures silent and efficient operation. After the wires are installed, they are ground to the finished diameter of the piston skirt.

The piston is cam-ground and slightly tapered. There are two compression rings and one taper ring. The top compression ring is chrome-plated and has a slight outside taper.

Impressions

There is no doubt that the Matchless "Sport Twin" is quite easy to handle. In some machines of comparable size and weight, one often has the feeling that in order to obtain a more suitable center of gravity for off-the-road running, some modification would have to be made. However, with the Matchless this was certainly not the case. On several occasions the machine was taken over some rather acute jumps and rough territory, and handling was always impeccable.

While the power plant did not deliver an over-abundance of power in the low range, it did give an ample amount when the speed was increased. Once the twin had progressed to around 20 mph., there was a definite surge of power. While in the dirt, the gearing used was not ideal and could have been somewhat lower, but the amount of power did prove to be ideal and adequate for any hill or obstacle.

The particular territory where the Matchless was tested was typical of the terrain used in many of the sporting events in Southern California. With us on the test run was the well-known motorcycle and sports car competitor, John McLaughlin, past winner of the National Championship Cross-Country Race, the Greenhorne Enduro Championship, the Catalina Grand Prix plus numerous other sporting events. This ex-air force pilot certainly agreed as to the conditions of the terrain: rugged. Hills, brush, rocks and mud were all encountered at one time or another during the testing period. The "Sports Twin" is thoroughly equipped with sports-type tires, which proved to be quite a help when the going got rough.

The only difficulty discovered during the test was a slight oil leak. The leak occurred on the return line to the oil tank. This leak was caused by an im-

Don Brown takes the Matchless around pavement corner.

proper or defective rubber fitting. While these fittings do not usually cause any trouble, the fact that they are not provided with clamps at the crucial spots could be the subject of some discussion.

All of the controls of the Matchless worked as would be expected. The Burman gearbox gave clean and positive gear changing, and it was always possible to engage first gear without any gear clash caused by clutch drag. The clutch itself was easy to operate, and little pressure was needed for disengagement. The clutch did have to be adjusted after repeated acceleration tests, but no slippage was noticed during the entire proceedings.

During the speed testing period, it became apparent that with a few more easy road miles on the machine, it would undoubtedly have attained a higher top speed. However, 93 mph is not dawdling along for a motorcycle of its size and engine capacity. We say this with the thought that the machine never actually attained its top rpm. during the speed trials, mainly due to engine tightness.

In the acceleration department the machine proved to be above average. On pavement, and considering the type cam employed, the Matchless got off the mark fast, but it was in the dirt that the machine really proved its worth. The Matchless steamed through the standing quarter mile in just 14.9 seconds. The standing tenth was covered in 9.75 seconds.

The braking power of the "Sports

John McLaughlin throws the Matchless into a severe slide in the dirt. Weight distribution seemed ideal and no tendency to "pogo" was noticed. While the suspension seemed to be ideally suited to the rough stuff, on high speed pavement corners some mushiness was apparent.

Front wheel shot illustrates the handsome design of the front section. Note the full width hub.

Twin" proved to be more than ample both in the dirt and on paved surfaces. It could be noted at this point that the tires fitted to our Twin certainly were not expressly meant for stopping on paved surfaces. In spite of this, the Matchless managed to turn in a commendable job of bringing man and machine to a screeching halt.

Using the front brake only, John McLaughlin was able to stop the Twin in just 18 feet and, with the rear brake only, the Matchless slowed to a stop in 33 feet. When both brakes were applied, the Twin came to a dead stop after covering only 13 feet. All of the afore mentioned braking distances were figured from 25 mph. to a full stop.

Rear Suspension

Top marks must be given to the Matchless Teledraulic front and rear suspension. The rear suspension rate is somewhat softer than on most makes, and it really does an excellent job of absorbing small road irregularities, while the hydraulic damping gives the desired progressiveness on bad bumps.

The rear wheel is mounted in a fork hinged just behind the gearbox. The fork hinge has plain bearings which are lubricated from a reservoir containing 42.6 cc. of heavy gear oil, which should last almost indefinitely. Provision for topping up, however, is made by removing the screw in the right-end cap of the bearing. Suspension is by springs in the "Teledraulic" legs, and spring movements are damped by dampers identical in design to those used in the front fork assembly. Each leg is filled with 85 cc. of hydraulic fluid of S.A.E. 20.

Rear Wheel

The rear wheel is of the quickly detachable type which, for 1955, is completely new. The pull-out spindle system permits the wheel to be removed with the hollow axle and bearings, leaving the rear drive and brake assembly completely undisturbed. Rear wheel driving torque is taken by five shouldered studs.

Inasmuch as the model we tested was provided with the new Amal Monobloc carburetor—described in this issue, we had a chance to find out just how this new addition would react under extreme conditions. At no time during the test period did this component let us down. The Monobloc was an easy starter, and it took full throttle from low speeds without the slightest sign of sputtering once the machine was thoroughly warmed up. On one occasion, when the machine was taken down a long and very steep hill, we thought that perhaps this would be the carburetor's downfall. Carburetion, however, remained faultless, and at no time was there any tendency to "load up".

The appearance of the Matchless "Sport Twin" leaves little to be desired. The dual exhaust pipes flow along the lower part of the frame and funnel into megaphone mufflers of handsome design. When the Matchless is fired up for the first time, one immediately notices the even, subdued exhaust note, which is carried throughout the entire speed range. The beauty of the machine is further enhanced by the special light alloy fenders fitted to this particular export series.

Summing up, we believe the Matchless G 9 twin is a potent and versatile machine of thoroughly proven design and due to its excellent suspension system, is equally at home on or off the beaten track.—*Hank Elfrink—Don Brown*

Here John McLaughlin flies over a series of bumps while testing the Matchless suspension. During a series of such tests, the machine was ridden down the middle of the railroad tracks which can be seen in the background. Tires with which the Matchless was equipped proved to be ideal for the dirt but not so suited for pavement riding.

Specifications

Engine: Vertical twin cylinder, ohv. 2.72 x 2.87 in. Cubic capacity 545 cc. (33.4 cu. in.). Push-rod operated overhead valves by two gear-driven camshafts. Light alloy connecting rods with Vandervell plain bearings. Cast alloy iron crankshaft supported in three berings (two journal roller bearings and a center Vandervell plain bearing). Compression ratio 8:1. Maximum power: 33 hp at 6800 rpm. Dry sump lubrication with two spur gear pumps. Amal Monobloc carburetor.
Electrical Equipment: Lucas magneto ignition (type K2F). Cut-out button on magneto contact breaker cover. Spark plug K.L.G. (FE 80) Corundite ¾ in. reach. 6 Volt 45 watt Lucas generator with automatic voltage control. 7 in. headlamp with twin pilot lights.
Transmission: Primary chain enclosed and oil-bath lubricated. Rear chain protected by deep section guard. Cam-type crankshaft shock absorber. Burman four-speed gearbox with foot control. Ratios: First 13.91:1, second: 8.91:1, third: 6.86:1, top: 5.25:1. (with standard 20 teeth engine sprocket). Multiplate clutch.
Frame: Duplex cradle frame, Teledraulic front and rear forks. Spring up central and side stands.
Brakes: 7 in. brakes front and rear. Full width front light alloy hub and straight spokes.
Tires: 3.00 x 21 front; 4.00 x 19 rear.
Fuel tank capacity: 4.5 U.S. gallons.
Wheelbase 55¼ in.
Dual seat height: 31½ in.
Weight as tested: 410 pounds.
Manufacturers: Associated Motorcycles Limited, Plumstead Road, London, England.

Performance Summary

MAXIMUM SPEEDS

1st gear	45 mph
2nd gear	56 mph
3rd gear	81 mph
4th gear	93 mph

ACCELERATION

¼ mile drag	14.9 sec.
1/10 mile drag	9.75 sec.

BRAKING DISTANCES

from indicated 25 mph to full stop—

Front brake only	18′
Rear brake only	33′
Both brakes	13′

GASOLINE MILEAGE

overall average	56 mpg.

SLOW RUNNING

High (4th) gear without chain snatch	15 mph

The 349 c.c. Three-valve A.J.S.

Exclusive Full Description of a Single-cylinder Power Unit with a Three-valve Head, and Triple Overhead Camshafts Operated by an Elaborate System of Bevels and Pinions

By VIC WILLOUGHBY

THE 349 c.c. triple-overhead-camshaft engine used by Rod Coleman to win the 1954 International Junior T.T. was, at that time, the latest of a 27-year lineage of A.J.S. racing engines employing chain-driven overhead camshafts. The three-valve engine which forms the subject of this article and on which the o.h.c. drive is by means of bevels and shaft instead of by sprockets and chain, is a more recent variant of Coleman's T.T.-winning power unit. Model A and Model B are the designations given respectively to the chain- and the bevel-driven o.h.c. versions of the three-valve engine.

Model B has not yet been raced. Its construction was not completed until July last year and little or no development work has been undertaken; and, in view of the subsequently changed outlook in British racing policy, the engine may never be seen in action.

It was in 1927 that the first A.J.S. chain-driven-o.h.c. racing engine made its début in the Isle of Man, and it was not long before victories in classic events focused attention on the design. The success of the concept in national and international competition led to the marketing of sports and racing machines powered by similar engines. Pre-war culmination of this policy was the 39/7R road-racing machine available to the public in 1939.

When the extremely popular post-war 348 c.c. 7R production racer was introduced in 1948, it was partly tradition and partly the simplicity and mechanical efficiency of chain drive which dictated its retention as the means of operating the single overhead camshaft.

In 1952 the late H. J. Hatch was responsible for the ingenious three-valve layout employing a separate camshaft for each valve. Most important of the basic aims of the design was to achieve lower exhaust-valve temperatures and thus increase volumetric efficiency by reducing heat transfer to the incoming charge. Much of the heat picked up by an exhaust valve is dissipated through its seat when the valve is closed. Because the ratio

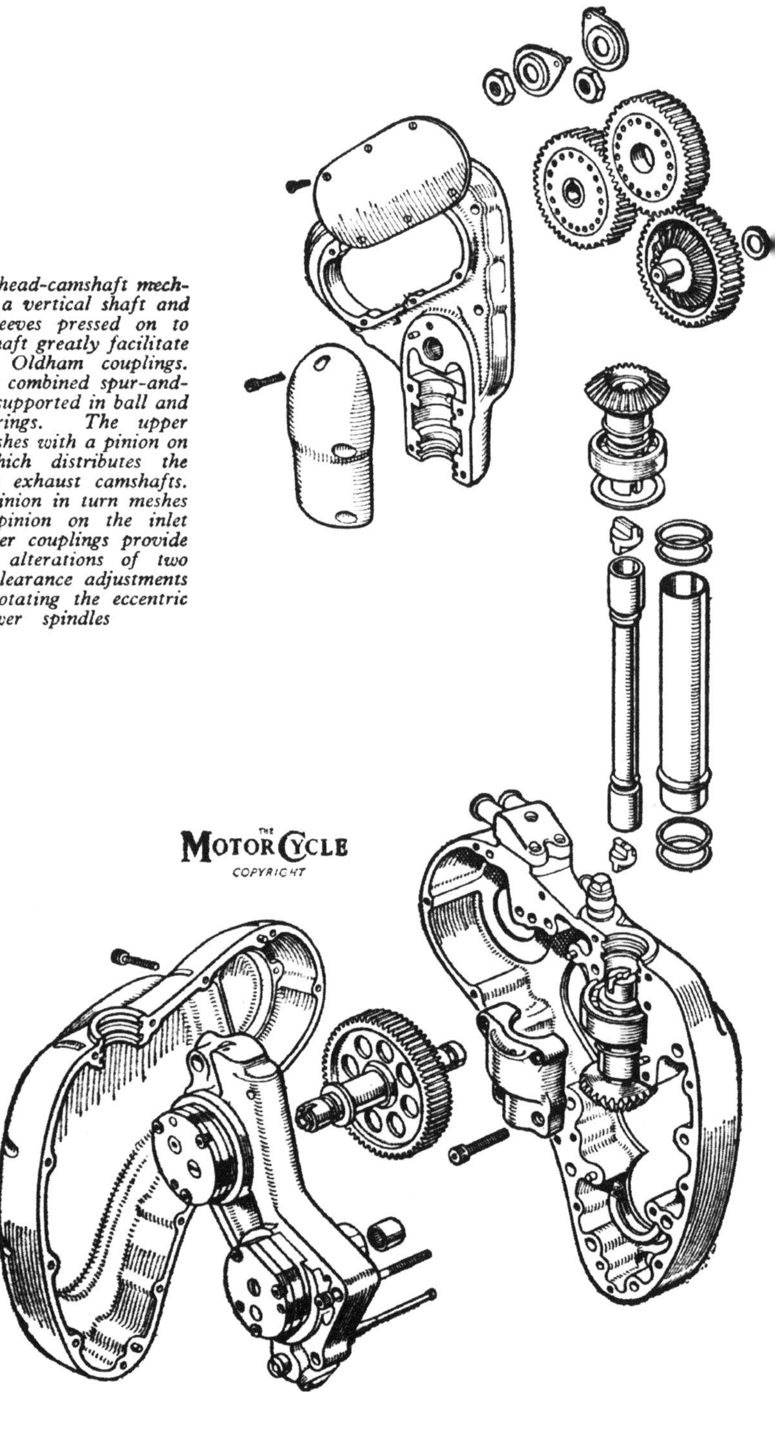

Drive to the overhead-camshaft mechanism is through a vertical shaft and gears. Short sleeves pressed on to the ends of the shaft greatly facilitate assembly of the Oldham couplings. Upper and lower combined spur-and-bevel pinions are supported in ball and needle-roller bearings. The upper combined gear meshes with a pinion on a cross-shaft which distributes the drive to the two exhaust camshafts. The cross-shaft pinion in turn meshes with a similar pinion on the inlet camshaft. Vernier couplings provide for valve-timing alterations of two degrees. Valve-clearance adjustments are effected by rotating the eccentric cam-follower spindles

RACING ENGINES

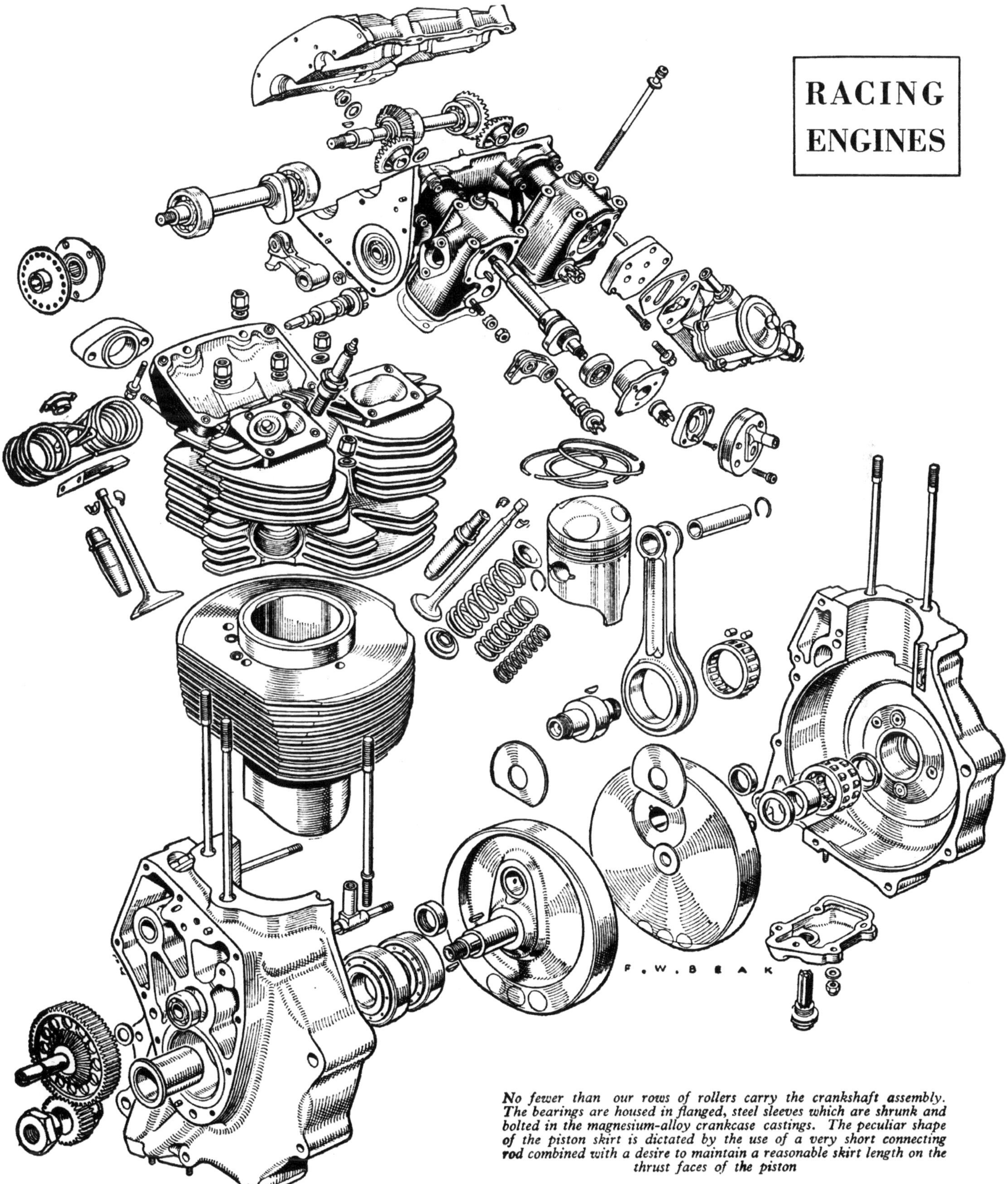

No fewer than our rows of rollers carry the crankshaft assembly. The bearings are housed in flanged, steel sleeves which are shrunk and bolted in the magnesium-alloy crankcase castings. The peculiar shape of the piston skirt is dictated by the use of a very short connecting rod combined with a desire to maintain a reasonable skirt length on the thrust faces of the piston

of seat area to valve-head surface is greater with a small valve than with a large one, two small valves can be made to operate at a lower temperature than one large valve while providing equivalent valve-opening area.

The bifurcation of the exhaust tract permits a good air flow round the ports with a consequent improvement in cylinder-head cooling. Other advantages of the three-valve layout are that the sparking plug can be located nearer to the middle of the cylinder head, thus shortening flame travel, and that valve-inertia forces are reduced by virtue of the lower weight of the individual valves.

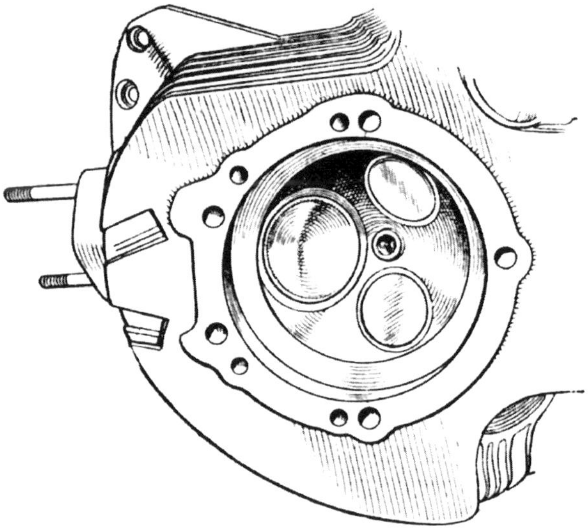

Primary reason for the use of two exhaust valves is to lower running temperature by increasing the ratio of valve-seat area to valve-head surface. Reduced heat transfer to the incoming charge results in higher volumetric efficiency. Position of the sparking plug approximates to the middle of the cylinder head

It will be appreciated that in a part-spherical cylinder head for a given bore dimension, the use of two small exhaust valves, suitably disposed, in place of one large valve enables a larger inlet valve to be accommodated. Exploitation of the potential advantages of a larger inlet valve was, however, left to subsequent development work. In the three-valve design under review, inlet-valve throat diameter remains at 1$\frac{1}{4}$in, as in the factory two-valve engines. Peak power output of the three-valve engine is higher than that of its predecessor though, initially, the low-speed torque of the two-valve unit was better. Compression ratio of both engines is approximately 10 to 1.

But on the three-valve engine, as on the single-o.h.c., two-valve unit, the necessity for maintaining fixed sprocket centres in the camshaft drive hampered development work. Even a minor alteration in compression ratio involved the making of a new piston. In 1954 Mr. Hatch removed this restriction from the layout by converting the camshaft drive to bevel gears and shaft. During experimental work, any required alteration to the effective length of the shaft can easily be made by substituting Oldham couplings of different thicknesses. An incidental advantage of the new camshaft drive is improved cylinder-barrel cooling. The shaft-enclosure tube is slimmer than the chaincase and thus offers less obstruction to air flow and necessitates less cutting away of the cylinder fins.

Throughout the design of the Model B three-valve engine there is evident an appreciation of the importance of rigidity and of minimizing frictional losses. Adequate shaft support and the extensive use of ball and roller bearings are significant characteristics of the engine.

The crankcase and crankshaft assembly bear a marked resemblance to those of the current 7R production racer. Magnesium alloy is the material used for the crankcase, which is chromated as a protection against corrosion. The same material and the same treatment are also employed for the castings which form the timing case and the housings for the bevels and camshafts. The two halves of the crankcase are webbed for stiffness and are located by a spigot. A small oil sump is bolted to the bottom of the case at the rear. Oil is directed into the sump by knife-edge scrapers which are bolted to the inside of the case and lie close to the periphery of the flywheels.

Because of the tendency for heavily loaded bearings to move when housed directly in magnesium-alloy castings which are subject to varying temperatures, the main bearings are shrunk in flanged steel housings which are themselves shrunk and bolted into the crankcase halves. The drive-side bearing comprises two rows of $\frac{5}{16} \times \frac{5}{16}$in rollers located in a common cage and running on a hardened-steel sleeve pressed on to the mainshaft.

The timing-side mainshaft is supported in two separate journal roller bearings carried on a sleeve pressed on to the shaft. Since the outer end of the sleeve is flanged and the two bearings are of lipped variety, lateral location of the flywheel assembly is effected by the timing-side bearing assembly.

Forged in a high-grade carbon steel (KE805), the flywheels are machined all over and have separate, flanged mainshafts (of nickel-chrome molybdenum steel) pressed in from the inside and keyed. A spring-loaded, diaphragm-type crankcase breather is screwed into the outer end of the hollow drive-side shaft. Diameter of both shafts is 1in. Fywheel diameter is 7$\frac{1}{8}$in, thickness $\frac{15}{16}$in. Balancing is effected by drilling three $\frac{15}{16}$in-diameter holes near the rim of each wheel on the side remote from the crankpin, and filling the holes with a special heavy alloy—a G.E.C. product.

The crankpin consists of a hardened sleeve of 1$\frac{3}{4}$in diameter and $\frac{13}{16}$in width, pressed on a tough, 1in-diameter pin which is a parallel interference fit in the flywheels. Hexagon heads of the crankpin retaining nuts are machined off after assembly. Sixteen $\frac{3}{16} \times \frac{9}{16}$in big-end rollers are carried in a light-alloy cage. As in the case of the drive-side main bearing, the bars of the cage are not relieved where they rub on the roller track. The large contact area which results reduces wear of the crankpin sleeve by the cage to negligible proportions.

At 5$\frac{3}{4}$in between centres, connecting-rod length is extremely short, with the object of achieving the most effective leverage as early as possible during the power stroke. The I-section rod, which is highly polished all over, is forged in KE805 steel and has two strengthening webs around both big-end and small-end eyes. The big-end bearing sleeve and bronze small-end bush are pressed in place.

Because of the shortness of the connecting rod, a compromise has been struck to achieve the necessary clearance between piston and flywheels at bottom dead centre without sacrificing too much piston-skirt area. In the first place the flywheel radius is reduced on the side opposite the big end and the wheels are heavily chamfered on their inner edges at that position. Additionally, the sides of the piston skirt are cut away to clear while the middle of the skirt at front and rear is left full length and extends well down between the wheels at the bottom of the stroke.

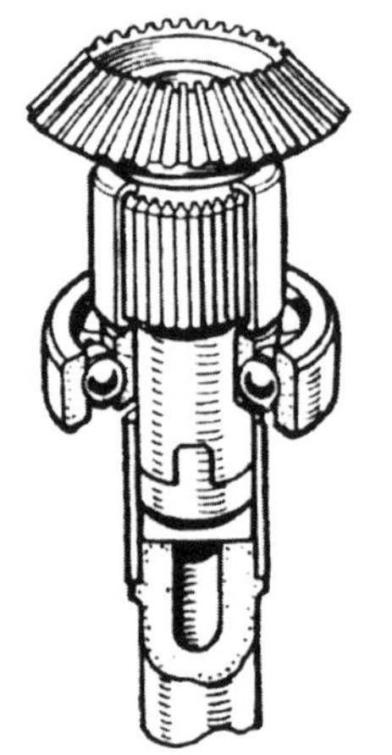

Axial loading in each vertical bevel gear is accommodated by a ball bearing. A relatively long, needle-roller bearing close to the head of the gear resists side thrust

An interesting feature of the piston is that its two compression rings are of Dykes type, that is, they resemble a shallow L in section. The rings remain free from flutter up to higher piston speeds than is the case with rectangular-section rings. A slotted oil-control ring is fitted. Gudgeon-pin diameter is $\frac{7}{8}$in, and the moderately domed piston crown has three cutaways for valve clearance.

Close-pitch finning characterizes the

The 349 c.c. Three-valve A.J.S.

Wellworthy Al-Fin cylinder barrel; the average distance between the fins is ¼in. Approximately half of the barrel is recessed into the crankcase. Indeed, at bottom dead centre the piston crown is level with the mouth of the case. Bore and stroke dimensions are 75.5 and 78mm respectively, giving a swept volume of 349 c.c.

Five long, waisted studs secure cylinder and head to the crankcase. Four of the studs are threaded directly in the case. The fifth, lying in the plane of the crank-case joint, is screwed into an inverted T-shape trunnion located in a tunnel in both halves of the case.

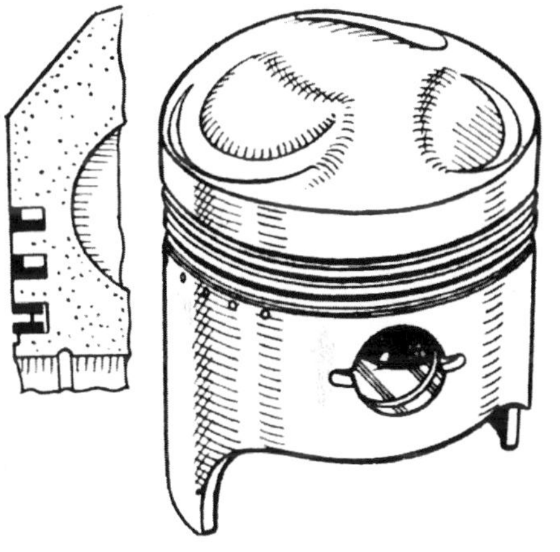

Left: Pumping losses in the crankcase are minimized by a spring-loaded diaphragm-type breather screwed into the end of the drive-side main-shaft. Right: Freedom from piston-ring flutter at high engine speeds is achieved by the use of Dykes-type compression rings which are L-shape in section. A conventional slotted scraper is employed for oil control. Sides of the piston skirt are cut away to clear the fly-wheels at bottom dead centre. Wire circlips retain the taper-bored gudgeon pin

The deeply finned cylinder head is cast in a high-strength aluminium alloy to specification D.T.D. 238 or similar. Of 1 3/16in diameter, the inlet port is inclined downward at 16 degrees. All valves are Stellite tipped, have chromium-plated stems and operate in phosphor bronze guides. In an annulus near the upper end of the bore of the inlet guide is a synthetic-rubber seal which prevents oil from creeping down the guide and thus finding its way into the combustion chamber. The exhaust valves are sodium cooled and have flat heads. Austenitic-iron inlet and aluminium-bronze exhaust valve-seat inserts are shrunk into the cylinder head casting.

Overlapping hairpin springs return the inlet valve to its seat, but the exhaust valves are so light that triple coil springs are adequate in their case. Respective poundages with the valves seated are: inlet, 90; exhaust, 70.

Drive to the overhead camshafts is straightforward. Meshing with a spur pinion keyed to the timing-side mainshaft is a combined spur-and-bevel pinion which runs at half crankshaft speed. The double pinion is supported in ball and needle-roller bearings and drives the lower vertical bevel gear. This gear, too, is very rigidly supported in similar-type bearings of larger dimensions.

Connection between the lower bevel gear and a similarly supported upper bevel gear is by means of a hollow vertical shaft and two Oldham couplings. A tubular shroud pressed on to each end of the shaft surrounds its coupling closely. A light-gauge steel tube, recessed in the timing case and the upper bevel case, encloses the shaft. Oil sealing of the tube is effected by two synthetic-rubber rings at each end trapped in annular grooves in the respective castings.

The upper bevel gear drives another combined bevel-and-spur pinion running in ball and needle-roller bearings. This pinion meshes with a plain spur pinion on the end of a cross-shaft which distributes the drive to the two exhaust camshafts. Engaging with the spur pinion on the end of the cross-shaft is a similar pinion on the inlet camshaft; both pinions are attached to their shafts by vernier couplings, by means of which valve-timing adjustments of two degrees may be effected. The inlet camshaft is carried in two ball bearings—one close behind the driving pinion, the other in a similar juxtaposition relative to the cam.

Two bevel gears are fixed to the cross-shaft. There is a ball bearing alongside each bevel and a needle-roller bearing close behind the driving pinion. Bevel gears on the rear ends of the exhaust camshafts take their drive from the cross-shaft bevels. Each exhaust camshaft is housed in a separate casting and carried in two ball bearings.

All three cams operate their valves through highly polished followers which are pivoted on plain bushes. Rollers of ⅝in diameter, supported on hardened-steel pins, are incorporated in the followers to transmit cam thrust. Valve lifts are not fantastic: that for the inlet valve is less than ½in; for the exhaust valves the figure is approximately 5/16in.

The pump which lifts fuel from the low-mounted tank to the carburettor is driven by the front end of the left-hand exhaust camshaft. Rev-meter drive is by the right-hand shaft.

Reverting to the timing chest, the lower combined gear drives an intermediate pinion which, in turn, meshes with the magneto driving gear. The gear-type oil-delivery pump is driven by the outer end of the live spindle of the combined gear; the scavenge pump is similarly driven by the intermediate-gear spindle. Mineral oil of S.A.E. 50 viscosity is circulated at a rate of 30 gallons an hour when the engine is running at its peak-power speed of 8,000 r.p.m. Oil consumption is negligible.

With the exception of an external pipe taking oil from the timing case to the housing for the inlet camshaft, both delivery and scavenge circuits are completed by means of internal drillings. A quill protruding into the timing-side mainshaft feeds the big-end bearing in the orthodox manner.

The feed to the inlet-camshaft housing is ducted first to the cam-follower pivot bearing, from which oil escapes to lubricate the inlet-valve gear. Drillways from the cam-follower pivot bearing conduct oil to wells surrounding the cross-shaft bevels, whence further ducts lead to the exhaust cam-follower pivot bearings. Overflow from these bearings lubricates the exhaust-valve gear.

Drillways in the cylinder head and barrel drain oil from the three valve wells to the timing gear and the crankcase sump, whence the scavenge pump returns the lubricant to the tank.

Though the Model B three-valve engine has not been raced, it is thought that its performance, at the present stage of the engine's development, would be indistinguishable from that of the T.T.-winning Model A. Coleman's machine, on Isle-of-Man gearing of 5.09 to 1, had a level-road maximum speed of about 120 m.p.h. Maximum power, as stated earlier, is produced at 8,000 r.p.m. The normally used engine-speed range is from 6,000 to 8,200 r.p.m.; the latter speed is registered when the rider is accelerating through the gears.

Maximum torque (i.e., most efficient cylinder filling) is achieved at about 6,500 r.p.m. Thorough pre-race warming up of the engine is vitally important because of the large area of heat-dissipating fins. Fuel consumption of the Model A during Coleman's victorious T.T. ride proved to be almost exactly one gallon per 37¾-mile lap.

BIGGER PARALLEL TWINS

Impressive 592 c.c. Models Introduced by A.J.S. and Matchless : New 347 c.c. Trials Machine : Extensive Cleaning-up and Numerous Practical Modifications

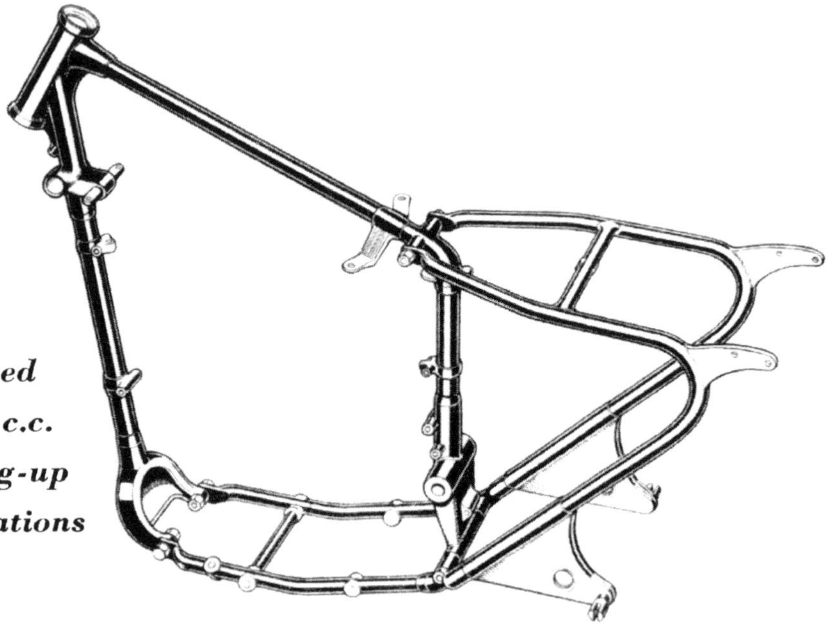

Above: All 1956 roadsters feature a new frame with vertical seat tube

The imposing 498 c.c. Model 20 A.J.S.

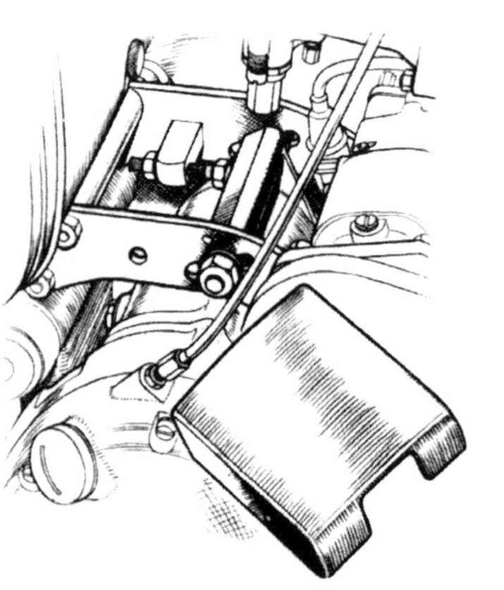

Removal of a snap-on cover reveals the accessible primary-chain adjuster

TWO 592 c.c. parallel twins—which are identical with the 498 c.c. models except for tank finish and engine capacity—are announced by A.J.S. and Matchless. The reason for the introduction of the new models will be obvious to all enthusiasts: the demands for increased power from important oversea markets, particularly the U.S.A., and from sidecar drivers at home who require greater torque than is available from 500 c.c. For the solo rider in Britain, too, whether he requires extra speed or not, there are advantages accruing from the use of increased capacity. Improved torque means better low-speed pulling; and reduced engine wear and tear is likely to be achieved since the power unit, for most of its life, is operating comfortably within its limits.

That no solid-frame machines are to be produced by the Plumstead concern in the coming year is hardly surprising in view of current trends. Rear springing is gaining popularity in trials just as quickly as it is for sidecar work. And, of course, production and servicing efficiency can be improved as the number of different models in a given range is decreased.

A.J.S. and Matchless machines are, of course, basically similar. Frames of the roadster and trials models have been redesigned; the roadsters have matching, pannier-type oil tanks and tool boxes. Brake operation is altered and there are other modifications to improve appearance or enhance practicality.

The 347 and 498 c.c. singles and the 498 c.c. twins are famous the world over for their mechanical quietness and excellent finish. Only the 347 c.c. power unit is employed for the new trials models which, not surprisingly, feature pivoted-fork rear springing and are similar to the machines used by members of the A.J.S. and Matchless factory teams in 1955 sporting events.

As already stated, the new six-hundreds closely resemble the smaller twins. The bore is 72 mm, as against the 66 mm of the five-hundreds; stroke in each case is 72.8 mm. These twin-cylinder engines are notable for their robust construction. Primarily responsible for their ruggedness is the three-bearing, cast-iron crankshaft; the middle main bearing is of the plain pattern and serves to distribute oil, from the gear-type pump, equally to the plain big ends.

Separate deeply spigoted cast-iron cylinder barrels are employed and the cylinder heads are in light alloy. Rocker spindles for the overhead valves are carried in pedestals integral with the heads and have eccentric ends for valve-clearance adjustment. The twin camshafts are gear driven and a separate dynamo and manually controlled magneto are fitted.

Cylinder dimensions of the 347 and 498 c.c. singles are respectively 69×93 mm and 82.5×93 mm. Each engine incorporates overhead valve gear and light-alloy cylinder heads. Ignition of the roadsters is by rotating-magnet magneto with automatic advance and retard; the trials three-fifty has the usual Lucas wading magneto.

Engine modifications on the 498 c.c. twins (and incorporated also, of course, on the new 592 c.c. models) are wider scraper rings and the replacement of the oil-tank fabric filter by a crankcase magnetic filter of the kind already used in the racing engines. Because of the general availability of high-octane fuels, standard compression ratio of the smaller engines has been raised from 7.0 to 7.8 to 1; the ratio on the six-hundreds is 7.5 to 1.

The roadster singles, too, have higher compression ratios: that of the 347 c.c. engines goes up from 6.5 to 7.5 to 1, while the five-hundreds now run on 7.3 to 1 in place of 6.3 to 1. On the trials unit the 6.5 to 1 ratio is retained. Low-compression pistons can be supplied for roadster models to special order.

To obviate any possibility of the exhaust guide on the singles moving when hot, circlip location was introduced during the current season and is specified again for 1956. These engines feature the same design of magnetic oil filter as do the twin-cylinder units.

Introduced last year for the twins, the improved forward exhaust-pipe stay has been adopted for the singles; it comprises a long pillar nut on the forward cradle bolt, in place of the double-cranked, strip-steel bracket which was employed previously. A ring of internal ventilation holes is provided near the outlet of each of the Matchless twin's silencers to overcome any possible tendency to corrosion.

On both roadster and trials frames, the seat member, instead of running diagonally from under the seat to the engine-and-gear-box plates, lies vertically behind the gear box. Lateral rigidity is improved in comparison with earlier rear-sprung frames by housing the rear-fork pivot in a sturdy, malleable-iron lug which is clamped between the cradle tubes and brazed to the lower end of the seat tube (the pivot spindle was previously carried in a light-alloy casting linking the seat tube with the rear of the cradle).

Conventional brazed- and bolted-up construction is employed for the roadster frame; by minor modifications to the rear sub-frame, the horizontal line of the fuel-tank base is continued through to the rear of the machine, thus enhancing the general appearance.

The main portion of the trials frame is somewhat similar to that introduced for the 1954 solid-frame competition models, which was of welded construction with a single tube forming the front-down and top members, and a wrap-round gusset plate at the steering head. However, the main tube now extends downward under the seat, to link up with the pivot lug.

Although the pivot lugs of the roadster and trials frames bear a resemblance to one another there is an important difference: the pivot axis of the roadster frame is offset ½in to the rear of the seat-tube centre line whereas that of the competition frame is offset ahead by the same amount, thus providing a 1in difference in wheelbase.

Ingenious strengthening has been applied to the fabricated sidecar-attachment brackets on each side of the rear sub-frame of the roadsters. The brackets, which also support the silencers and pillion footrests, were previously fillet-welded to the sub-frame tubes and instances of distortion had been experienced. A one-piece pressing, each of the new brackets is wrapped round the tube and brazed to it, and the two meeting edges are subsequently welded together.

A much sleeker aspect has been given to the roadster models by the integration of the oil tank and tool box. Increased in capacity from four to five and a half pints, the oil tank is of pannier type and fits neatly into the sub-frame loop. Balancing it on the left is a tool box of similar shape, and the two are bridged at the front by a detachable cover. Should an air filter (available at extra charge) be

The toolbox and oil tank on the roadsters are blended together by a detachable fairing

Stronger sidecar lugs are fitted and the rear brake has a modified pedal and adjuster

A rear-sprung trials model —the 347 c.c. A.J.S. 16MC

Below: View showing the inbuilt speedometer and the neat and practical Lucas dipper-and-horn-push unit

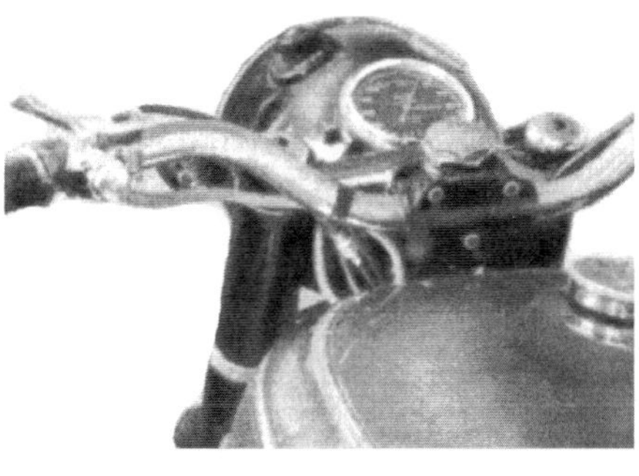

The new 592 c.c. G11 Matchless, one of the two larger-capacity twins, appears trim and powerful

fitted, it is concealed behind the cover. Because of the relatively high oil temperatures attained with the twins, a shield is mounted to the outside face of the oil tank and is separated from it by a ¼in air gap.

Having a lid hinged at its lower edge, the tool box houses the battery in its forward end. The battery is secured, continental-fashion, by a quick-release rubber strap and is separated from the tool compartment by a bulkhead. At the upper-rear corner of the tool box is housed the automatic voltage-control unit, which sits in what might be described as an anti-vibration, sponge-rubber nest. There is rubber behind, above, below and in front of the unit. In fact, the system is similar to that adopted for the A.J.S. and Matchless International Six Days' Trial models. The horn is concealed under the twin-seat, the length of which has been increased by 2½in to provide additional accommodation for adults of above-average stature.

A detail improvement on the electrical side is the introduction of a Lucas combined horn button and dip switch, mounted close to the left grip. The horn push is of large diameter and both it and the dipping lever are within easy reach. The cables of the wiring harness are plastic covered.

Although the primary-chain adjuster used previously was accessible, it was felt to be rather unsightly and hence it has been moved to an equally accessible location between the gear-box plates where it is concealed by a snap-on cover.

The full-width hubs, with their straight spokes, and the internals of the brakes remain unchanged but there are differences in the brake-operating mechanism. In order to achieve a cleaner appearance, the front brake has, in effect, been rotated through 180 deg, so that the cam lever lies above the wheel spindle and behind the fork leg. At the rear, the brake adjuster has sensibly been transferred to the forward end of the operating rod so that it is accessible from the saddle. All A.J.S. and Matchless roadster models feature a quickly detachable rear wheel, removal of which leaves the brake undisturbed.

The left of the machines is brightened by a polished-aluminium rear-brake shoe plate which replaces the previous black plate. A modified rear-brake pedal is fitted: it has a straight shank and the shorter arm is inclined at approximately the same angle as the sub-frame tube. The cotter-type retention of the pedal to its spindle and the concealed return spring, introduced last year, are still employed but the adjustable stop is clamped between the pedal spindle and the frame.

Since all 1956 roadster models have a centre stand a front stand is no longer necessary and is replaced by a simple, tubular stay. The use of mudguard forward stays was, of course, discontinued for 1955.

Another distinctive feature to be retained is the use of twin, streamlined pilot lamps flanking the headlamp. As before, the lower rear of the lamp shell is cut away to provide easy access to the head of the inbuilt speedometer.

The front-brake cam lever, located above the wheel spindle, is cranked round the fork slider

The front Teledraulic fork is modified in detail only. The wheel-spindle retaining caps at the lower ends of the sliders have been made of cleaner design; they are now cast in one with the slider and milled off later. At the top of the legs, the deeply domed, push-in caps are replaced by shallower domes incorporating socket hexagons; the caps are screwed in by the key used also for the fork lower-yoke clamp bolts. The sockets normally are filled by rubber plugs and the result is both neat and practical.

An unusually clean "rider's eye view" results not only from the features already mentioned but from two other detail changes. Short plastic sleeves hold together air and front brake cables on one side and ignition and clutch cables on the other. Moreover, on each side of the malleable-iron fork top yoke is an ear having a hole through which those cables pass, so that they are neatly located without constraint. All cable casings are plastic covered, for durability and waterproofing, and lubrication nipples are fitted to clutch and throttle casings.

Normal finish on all machines save the two Matchless twins is black and chromium. The two exceptions have dark-red and chromium petrol tanks. To distinguish between the five-hundred and six-hundred twins, petrol tanks of the larger models have chromium-plated side panels. On the A.J.S. models the three-dimensional tank motif is of ¼in larger diameter than before.

Those two popular racing machines, the 348 c.c. 7R A.J.S. and the 498 c.c. G45 Matchless, will again be produced next season in strictly limited quantities. Modifications to the present specification will be made in the light of experience gained with the current season's factory models. Two scramblers in each range will be marketed but here, too, certain details have still to be settled; racers and scramblers will therefore be the subject of separate descriptions later.

A.J.S. and Matchless machines are manufactured by Associated Motor Cycles, Ltd., Plumstead Road, London, S.E.18. Prices will be announced shortly.

The 347 c.c. A.J.S.—noted for its mechanical quietness

498 c.c. of sturdy simplicity—the G80S Matchless

A.M.C. Competition Models

Redesigned Scrambles Engines of 348 c.c. and 497 c.c. : Latest Developments on the 7R A.J.S. and G45 Matchless

REDESIGNED engines are employed for the latest A.J.S. and Matchless scrambles models. The five-hundreds (A.J.S. 18CS and Matchless G80CS) have bore and stroke dimensions of 86 × 85mm, giving a swept volume of 497 c.c. Engine dimensions of the three-fifties (A.J.S. 16MCS and Matchless G3/LCS) are 72mm bore × 85mm stroke, swept volume 348 c.c. As suggested by the similarity of stroke measurements, identical crankshaft assemblies are used in both sizes of engine. With the obvious exceptions of tank transfers and the shape and lettering of magneto-chain covers, there is no difference between the equivalent-capacity models of the two marques.

The flywheel discs are machined from steel forgings and the assembly has been stiffened by an increase in crankpin diameter. A new and robust connecting rod has a single web surrounding the big-end eye; a single row of long, caged rollers is used in the big-end bearing. As on earlier competition engines, the drive-side mainshaft is supported in two ball bearings. The plain bronze bush supporting the timing-side shaft (and acting as an oil distributor) is retained, but is supplemented by a roller bearing close up to the flywheel.

Most noticeable engine modification at a casual glance concerns the deeply finned light-alloy cylinder barrel which now incorporates integral pushrod tunnels. The cylinder has a bonded cast-iron liner. A full-skirt piston, carrying two compression rings and a slotted oil scraper, gives a compression ratio of 8.7 to 1 in the case of the five-hundred and 9.9 to 1 in the 348 c.c. engine.

No alteration has been made to the die-cast light-alloy cylinder head. Austenitic-iron valve seats are cast in position and the ports are polished. Valve operation is by flat-base tappets and tubular, light-alloy push-rods. The cams provide exended valve-opening periods; measured with running valve clearances the timing is: inlet valve opens 59 degrees before top dead centre and closes 69 degrees after bottom dead centre; exhaust valve opens 74 degrees before b.d.c. and closes 48 degrees after t.d.c.

Amal Monobloc carburettors are standardized on both sizes of engine. Choke diameter on the larger unit is $1\frac{3}{16}$in, while on the three-fifty it is $1\frac{1}{16}$in. The Lucas waterproof magneto has manual control. Exhaust-pipe diameter has been increased by $\frac{1}{8}$in to $1\frac{3}{4}$in. Standard equipment includes a silencer, which is replaced by a plain extension for competition use.

Crankshaft layout of the latest A.J.S. and Matchless scrambles models

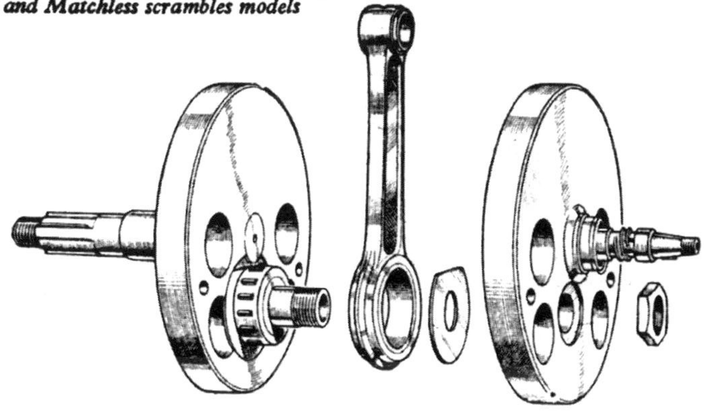

Frame of the scramblers is similar to the new roadster pattern described in *The Motor Cycle* for September 8, except that the upper attachment lugs for the rear shock absorbers are braced together by a welded strut. The suspensions are heavier than on roadster machines.

Many alterations have been made to the production racing machines—the 348 c.c. 7R A.J.S. and the 498 c.c. G45 Matchless twin—in pursuance of the firm's policy of passing on to customers the benefits of development work carried out on he factory racing models. On the 7R the cylinder bore is increased to 75.5mm and the stroke shortened to 78mm. The piston has only two rings—one Dykes-type compression ring and a slotted oil scraper.

Cylinder filling is improved as the result of an alteration to the shape of the inlet port. A $1\frac{5}{32}$in-choke Amal GP carburettor is standardized, though an instrument with a $1\frac{3}{16}$in-diameter choke can be specified if desired for circuits involving almost continuous full-throttle running, such as at Spa, in Belgium, for example. The carburettor is insulated from the heat of the engine by a fibrous packing. Flywheel weight has been reduced slightly and the housing for the timing-side main bearing (of double-row roller pattern) is stiffened.

A rotating-magnet magneto replaces the earlier-type instrument and the exhaust megaphone has a reverse-cone outlet. Other improvements concern crankcase breathing and oil scavenging.

Many of the 7R engine modifications are duplicated on the G45, including improved inlet-port shape, heat insulation of the carburettors, the new piston-ring layout and the use of a rotating-magnet magneto and reverse-cone megaphones. In addition, triple-coil valve springs of an improved type are used and the diameter of the big-end studs has been enlarged.

Alterations to the cycle parts are common to both models and include the use of a twin-feed primary chain oiler, a magnetic rev-meter and new rear shock absorbers. Bowdenex front-brake and clutch cables are standardized, as are a front-brake lever with built-in cable adjuster and a positive-stop twistgrip, i.e., a grip which, when held hard in the fully open position, relieves the throttle-cable nipples of undue strain.

Ventilation of the brakes is increased by alterations to the air scoops and hub castings, and the latest Ferodo friction material is fitted. Slight frame strengthening has been carried out and the layout of footrests, brake and gear pedals is identical with that on the factory machines. Other developments include a slight variation in gear-box internal ratios and modification of tank shapes.

The makers are Associated Motor Cycles, Ltd., Plumstead Road, London, S.E.18. Prices (in which total price includes purchase tax, payable only in Great Britain) are as follows:—

	Basic Price £	Basic Price s	Total Price £	Total Price s	Total Price d
A.J.S. 16MCS and Matchless G3/LCS, 348 c.c. ..	182	0	225	13	7
A.J.S. 18CS and Matchless G80CS, 497 c.c. ..	196	0	243	0	9
Electric lighting extra ..	8	5	10	4	7
A.J.S. 7R, 348 c.c.	325	0	403	0	0
Matchless G45, 498 c.c. ..	325	0	403	0	0

The 497 c.c. A.J.S. scrambler for 1956

ROADSTERS, COMPETITION MOUNTS, RACERS, FOR NEXT YEAR : MEASURES TO ENSURE OIL TIGHTNESS : ENGINE PERFORMANCE INCREASED : LATEST GEAR BOX AND CLUTCH SPECIFIED : NO PRICE INCREASES

A.J.S. and Matchless Modifications

Above: The 498 c.c. A.J.S. model 20 fitted with chromium-plated tank panels

Right: Largest - capacity Matchless is the 592 c.c. model G11

TOURISTS, racing men, trials riders and scramblers can all find models to whet the appetite in the A.J.S. and Matchless ranges which are, of course, basically similar. All models have pivoted-fork rear springing and telescopic front forks. The roadsters—comprising singles of 347 and 498 c.c. and parallel twins of 498 and 592 c.c.—are renowned for their comfort, high-quality finish and mechanical quietness. A unique, three-bearing crankshaft layout is a feature of the twins which have an enviable reputation for robust construction and tireless performance.

The wide popularity with amateur sportsmen of the trials and scrambles models furnishes eloquent testimony of their suitability for competition riding off the beaten track. Although the scrambler is available both as a three-fifty and a five-hundred, the trials model is marketed only with a 347 c.c. engine; all the power units are singles. The racing machines, too—the 348 c.c. 7R A.J.S. and the 498 c.c. G45 Matchless twin—are well to the fore in sporting circles and may be seen in appreciable numbers at most national and international road-race meetings.

For 1957 no sweeping changes are announced for any of the models. Nevertheless, progress has been made in several directions. All models are fitted with the new gear box and clutch described in *The Motor Cycle* for May 24 last. Meticulous attention to oil sealing is a notable characteristic of the design of the box, while the clutch is claimed to be unaffected by the presence of oil on the friction faces. It is claimed that the performance of all the roadsters has been stepped up without loss of tractability. Appearance of the twins is brightened by a new method of petrol-tank decoration. For the first time since rear springing was introduced in 1949 proprietary shock absorbers are used in place of the A.M.C.-manufactured units. In addition to the foregoing changes there are several minor modifications aimed chiefly at cleaning up appearance and eliminating oil leakage.

A redesigned frame for the trials model is of real significance, since it results in a 3in increase in ground clearance. Cross-country racing men will be delighted to learn that the power output of both sizes of scrambles engine has been increased. As far as the two road-racing models are concerned there is, of course, a continuance of the firm's policy of incorporating the benefits of experiments carried out on the factory racing machines during the previous season.

Liberal use of synthetic-rubber O-rings and a special mainshaft oil seal behind the final-drive sprocket are the means adopted to ensure oil-tightness of the new four-speed gear box. Wedge-shape rubber blocks in the clutch centre serve as a transmission shock absorber. The unusual method of clutch operation employs a ball and roller between which works a suitably profiled floating lever. Adjustment is effected by a screw and lock nut in the centre of the clutch pressure plate. Access to the adjuster and those for the three clutch springs is obtained by removal of a dome from the primary chaincase.

General liveliness of A.J.S. and Matchless roadsters has never been in doubt. But it is the makers' experience that many buyers attach considerable—and, perhaps, undue—importance to speed and acceleration. Hence the quest for more engine power which has led to the adoption of a new inlet cam for the singles and modified inlet and exhaust cams on the twins. For those readers who are unfamiliar with the general layout of A.M.C. roadster engines it may be said that the singles have light-alloy cylinder heads, pushrod-operated overhead valves and wire-wound pistons. A chain-driven Lucas rotating-magnet magneto is mounted in front of the cylinder base and a separate dynamo behind the crankcase. On the twins, separate cylinder barrels and light-alloy heads are used and valve-clearance adjustment is effected by means of eccentric rocker spindles; the third crankshaft bearing is a split, white-metal-lined bush in the middle of the shaft.

The new method of petrol-tank embellishment on the twins consists of separate, chromium-plated side panels fixed to the tank by the two screws which retain the plastic motif and by two more screws which hold the kneegrip plates in position. Plastic beading round the edge of the panel seals the gap between it and the side of the tank. Colours of beading

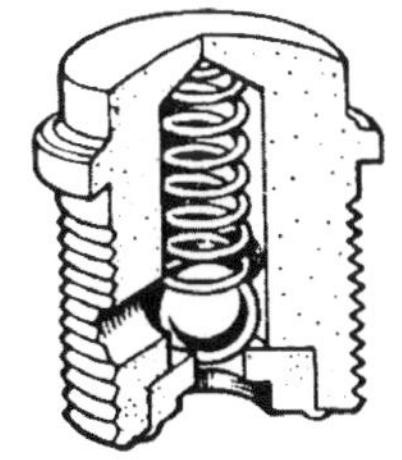

A modification to the twin-cylinder machines concerns the oil-filter end cap which is now of one-piece construction as shown

and tank enamel are very well blended so that the overall impact is most attractive. The 592 c.c. A.J.S. Model 30 has a royal-blue tank and light-blue beading, while the six-hundred Matchless G11 has a dark-red tank with black beading. On both five-hundred twins (A.J.S. Model 20 and Matchless G9) the tank is enamelled black; the beading on the A.J.S. is light blue and that on the Matchless is red.

Tank finish on the single-cylinder models is unaltered. In other words, the tanks are enamelled black on a Bonderized surface and subsequently lined by hand. Lining in the case of the Matchless models is in silver and for the A.J.S. machines in gold. Plastic tank badges similar to those on the twins are fitted. When output permits, tanks with panels will be offered on the singles at extra charge.

In adopting Girling shock absorbers for the rear springing, A.M.C.s are one of the last manufacturers to take advantage of the specialized products available in this field. Incorporating a three-position adjustment for load, the units are similar to those fitted to several other makes of machine, but the upper and lower attachments are specially made for the Plumstead models.

At the bottom of each shock-absorber unit is a rubber-bushed, cast light-alloy yoke. The top fixing takes the form of a single eye containing a rubber sleeve. As the A.M.C. shock absorbers embodied yokes at both ends, the plain upper-attachment ears formerly welded to the frame loops have been replaced by box-section lugs. Thus the new units are not interchangeable with the old. Another minor departure from normal Girling practice is the use of polished light alloy for the split collets which retain the upper spring shroud.

Incidentally, to provide added frame stiffness for sidecar duty on the twins and 498 c.c. singles, the bolt which clamps the rear-frame loops to the top of the seat tube is increased in diameter to $\frac{3}{8}$in. Also the attachment point has been moved forward slightly so that the lug on the main frame member is formed integrally with the rear tank-support lug, whereas previously two separate lugs were used.

A mid-season alteration to front-fork damping is continued for 1957. On both shock and recoil movements the hydraulic check is more progressive in action, with the result that front-end pitching is said to be eliminated.

To prevent the possibility of chatter occurring after an extensive mileage has been covered, and to provide an added degree of resilience in the transmission, a considerable alteration has been made to the five-pin drive of the quickly detachable rear wheel on all the roadster models. Previously the driving (and braking) torque was transmitted to the solid pins bolted in the full-width, light-alloy hub by their engaging with five holes drilled in the malleable-iron casting which serves the dual purposes of chain sprocket and brake drum. The radius at which the pins are disposed from the hub axis has been doubled, while the pins themselves are hollow and of much larger diameter than before; each pin is rubber sleeved externally. In the brake-drum casting the holes with which the pins engage take the form of tubular bosses, about $\frac{7}{8}$in long, which are connected by an annular rib.

A year ago the tool kit, battery and voltage regulator were grouped in a large box which fits snugly in the left-hand rear frame loop and matches the oil tank on the other side of the machine. Hinged along its bottom edge, the lid of the box could not be opened fully if the model was fitted with a sidecar because of the proximity of the upper rear sidecar connection. To overcome the difficulty the lid is no longer hinged to the box. Instead, two ears are formed at the lower edge of the lid which engage with slots in the box. Hence the lid can be completely detached from the box irrespective of the nearness of a sidecar connection.

In place of the endless red rubber band which held the battery in place in the front of the box there is now a black moulded-rubber strap; the lower end of the strap carries a transverse steel peg which hooks into slots in the battery platform. Disengagement of the peg from the slots releases the battery for maintenance purposes.

On the twins a separate cover is attached to the side of the oil tank to match the tool-box lid. Both cover and lid have three horizontal ribs embossed on them to minimize the possible effects of chafing by a pillionist's leg-wear and to enhance the appearance of the models. The singles do not have oil-tank covers and in their case the ribs are embossed on the tank.

Prevention of oil leakage is the aim behind a number of changes which have been made to the rear portion of the pressed-steel primary chaincase. First, the edge of the hole through which the engine shaft protrudes is turned outward and a composite cork washer is interposed between crankcase and chaincase at that point. Secondly, a sliding oil thrower, consisting of two dished plates riveted together one on each side of the slotted hole in the case, locates on the gear-box mainshaft and moves backward or forward with the gear box during primary-chain adjustment. Finally, in the case of the singles, an oil deflector

Stark scrambler—the 497 c.c. G80CS Matchless

The 347 c.c. trials-equipped A.J.S. 16MC

is welded into the rear portion of the chaincase where the dynamo shaft passes through. Additionally the rear face of the dynamo sprocket is profiled to fling oil away from the hole.

Incidentally, because it is considered that the rubber vanes in the clutch centre and the rubber sleeves on the rear-hub pegs make adequate provision for smoothing out transmission shocks, the face-cam type of shock absorber previously fitted to the engine shaft has been dropped. In consequence the dome at the forward end of the chaincase outer half is much shallower than before.

Shielding of the rear chain has been increased by extending the chainguard farther downward behind the rear-wheel sprocket and by fixing a curved guard to the back of the primary chaincase to cover the rear chain where it runs on to the gear-box sprocket.

The desire to eliminate another possible source of oil leakage is responsible for a slight modification to the end cap of the pressure oil filter situated in the crankcase left-hand half on the twins: the cap is now made in one piece instead of two, thus eliminating a joint which was subject to oil pressure. Though the spring retaining the filter element is different from that for the by-pass ball-valve, it was possible with the old end-cap layout to interchange them with disastrous results. The end-cap modification has rendered the interchange impossible.

Several random and minor changes contribute towards a general cleaning up in appearance. The adoption of Neoprene push-on oil pipes dispenses with the need for threaded unions. A more streamlined form has been given to the yoke at the rear end of the brake rod. The depth of the rear-mudguard valances has been increased slightly at the rear end. Plastic handlebar grips are standardized. On the twins there is a restyled end cover for the Lucas dynamo.

Known as the A.J.S. 16MC or the Matchless G3LC, the trials model is the only machine in the range to utilize a welded frame. For 1957 the integral tubular structure comprising duplex engine cradle and rear frame loops is raised considerably with the result that ground clearance (with the machine unladen) is increased from 7in to 10in. The altered sweep of the tubes has permitted the exhaust pipe to be tucked in more closely with the result that it is no longer necessary to use an exaggerated outward crank for the kick-starter pedal. No alteration has been made to the engine; with moderate valve timing and a low compression ratio it delivers smooth, lusty, controllable power down to ultra-low engine speeds and is thus particularly well suited to its purpose.

Conversely, on the scrambles models of both capacities it is the engines only which are the subject of slight alteration. Introduced during the past season the models have met with considerable success and the power increase now achieved should enhance their popularity still further. Both engines have modified inlet ports and larger inlet valves. Choke size of the Amal Monobloc carburettor on the 348 c.c. unit has been increased by $\frac{1}{16}$in to $1\frac{1}{8}$in. On the 497 c.c. engine the carburettor bore remains at $1\frac{3}{16}$in.

With the exception of the new gear box, no details are yet available of the modifications to be incorporated in the racing models as a result of development work carried out on the factory machines this year.

The makers of A.J.S. and Matchless machines are Associated Motor Cycles, Ltd., Plumstead Road, London, S.E.18. Prices (in which total price includes purchase tax, payable only in Great Britain) are as follows:—

Model	Basic Price £	s	Total Price £	s	d
A.J.S. 16MS 347 c.c. .... / Matchless G3LS 347 c.c.	170	10	211	8	5
A.J.S. 16MC trials ...... / Matchless G3LC trials ..	177	0	219	9	7
A.J.S. 16MCS scrambles .. / Matchless G3LCS scrambles	182	0	225	13	7
A.J.S. 18S 498 c.c. ...... / Matchless G80S 498 c.c. .	180	10	223	16	5
A.J.S. 18CS scrambles .. / Matchless G80CS scrambles	196	0	243	0	9
A.J.S. 20 498 c.c. twin .... / Matchless G9 498 c.c. twin	205	0	254	4	0
A.J.S. 30 592 c.c. twin .... / Matchless G11 592 c.c. twin	210	0	260	8	0
A.J.S. 7R 348 c.c. o.h.c. (racing model) / Matchless G45 498 c.c. o.h.v. twin (racing model)	325	0	403	0	0
Lighting extra, competition models	8	5	10	4	7

Ever-popular 347 c.c. 16MS A.J.S. roadster

A sturdy single—the 498 c.c. G80S Matchless

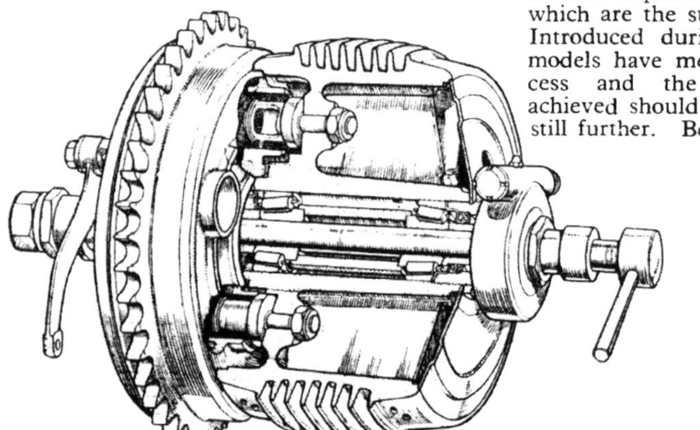
In the latest quickly detachable rear wheel the improved driving pins are disposed at a larger radius, are rubber bushed and locate in long bosses in the brake-drum casting

347 c.c. G3LC Matchless

Increased Ground Clearance for Famous Trialster By GEORGE WILSON

DURING a Scottish Six Days' Trial a few years ago a journalist was riding a five-hundred dingbat-wise down Glencoe. In the course of his dramatic dash, the speeding scribe overtook a three-fifty—a product of a certain Plumstead factory—with High Viney up. That transient episode had its repercussion for thereafter, on the road to Mamore, Viney demonstrated that his three-fifty had plenty of top-end power—equal, easily, to that of many a five hundred. Later, on the climb of Mamore's rock-strewn surface, he gave an exhibition of plonk to take the breath away.

Buzz or plonk—either characteristic can be easily achieved, but producing both from the same power unit is much more difficult. In terms of engine efficiency the latest Associated Motor Cycles' product tested—a Matchless G3LC—showed that nothing has been lost in the years between. From the pilot-jet supply upward, breathing is as clean as a razor's cut. Engine response to sudden throttle opening is instantaneous and full of lusty power. On a full day spent riding the machine over rough stuff, the carburettor didn't spit once.

The 1957 G3LC is bigger than its predecessors. In the search for ground clearance the rear sub-frame has been modified to provide a lower mounting for the top ends of the suspension units. The result is that the rear of the machine is forced upward to provide 8½in ground clearance when the suspension is in the position of static load. When the springing is extended the clearance increases to approximately 10in. The result is a machine that can probably tackle every known observed hill between Dartmoor and Aviemore without once marking its crankcase shield.

Two allied questions immediately arise when a competition machine with so much air under the engine is being considered. First, does the advantage outweigh the disadvantage of effective increase in top hamper? Secondly, has the saddle been raised to such an extent that the rider cannot foot without difficulty? Obviously, you can't have it both ways. One of the photographs shows the model topping a rock step that would certainly have stopped the 1956 model: It is also, probably, the highest step that could be surmounted by any trials machine. Numerous other rocky hills were tackled during the test and everywhere the result was the same. There is simply no gainsaying the fact that in the quest for ground clearance the design aim has been achieved.

For short riders such as I, it has not, however, been achieved without cost. I found the saddle-height increase and top hamper combined to make my day somewhat taxing. Saddle height is 33in with the machine unladen and about 32in with my 10-stone weight in the saddle. When my tailor runs me up a pair of trousers he makes my inside leg measurement 29½in. I simply couldn't foot at all! As for top hamper, when the machine became frisky it was not easy for me to bring it to heel. Total weight, with the lighting set fitted and a gallon in the tank, was 372 lb. In short, the machine is at its best when being ridden by citizens of Viney build.

I was thrilled by many aspects of the Matchless's performance. With the tyre pressures reduced to the extent that the gauge wouldn't register, the rear wheel found bite to an astonishing degree on loose, sandy going and greasy tracks. Excellence of delivery tune is a feature of all Matchless products. The G3LC was no exception. Handlebar control positioning, cable operation, the degree of finish were all of the very highest standard. The clutch took up the drive positively—so positively that if you did want to slip it on a hairpin you could do so with perfect peace of mind.

As well as being high the G3LC is very narrow downstairs and, indeed, the width across the footrests is a mere 21½in. The narrowness paid off handsomely. It meant that deep gullies could be ridden without fear of the rests digging into the sides. It meant that where a gap appeared ahead you could go for it with all the zest displayed by a spring lamb. Up top, the model is narrow, too—so narrow that after being on its side all of a dozen times in the course of our day out, it displayed no evidence of our misadventures. Maybe I had suffered—but no one could guess it!

DATA.—*Capacity:* 347 c.c. *Bore and stroke:* 69 x 93mm. *Compression ratio:* 6.5 to 1. *Carburettor:* Amal Monobloc 376/5. *Choke size:* 1 1/16in. *Gear ratios:* 21.3, 15.5, 9.6 and 6.5 to 1. *Fuel capacity:* 2 gallons. *Wheelbase:* 54in. *Seat height:* 33in. *Ground clearance:* 8½in. *Weight:* 372 lb with full lighting set and one gallon of fuel. *Price:* £177 (with purchase tax, payable only in Great Britain), £219 9s 7d. *Manufacturers:* Associated Motor Cycles, Ltd., Plumstead Road, London, S.E.18.

Two shots showing (left) the Matchless about to clear and (right) clearing unusually high rock steps

NEW AJS & MATCHLESS SCRAMBLER TWINS

Aluminum alloy primary chaincase and 19-inch front wheel are features of the AJS Scrambler twin. Note full-chromed forks.

Frank Cooper, U.S. Distributor for AJS and Matchless motorcycles announces that the 1958 Scrambler twins will sport several major changes and a great number of minor ones calculated to make the popular models still more appealing to sports riders. The 592 cc parallel twin AJS 30CS and Matchless G11CS have new cast and polished aluminum primary covers secured by cheese-head screws replacing the stamped steel covers of previous years. The new alloy chain housing has two inspection plugs for checking chain tension, topping up the chaincase with oil, and for adjusting the clutch throw rod. These plugs are flush, with milled screwdriver slots, and lend themselves smoothly to the streamlined shape of the chaincase.

Other components have returned to the drawing boards of Associated Motorcycle Company engineers to be improved or restyled. Among these is the rear fender of the Scrambler twins which has been made wider, deeper, and longer to provide greater protection. The rear frame section has been modified to accommodate new heavier rear shocks with longer movement that now slant forward to give better shock absorbing action and consequently a smoother ride.

The seat covering has been changed from black to a heavier duty tan material and the control levers now have ball-end rubber covers. Introduced last year, an A.M.C. four-speed box is fitted. The gear change has been lightened by use of a lower-rate selector spring.

To give the Scramblers a narrower width and thus permit closer tucking in, the oil filler cap has been relocated on the top of the tank where it will not interfere with the rider's leg. The new models are also fitted with the 19-inch front wheels that have been proved to be so effective in competition, and the gas tanks, which retain their familiar contour, have stronger, rubber-cushioned rear mounting bracket, and new barrel type petcocks.

A distinctive feature of the twin cylinder engine is the use of a third bearing in the middle of the crankshaft; the bearing is plain and carried in a separate cast-aluminum plate. The big-end bearings are also plain and the connecting rods are light-alloy forgings.

The models are prettied up by completely chromed forks and rims, new light Mediterranean blue paint on the gas and oil tanks and tool box of the AJS, and a brighter red paint on the tanks and tool box of the Matchless. Pillion footrest lugs have been relocated and the crossover exhaust pipes and muffler have been raised to give more ground clearance.

Virtually "convertible" machines, the Scramblers can be purchased with full quickly-detachable lighting equipment, which has a new voltage regulator of the unitized type, and used as roadsters. Wherever they are used, on the road or in the rough, the AJS and Matchless Scrambler twins will perform with the same flawless precision that has earned them a high place in the regard of American motorcyclists. Other AJS and Matchless models will be announced in CYCLE as they are released. ★

Redesigned exhaust pipes and muffler are shown in this view of the 1958 AJS Scrambler.

Ready for ROUGHIN'

ROAD TEST NO. 91

by the CYCLE Staff

By the time this road test reaches its readers cycle shops and bench racing sessions will be buzzing with the news that Matchless has a forty cubic inch twin for 1958. Previous Matchless (and its sister make AJS) twins had a 600 cc displacement, or roughly thirty-seven cubic inches. Though they performed well, Frank Cooper, the U.S. Matchless - AJS distributor, felt that instead of spotting their competition three inches, an additional model should be added to the line with the extra cubes needed to bring them right up to the top of their class.

Cooper let CYCLE in on the displacement boost before he told anyone else about it and he gave us one of the first forty inch Matchless Super Hurricane Scramblers to test.

We picked up the bike at Cooper's distributorship and spent the better part of an hour just plain admiring it. The '58 Matchstick shines with enough polish and sparkle to dazzle a Detroit designer. From its full-chrome 19 inch front rim, past its polished alloy front hub, up the chromed forks and all the way back to the tip of its polished alloy rear fender the Matchless is a beautiful and businesslike piece of machinery. All the painted parts (and there aren't many of these, most of the bike is either polished or chromed) are triple dipped in excellent quality black enamel and then baked to a glossy hardness that resists scratching and wear. To illustrate the toughness of this finish, after nearly 300 miles of hard riding, the footpegs on the test bike, which are painted by this process, were only beginning to show metal through the paint over most of their length.

GEARED TO THE GROUND—Heavy rear knobby digs in even in loose dirt. Here Assistant Editor Chuck Clayton tackles hill from dead stop. Made it, too.

There is a new shade of red on the tanks and tool box of the Matchless that immediately catches the eye. It's a brighter, more attractive red that should stay new looking for a long time. The red painted cover on the oil tank keeps your leg away from the hot tank on a long road ride and can be removed for scrambling. That way even if the tank gets dented (which should be pretty hard to do since it is well tucked in) the cover can be screwed back into place and the bike will still look cherry.

The seat is also a new color, a sort of imitation pigskin brown. Somehow this new shade just doesn't go with the rest of the bike. We'd like to see Associated Motor Cycles Ltd. return to the traditional black seat with white piping for scramblers and leave the cute colors to the ladies' machines.

The rear fender is broadened and lengthened to give more weather protection. We got a chance to test this feature the first night that we had the bike. The Los Angeles Chamber of Commerce had ordered some rain to make the tourists feel at home and one of CYCLE'S testers found that while the cars ahead of him on the freeway had no trouble keeping his face drenched, the rear mudguard caught all of the water and road slush thrown up by the back knobby, keeping the back of his jacket as clean as an atheist's Bible.

We cut our admiration of the Matchless' appearance short because we wanted to see how that brute engine performed. The motor was warm and fired up with one easy kick. We later found that starting on cold mornings was more difficult. The twin has so much compression when cold that it feels like you're breaking a weld every time you kick it over. After a few kicks it starts, though, and from then on everything goes smoothly.

The engine idles with a steady drumming exhaust note and a turbine-like whirr that gives promise of great things to come. And it keeps its promise. You lift the gearshift until it nicks smoothly into low, feed the horses to the rear wheel and something

ATTRACTIVELY TRACTABLE — Was Editor Jack Snyder's description of the Hurricane Scrambler. Hill in background is the one navigated in photo at left.

SPORTY FORTY incher took lots of starting and stopping abuse without a whimper.

moves — It's the motorcycle and you with it. Since it is made for TT and scrambles competition the Hurricane engine is tuned to spread its power over a wide rpm range. Even at that it can break an honest 106 mph at the top end (with road gearing and fully muffled), but where the power is really felt is in accelleration.

The extra inches are gotten by stroking the engine. The stroke in this case has been increased from 2.866 to 3.13 inches. Lengthened stroke means added torque and added torque means added acceleration. Added acceleration means getting ahead of the pack at the start and getting head of the pack at the start means having a dust-free, uncrowded lead in an event, and that means — if you're a reasonably good rider—a new trophy to add to the shelf.

Check the performance chart and you will find that the scrambler twin accellerated through the 1/10th and 1/4 mile drags in exceptionally good time. These tests were conducted on smooth macadam with road tires and 5:1 gear ratios. The bike was supplied out of the crate with knobby tires, of course, but we switched to road specs for the speed and braking tests because a scrambler is poorly set up for road work and our tests of this type have to be made on pavement if they are to be standardized. No changes were made on the engine except to remove the air cleaner and swap the plugs for better high speed performance.

The Matchless performed well on the road except for vibration at about 60 mph in high with road gearing and about 50 mph with dirt cogs. The tough, 3-bearing lower end of the Hurricane engine is basically the same as the famous Matchless G45 road racer, and the intake valves are identical diameter, but since this is a road test of a scrambler let's get those knobbies and scrambles gears back on it and take it to the boondocks.

In its favorite trim with a heavy hand ot the throttle the Matchless once and for all answers the old question, "what's in a name." Through the years this make's handling, ride, and performance have truly been Matchless. Now, with its new larger engine, it is just about catchless. The teledraulic front and the newly modified Girling rear swinging arm suspension can only be described as excellent. With its terrific engine and traction combination the Matchless scrambler climbs like a starving chimpanzee in a banana grove. It barrels smoothly around curves with no hint of breaking loose until you want it to and plows confidently through territory where experts fear to tread.

ON THE DOWNGRADE—For the faint of heart the engine's compression would hold the machine to a reassuring speed on downgrades. More fun to give 'er some gas, though.

The brakes on the machine are excellent. After being impressed by their stopping power in actual use and the braking tests, we decided to give them the acid test. At the end of each of ten high speed runs we applied the brakes hard to stop the bike as quickly as possible. This was bringing the machine to a halt from 100 miles per hour in a few hundred yards, several times without using engine compression. At the end of each of these stops the rear brake would be smoking, but the brakes never did fade completely and always returned to their original power after a few seconds cooling period. The amazing part was, during the whole test we only adjusted the brakes once! Which brings to mind another good feature. The rear brake adjuster was changed in 1957 to the forward end of the brake rod near where the brake pedal pivots. This

DIGGING FOR GO—Clayton pulls the trigger that shoots the Matchless through the 1/10 mile accelleration test. Source of light for electric eye timer is at left.

POTENT PACKAGE OF POLISHED POWER describes the 1958 Matchless twin. Note new slanted rear shocks, tucked-in crossover pipes, 19 inch front wheel.

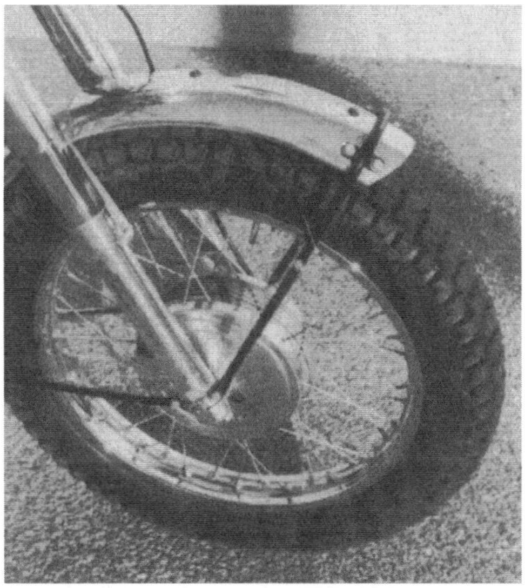

BRIGHT SIGHT—The front of the Matchless is nearly all chrome and polished alloy.

puts it up out of the way and makes adjustment easier.

The new forty incher should be a real trophy-taker when the nation's hotshoes get ahold of it. We could wind the throttle open in the rough and there'd be a sudden surge forward no matter how fast we were already going. Several detail features help make scrambling with the Matchless easier such as rubber ball-ended clutch and brake lever covers that keep fingers from slipping and help prevent hand fatigue; a tucked-in oil tank that doesn't interfere with the rider's leg; and heavy duty components that are built to take plenty of abuse and then come back next Sunday for some more.

The kickstarter folds way in to keep it out of the way, but sometimes the ball snap that holds it in position became stuck, then it was very difficult to unfold for starting.

Pillion passengers liked the high rear peg arrangement and the comfort of the dual seat and rear suspension. Although passengers were not aware that the engine displacement had been increased, they remarked on the extra amount of urge that the Matchless seemed to have.

The model is supplied with full lighting equipment which is quickly detachable. The lights were adequate. The tool kit is of average quality and the instruction manual is good.

The gearbox, while not fully broken in, was a joy to operate except when down-shifting. Sometimes we held up traffic trying to get the machine into low gear, but the trouble seems due only to the stiffness that all new machines have and will no doubt go away in a few hundred more miles.

Oil consumption was very low despite plenty of running at full bore. Much of this is due to the fact that the engine leaked practically no oil at all, just a little weeping out of the rocker covers. The gearbox leaked a little more, but not enough to warrant a gunking. Just about the most oil-tight component of all was the new alloy primary chaincase, which didn't let enough lubricant leak to show us where the joint was. This chaincase is indeed a welcome improvement on A.M.C. motorcycles, whose old-style pressed steel chaincase was sometimes prone to problems.

The scrambler twin is equipped with jiffy and center stands, both of which were stiff to operate, and crossover pipes plugged with a very efficient muffler.

The performance figures and our handlebar impressions indicate that the new Matchless Hurricane scrambler promises to blow up a storm on American scrambles and TT courses. It will remain for U.S. competition riders to prove whether this promise will be fulfilled or not. But in any case, this is one machine to keep your eye on in any starting lineup.

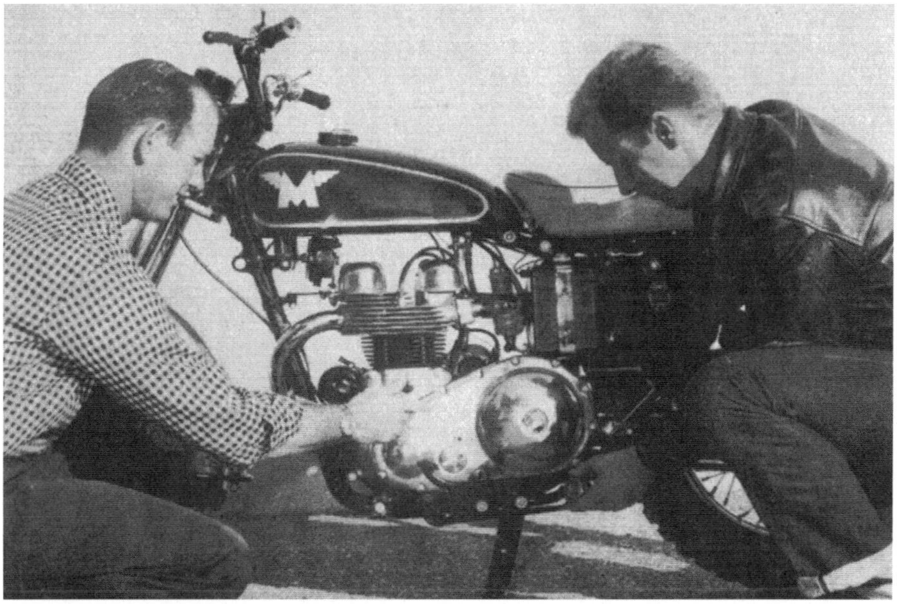

POLISHED PRIMARY of aluminum alloy is a feature of new Matchlesses. Cooper Motors Sales Manager Bill Adams points out the clutch rod adjusting aperture.

MATCHLESS SUPER HURRICANE

PERFORMANCE SUMMARY

TOP SPEEDS:	
(With scrambles pipes and muffler)	
Third Gear	92.02 mph
Fourth Gear	106.84 mph
ACCELLERATION:	
Tenth Mile	8.05 sec.
Quarter Mile	14.17 sec.
BRAKING DISTANCE:	
(From 25 mph)	
Front Brake Only	19′ 2″
Rear Brake Only	26′ 8″
Both Brakes	13′ 8″
WEIGHT:	
With Lights	389 lbs.
GASOLINE MILEAGE:	
Average with Road Gearing	50 mpg
Aver. with Scrambles Gearing	38 mpg

SPECIFICATIONS

Engine: Vertical twin cylinder ohv. Bore 2.834, stroke 3.130, 40 cu.in. Light alloy connecting rods with Vandervell plain bearings. Cast alloy iron crankshaft supported in three main bearings, journals roller, center Vandervell. Compression ratio 8.5:1. Dry sump lubrication with two spur gear pumps. 1 1/8″ Amal Monobloc carburetor.

GEARBOX: A.M.C. four-speed, foot-operated, positive stop type. Multiple plate clutch, engine shock absorber. Primary chain enclosed in polished alloy housing.

ELECTRICAL EQUIPMENT: Lucas magneto ignition and lighting equipment. Quickly detachable headlight and taillight. Gear-driven generator and magneto.

FRAME: Duplex crade frame, skid plate under crankcase, teledraulic front and Girling rear shocks. Swinging arm rear suspension. Spring-up side and center stands.

BRAKES: Seven inch front and rear brakes. Full width hubs and straight, heavy-duty spokes.

TIRES: 3.25x19 front, 4.00x19 rear. Sports knobbies. Road tires optional.

TANK CAPACITY: 2.2 U.S. gallons.

MANUFACTURERS: Associated Motor Cycles Ltd., Plumstead Rd., London, England.

U.S. DISTRIBUTOR: Cooper Motors, 5832 So. Broadway, Los Angeles, Calif.

New Two-fifties From Woolwich

Exciting Overhead-valve Singles Added to A.J.S. and Matchless Ranges: Technical Soundness Allied with Stylish Appearance

"We have tried," said Jock West, sales director of Associated Motor Cycles, "to achieve that indefinable modern appearance without resorting to the excessive bulk and extreme 'one-ness' that cost money and often result in poor accessibility." He was talking about the new 248 c.c. four-strokes which make their début today at the Geneva Show in two forms—the Model 14 A.J.S. and the G2 Matchless. That those efforts have been highly successful is clear from the illustrations on these pages. Here, indeed, are worthy additions to the ranks of modern British two-fifties, models in which technical soundness and functional layout are united with unusually happy styling.

Both models are essentially similar. Though design is largely orthodox, there is evidence of a healthy willingness to accept unorthodoxy, but only if it offers worthwhile advantages. Thus, the over-square engine has a *désaxé* cylinder (i.e., its axis is offset from the crankshaft axis) and an obliquely mounted cylinder head. Again, whereas the gear box appears to be in unit with the engine, it is actually a separate component and the primary chain is adjustable. The frame is mainly tubular in construction, but ingenious use is made of steel pressings, and full enclosure of the rear chain is available as an optional extra. The chief impression is of an up-to-the-minute machine in the classical tradition, compact and efficient.

Bore of the deeply finned cast-iron cylinder is 69.85 mm and the piston stroke is 64.85mm—not quite the same as twice the crank throw because of the offset cylinder axis which lies ¼in ahead of the crankshaft axis. The *désaxé* arrangement is not uncommon in car power units and two advantages are claimed for it: the movement of the piston across from the non-thrust to the thrust side of the cylinder at the top of the stroke is more gentle, so that slap is minimized, and the connecting-rod angularity at maximum combustion pressure is reduced. This second point means a slightly better use of that portion of the power stroke together with a small reduction in piston friction.

The degree of offset used, however, is

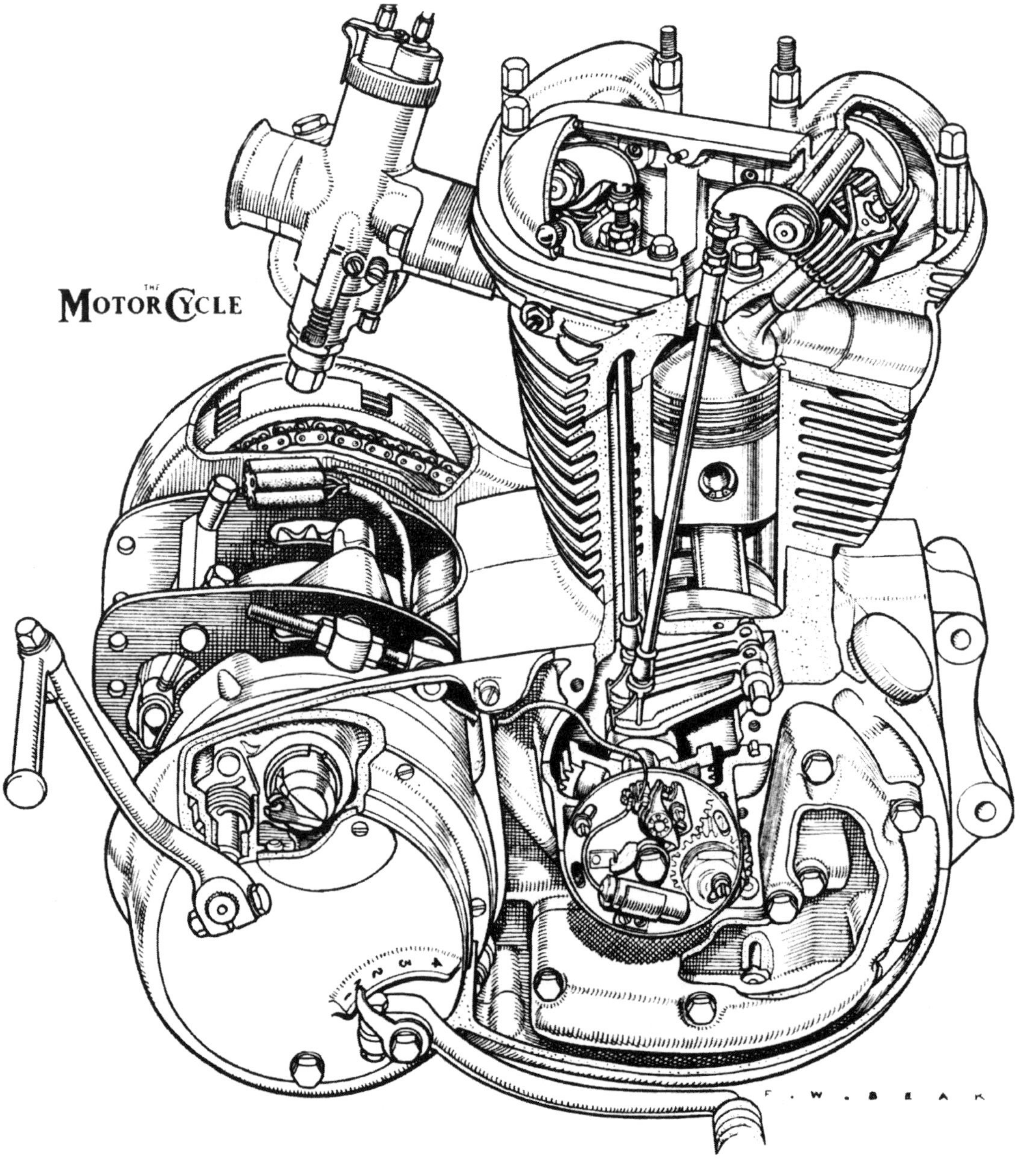

Clearly shown in this part-sectioned drawing are the oblique positioning of the cylinder head, the simple timing gear and the cylindrical gear-box shell. The securing straps and adjuster are normally concealed by a sheet-steel cover

The Model 14 A.J.S. and the G2 Matchless (lower photograph) are remarkably handsome yet essentially British

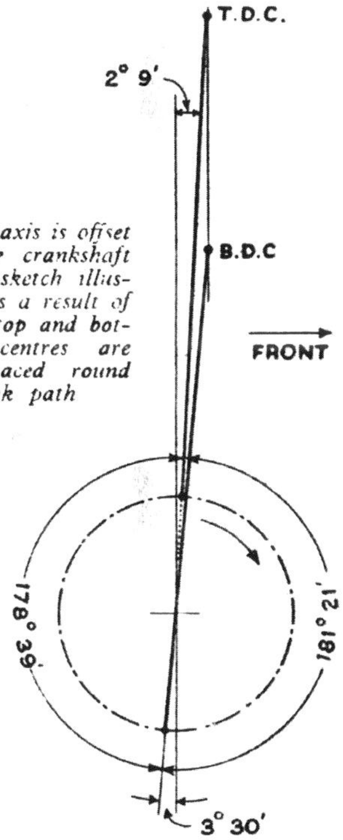

The cylinder axis is offset ahead of the crankshaft axis. This sketch illustrates how, as a result of that layout, top and bottom dead centres are unequally spaced round the crank path

not great and the system was chosen not just for those minor benefits stated but also because it fitted in with the general geometry of the engine. A single camshaft with two trailing lever-type followers was one of the fundamental design requirements and by positioning the cylinder forward and the camshaft rearward of the crankshaft axis, more room was obtained for the cams and followers.

A further requirement was a minimum of side thrust on the followers, which meant that the pushrods had to be as near as possible vertical in front elevation. Rotation of the plane of the valves (through $21\frac{1}{4}$ degrees clockwise, seen from above) made this practicable without the use of rocker spindles of different lengths and had the additional advantage that it brought the exhaust port and sparking plug more directly into the air stream, to ensure maximum cooling.

The cylinder head is in DTD 424 aluminium alloy and has a part-spherical combustion space with cast-in valve seats of austenitic cast iron. In addition to the deep horizontal finning there are diagonal vertical fins on the left to direct air over the sparking plug and combustion chamber. The silicon-chromium inlet valve is considerably larger than the exhaust valve, which is of Jessops G2 steel (respective head diameters are $1\frac{15}{32}$ and $1\frac{3}{16}$in). Because of this difference, and the wish to avoid too sharp a bend in the exhaust tract, the valve included angle of 75 degrees is split unequally: the inlet valve is at 35 degrees to the vertical and the exhaust valve is at 40 degrees.

Swirl is imparted to the charge and adequate scouring of the sparking plug achieved by offsetting the inlet tract 10 degrees towards the right of the centre line of the head. The exhaust tract, too, is offset to the right—by $11\frac{1}{2}$ degrees—and is counterbored at its forward end to take the push-in $1\frac{1}{2}$in-diameter exhaust pipe. Choke diameter of the Type 376 Amal Monobloc carburettor is $1\frac{1}{16}$in and the downdraught angle is 12 degrees. There is a thick heat-insulating spacer between the carburettor and cylinder-head flanges.

Valve guides, of cast iron, are located axially by circlips and crossover hairpin valve springs are employed. Four long studs screwed into the crankcase hold down the cylinder barrel and head; the separate die-cast rocker box is attached to the head by nine bolts. Rockers are of typical A.M.C. built-up construction: on each, the arms are splined to the ends of a "live" spindle and are separated by a hardened sleeve to which they are clamped by nuts on the spindle ends. Each rocker is supported by two bronze bushes with an oil-retaining felt sleeve between them. Owing to the unequal valve angles in the head, and the desirability of utilizing interchangeable pushrods, the exhaust rocker is mounted slightly lower in the box than the inlet rocker.

For simplicity and quiet running, the single camshaft is driven directly by the crankshaft pinion. Ends of the camshaft are supported in bronze bushes, one in the

There are no external oil pipes and the fabric filter is on the left of the engine. Two ball bearings support the drive-side mainshaft which carries the rotor of the A.C. generator—which is of Wico-Pacy manufacture

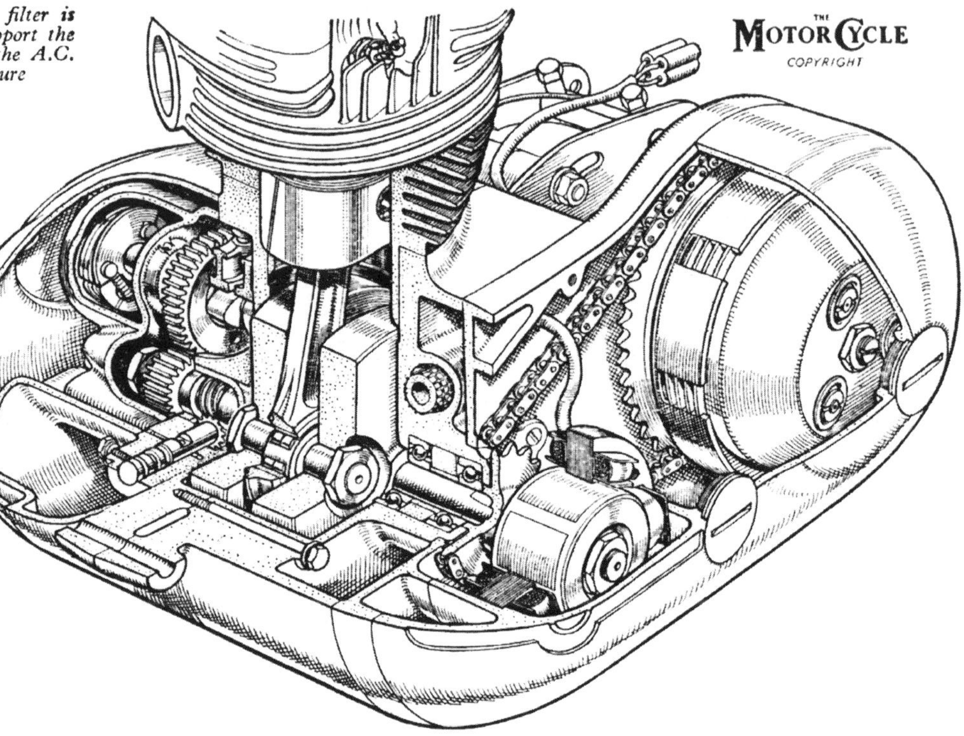

crankcase wall and the other in a light-alloy bolted-on case which surrounds the gears. On the outside of the case is the contact-breaker unit, actuated by an extension of the camshaft and embodying centrifugal automatic control of the ignition timing.

The cam followers have curved rubbing faces and oscillate directly on a common spindle ahead of the camshaft. Of light-alloy tubing, the pushrods are fitted with steel ends, the upper of which embody the valve-clearance adjusters. The rods are appreciably splayed and operate in a single cored tunnel of figure-eight section in the barrel; in the head each has its own tunnel.

Following normal A.M.C. practice, the piston is wire wound above the gudgeon pin and has a split skirt. It carries a scraper ring and two compression rings, the upper of which is chromium plated (to obviate scuffing) and has a taper face for quick bedding-in. The piston crown features a shallow dome and flats to provide valve-head clearance during the overlap period. Standard compression ratio is 7.8 to 1.

The connecting rod, or 1 per cent chromium steel, measures 5⅜in between the centres; within the ribbed big-end eye is a hardened-steel sleeve which forms the roller track; the small-end bush is of aluminium alloy. Comprising the big-end bearing are two rows of ¼in × ¼in rollers, 10 in each, spaced by a Duralumin cage. Nominal crankpin diameter is 1.2in and En. 351 is used for the hardened roller-track sleeve, on a body of KE805. The pin is a parallel press fit in the high-grade cast-iron flywheels (which measure 6½in diameter by ⅞in rim width) and is secured by nuts.

Of Ubas steel and with a journal diameter of ⅞in, the mainshafts have flanged inner ends and, too, are a parallel interference fit in the flywheels; each is located by a Woodruff key. Two ball bearings support the drive-side shaft, their inner races spaced by the keyed-on sleeve of the timed crankcase breather which discharges to the rear of the case. Timing-side bearing is a long bronze bush, cut out on its underside to accommodate the oil-pump worm drive.

The engine sprocket is splined on to the drive-side mainshaft and outboard of it is the rotor of the Wico-Pacy A.C. generator; the stator is carried in the outer half of the primary chaincase and both sides of the inner half of the case are dowelled to ensure accuracy of alignment. The drive-side crankcase half mates up with the primary chaincase and the timing-side half blends into a domed cover which carries a circular plate giving access to the contact breaker.

The reciprocating plunger oil pump, worm driven from the timing-side mainshaft, is similar in design and operation to those fitted to the other A.J.S. and Matchless singles. It draws oil from a 2½-pint oil container of kidney shape which is bolted to the outside of the timing-side crankcase half, ahead of the timing case; the inboard wall of the container is formed by the crankcase and the filler neck is in the case.

Some of the oil picked up by the supply side of the pump is fed to the worm chamber and passes through holes in the root of the worm thread into the mainshaft whence it travels to the big-end bearing through drillways. The balance of lubricant travels to the overhead rocker shafts via passages in the barrel and head. Grooves in the sides of the valve-operating arms of the rockers lead some of the emerging oil to the valve-stem ends—and there is an adjustable bleed to the inlet valve guide.

Lubricant drains from the valve compartment down the pushrod tunnels to the timing gear; from the timing chest it overflows into the crankcase to join that thrown out from the big end. The scavenge pump picks up the oil from the bottom of the case and passes it across the rear of the engine to the felt-cartridge oil filter, mounted longitudinally in the left half of the crankcase and removable from the front. From the filter the oil returns across ahead of the cylinder and, with the filler cap open, can be seen re-entering the tank.

Although no official claims are made for the power output and torque of the power unit, it is stated that prototypes have been running up to 7,400 r.p.m. and producing over 18 b.h.p. on the bench.

The most ingenious single feature of the new models is the combination of the neatness of an integral gear box with the ease of dismantling or unit replacement afforded by a separate box.

Orthodox adjustment of the primary chain, too, is achieved. The gear box has a cylindrical shell which butts against substantial faces of matching curvature on the rear of the crankcase; the right-hand end cover of the box fits snugly within a hole in the engine side cover. Since the mainshaft lies appreciably above the shell axis, rotation of the shell alters the position of the primary-chain centres. The mounting is identical in principle with that commonly used for belt-driven dynamos on cars.

The gear box is held on its seating by two substantial steel straps attached under the rear of the crankcase. At the upper ends of the straps are eye blocks threaded to take tensioning bolts passing through a bar bridging the two steel plates which embrace the gear box and are bolted to the crankcase and frame seat tube to form the rear mounting of the power unit. A draw bolt of normal pattern rotates the gear box after the straps are slackened. The drawbolt and the strap bolts are readily accessible on removal of a pressed-steel cover plate which is secured by two screws and, curving well round the back of the gear box, gives an extremely neat appearance with no indication of what lies beneath.

In the interest of standardization, the four-speed gear cluster is identical with that of the two-fifty A.M.C. two-stroke and provides internal ratios of 1, 1.30, 1.85 and 2.95 to 1. The foot-change and selector arrangements, too, are similar but the clutch thrust operation resembles that of the larger gear box for four-strokes in that the floating cable arm pivots on a thrust ball and has on its other face a cam profile bearing on a fixed-centre roller.

Primary drive is by ⅜ × 0.225in simple chain running on sprockets having 21 and 50 teeth. The clutch employs four friction plates with bonded-on oil-resisting material and embodies in its centre a vane-type transmission shock absorber employing rubber blocks as the resilient medium. In the outer half of the primary chaincase are two large holes closed by screw-in plugs; the forward hole serves as a combined inspection, filling and level orifice and the other gives access to the adjuster at the centre of the clutch pressure plate. Final drive is by $\frac{1}{2} \times 0.305$in chain on 19- and 55-tooth sprockets. Overall top gear is thus 6.9 to 1.

The malleable-iron steering-head lug of the frame is gusseted for extra stiffness and into it is brazed the $1\frac{1}{2}$in × 14-gauge top tube which curves downward at the rear of the petrol tank to become the seat tube. Part-way down it passes through another malleable lug in the front of which are two bosses for the attachment of the gear-box plates already described. This lug also incorporates the pivot bearing of the rear-suspension fork; the spindle runs in bronze bushes and the seat tube is scalloped out within the lug to provide spindle clearance. Further bosses on the lug form the attachment points for the diagonal tubes of the rear sub-frame.

Dimensions of the single front-down tube of the frame are $1\frac{1}{4}$in diameter and 14 gauge. At the top the tube is brazed into the steering-head lug and its lower end carries a brazed-on sleeve to which is welded a fabricated pressed-steel box member in the form of a tilted pyramid. This box member supports the front of the crankcase at two points.

In addition to the previously mentioned seat-tube attachment through the gear-box plates, the crankcase is linked to the bottom of the seat tube by a 12-gauge steel pressing of inverted channel section. The pressing is held to the tube by the mounting bolt for the pillion footrests; it extends behind the tube to the rear mudguard and to its rearward end is welded the tubular silencer-support stay.

A second pressing, of 14-gauge channel section is also clamped to the bottom of the seat tube. It lies beneath and within the 12-gauge channel and extends forward beneath the crankcase, turning upward at the front to mate with the engine-mounting pyramid. The result is a frame cradle of remarkably neat appearance, free from nooks and crannies to harbour dirt.

To the upper of the two pressings are welded bosses which carry the square-section bar supporting the rider's footrests. The bar passes through a spacer tube between the flanges of the channel. Outboard of the bosses are short sleeves on which the centre stand pivots. In the footrest hangers are triple-square (i.e., twelve-point-star) holes which permit adjustment in steps of 30 degrees.

The return spring of the tubular centre stand is cleverly concealed within the channel pressings Welded-on ears behind the pivot eyes of the stand are linked by a double cranked rod passing through slots in the lower pressing. In the middle the rod is slightly "joggled" to locate one end of the spring; the other end hooks over the footrest spacer tube. The slotted forward attachment of the lower pressing permits it to be hinged down (after the slackening or removal of the various cradle bolts) for access to the spring in the event of breakage.

Of $\frac{7}{8}$in tubing the horizontal stays of the rear sub-frame are bolted at their forward ends to bosses on the bend of the top tube. At the rear they are welded and gusseted to the $\frac{7}{8}$in diagonal tubes and the joints are bridged by a transverse tube carrying studs which support the upper ends of the adjustable Girling rear legs. The sub-frame is unusually narrow and is almost completely hidden by the dual-seat and the matching side cases.

The arms of the pivoted rear fork are of $1\frac{1}{4}$in × 14-gauge tubing and their forward ends are brazed into malleable-iron lugs. Pressed into the right-hand lug, and welded to it on the outside for security, is the pivot spindle; the left-hand lug is split and is fitted with a cotter. The sleeved and trapped rear ends of the arms are slotted to take the wheel spindle.

Of pressed steel, the side cases mentioned earlier are cleverly shaped to blend with the lines of the machine and conceal the ignition coil, rectifier, horn and 12-ampere-hour battery which are stowed beneath the seat. The battery sits on a platform welded to the seat tube and is secured by a rubber strap. It is readily accessible when the right-hand lid is hinged down. Within the lid, which is of deep section, is the tool kit in a compartment formed by a second, internal lid. When fitted (it is an optional extra) the air filter is located within the left-hand case.

The telescopic front fork is hydraulically damped and its pressed-steel upper yoke has extensions to carry the 6in-diameter Wipac headlamp. The yoke fits closely round the rear of the headlamp

Continued on page 101

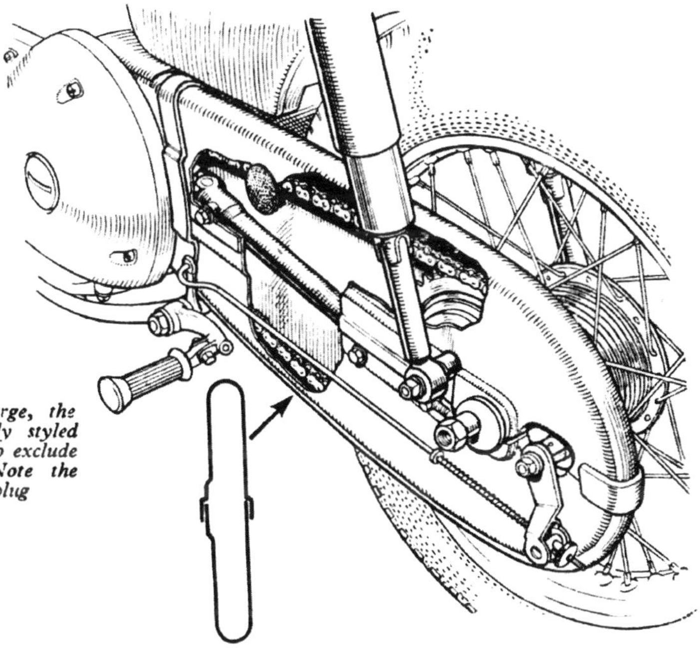

Available at extra charge, the rear chaincase is cleanly styled and the halves overlap to exclude dirt and moisture. Note the rubber inspection plug

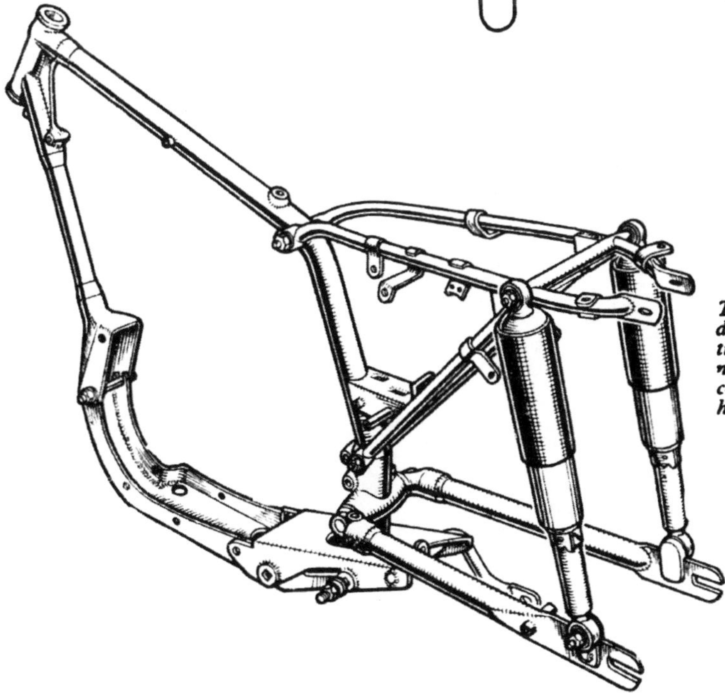

The frame is of simple design and the lower portion comprises two channel-section pressings. A cotter locates the left-hand arm of the rear fork

Hookey to

TALE OF A DAY'S OUTING

can be made as regards appearance. Confucianism creeps in and you find yourself recalling all sorts of wisecracks about beauty being in the eye of the beholder. That rather stops you in mid-sentence!

The weather on Wednesday of last week was poor. An uncompromising curtain of fog cut visibility to a few yards; the air felt waterlogged; the road verge, buildings, trees and other vehicles—all were apparent as vague blurs, becoming more vague between goggle wipes. Of course, I could have raised the goggles and persevered without them. But that for me would have meant streaming eyes and a headache later. So I kept them on.

My route to the coast was based on planning forced on me in my youth when I rode very old machines. The dogma then was that you always travelled on the outward run so as not to break anything; the fast work would keep for the return trip. I have never been able to break myself of the habit. So quitting the Dorking pylons, I took the little-used, high-hedged Newdigate road and on to Capel and Horsham. (Those living in London's south-west should note that route well,

Above: Fog enshrouds the lower part of Box Hill, famous Surrey landmark encountered on the road south. Right: The Model 14 is given a breather below the turrets of Arundel Castle, rising Baskerville-Hall-like above the lingering mist

LEFT glove, left eyepiece; right glove, right eyepiece . . . visibility, I suppose, was not too bad, but my arms were being kept on the move. They brought one hand and then the other up to my goggles as regularly as if they belonged to a Guardsman doing arms drill. Why my goggles were proving so troublesome I couldn't imagine. Car windscreens, in the main, appeared to be keeping clean and the insides of my lenses were not steaming up. At that famous roundabout near Box Hill, I stopped, dried the goggles and applied anti-dim compound to the outside. For all the good that did, I might have saved myself the bother.

Strictly speaking, I should have been at the desk. But the *Practical Handbook* discussed in the preceding two pages of prose had kept my nose to the office grindstone for nearly a month. The back of the job, I reasoned, was well and truly broken. A niggling voice from the region of my boots urged that the index was still to be completed and time was getting short. I pretended not to hear. A two-fifty Model 14 A.J.S. was burning a hole in the garage doors and had been giving me hang-dog, when-*are*-we-going-for-a-dice looks every time I'd put my head round the corner. In the end, just a week ago yesterday, I gave the looks best and pointed the job at the south coast. There, the weather men foretold, I could expect "warm sunshine in the afternoon." They were dead right, which made my delicious feeling of playing hookey all the more delicious.

This A.J.S.: since its introduction in March last year, I have looked on it with mixed feelings. Technically the design is undoubtedly right out of the top drawer. But it has always seemed to me to be a modern machine in which the stylist had not gone far enough. But to pontificate over styling is perhaps presumptuous. You can take most design features and say of them, in the light of experience, I know this to be right or I know that to be wrong. But no such cut-and-dried statement

the Seaside

By GEORGE WILSON

THROUGH FOG TO SUNSHINE WITH A 248 c.c. MODEL 14 A.J.S.

Above: At Shoreham-by-Sea the sun shone and riding gear was discarded for a brief spell. Below: The model is heeled through a roundabout on the Chichester bypass; the handling was exemplary at high and low speeds

for, as I say, it is very quiet and it runs parallel to the main coast road on which so much traffic-crawling is done in summer.)

First impressions of the A.J.S. were good—very good. There was a fair bit of mud on the road. Here and there "works" were going on and on two occasions I rounded corners flanked with roadworks signs to find, just round the apex, the road virtually blocked. These factors, combined with the poor visibility, meant that on occasion things had to be done quickly. From all these minor adventures the Ajay emerged with full marks. Both brakes were good (and the back rather better than the front). They were smooth and my feeling is that even used muttwise they would never lock the wheels. Then, too, the navigation was magnificent. The back end twitched once or twice on mud but there was never any tendency for the machine to get out of hand. There is just the degree of heaviness in the steering that I want—not so much as to make riding hard work but enough to give that feeling of true positiveness; of letting one know precisely where the front wheel is and what it's doing.

The suspension characteristics, too, play an important contributory rôle. The springs are softish and damping only moderately powerful. The result is that minor road irregularities are ironed out even at low speeds, yet at higher knots there is a very high standard of comfort and next to no pitching.

All the way to the coast the fog persisted in varying degree. Sometimes, briefly, I could read 50 m.p.h. on the speedometer, but for the most part I was groping at between 30 and 40 m.p.h. When I first reached the coast (approaching it from B2135 by way of Steyning and Bramber), it seemed that the weather prophets had let me down. Thin, wispy mist covered the area. Out to sea I could identify the foghorns of three or four vessels though I could see nothing.

A cup of tea was clearly indicated. I knew that would do the trick; and that satisfying break completed, the sun pierced through. Very soon afterwards Shoreham-by-Sea lay bathed in warmth. A beaming sun gleamed frankly from a bright blue sky.

When the engine is viewed from the timing side the oblique mounting of the light-alloy cylinder head is clearly evident. Though the gear box appears to be in unit with the engine it is in fact a separate compartment. There are no external oil pipes. The engine cover in the right-hand picture conceals the clutch and primary drive and the Wipac alternator

The quiet, deserted little township yawned in summer-in-winter. Flavourless town life seemed a long way off.

The A.J.S. is the perfect beach-potterer. The oversquare (69.85 × 64.85 mm) engine is beautifully quiet mechanically—one hears but a muted hum—and far more quiet on the exhaust than I had expected when the trip began. There was no audible piston slap. There is powerful punch low down on the r.p.m. scale so that one can trickle in a highish gear. If you ever hear a Model 14 being ridden obtrusively in a town you can label the rider "mutt" right away.

I lounged, perhaps, longer than I should, so that I hurried on my westward route to Arundel, family seat of the Duke of Norfolk, and so along the high-speed Chichester bypass to Portsmouth. If the A.J.S. had impressed earlier it did so to even greater extent then. The compression ratio is 7.8 to 1 and on the ordinary premium-grade fuel I was using there was no pinking when the grip was twisted hard. No official claims are made as regards power output, though I have heard mention of 18 b.h.p. It would appear to be all of that.

Inspired by an empty stretch of the Chichester bypass I tucked in a bit and held 70 m.p.h. With me sitting up (and I was wearing just about as much as I could get on) the Model 14 would get in the sixties quickly and was not being overdriven at that. Power delivery could not have been more smooth. There was no vibration. The lower three gears are well chosen and combined with the engine punch provide the sort of off-the-mark acceleration that makes Ford Consul drivers give up. The overall ratios are 20.12, 12.75, 8.96 and 6.89 to 1. Gear-pedal movement is just right; light without being over-light and with a range of movement that permits upward or downward changes to be made merely by the rider pivoting his foot about the right rest.

By one o'clock the sun was really doing its stuff. The temperature rocketed and I was able to lower the scarf off my chin. My hookey was really paying off. Affection for the A.J.S. grew with every mile. The last two-fifty I owned was a 1936 New Imperial—a beautiful little job for which I still harbour a deep affection. But where it was fragile the A.J.S. is robust. Maybe you could overdrive a Model 14. I just couldn't.

Between Arundel and Portsmouth the route is punctuated with built-up areas so that the bursts of full-chat were brief, giving the model time to recover its breath, as it were, between spells of fast going. We would see what would happen on A3, the main road back to London, with its long straights, sweeping, high-speed swerves and very little need for let-up.

The answer is that nothing happened. There was no wind. I never adopted any position more unorthodox than what Vic Willoughby calls my I.S.D.T. crouch. The road was dry. There was very little traffic. The 64 miles were reeled off in an hour and 20 minutes, an average speed of 47 m.p.h. Speed limits were observed throughout—and I'm much too long in the tooth to ride in any way that may court disaster. In other words, pretty well anyone could have done it, and many would have lowered that time by five minutes and some even by ten.

The standard of comfort provided by the suspension is, as I have said, high throughout. I liked the relationship between the seat and handlebar and, at speeds in the forties, that between the seat and the footrests. The riding position is truly "Viney" or "Manns." But batting in the fifties and lower sixties I would have preferred the rests an inch or so farther back, so that my legs could have relieved my arms of some of the load imposed by the increased wind pressure.

My early impression of the brakes was confirmed. Both are a treat to use. The rear-brake pedal could be felt moving underfoot with up and down movement of the rear wheel, but that never once worried me a jot. All the controls were light and smooth in operation—right in the A.J.S. tradition.

At the end of the day I had clocked 166 miles. I drew in at my local garage where I had filled up in the morning and looked interestedly at the gauge as the fuel went in: two gallons as near as might be—an average consumption of about 80 m.p.g. But I believe the feature that impressed me most was the utter cleanliness of the engine and gear-box side covers. They were free from oil in all respects and not much mud-dappled either—so that deep-section front mudguard really does its stuff as a mudguard should.

A famous designer, discussing motor-cycle trends a few years ago, prophesied that the machine of the future would be a lightweight. Not a lightweight in the utility sense, but a small-capacity model with a performance that would satisfy most of us. My experience with the A.J.S. bears this out. Low weight spells good steering and roadholding and a high standard of braking, and zipping acceleration from a relatively modest power output. My first new machine after the war was a single-cylinder five-hundred A.J.S. with which I all but wore a weekend groove between London and my home town in Fife. The two-fifty is very nearly that model's equal in terms of performance and it is superior by far in terms of comfort and roadholding.

By the way, should you be a Model 14 owner, you will be interested to know that the job I rode was fitted with a prototype prop stand soon to be made available. Mine was incomplete—less an extension to make it easily operable—but it promises well. Certainly it held the machine securely on all but one very steep camber in Arundel.

P.S. Don't worry overmuch about the index to the "Practical Handbook." We burned a gallon or two of midnight oil and, as you will find, when you receive the book with our special Spring Number and Buyers' Guide next week, it will all be there—all seven pages of it. Whew!

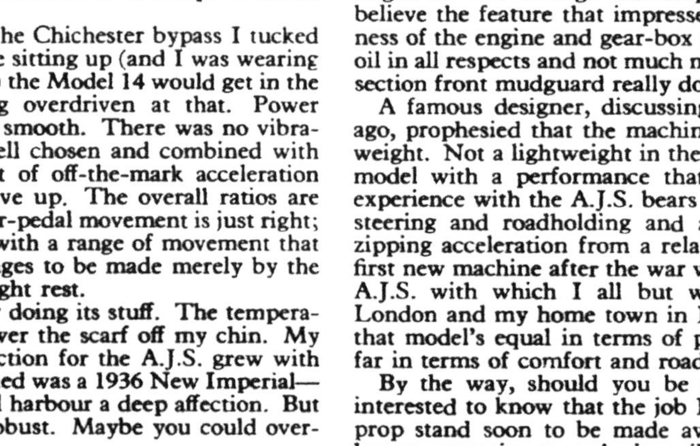

Neat handlebar assembly: the front-brake and clutch levers are welded to the bar and conveniently positioned for ease of operation; they lie below the horizontal plane of the grips by just the right amount

Built for Speed No. 2

JOHN GRIFFITH describes

Grown-up "Boy Racer" for Cross-country

Harold Lines' 500 c.c. o.h.c. A.J.S. Scrambler

EXCLUSIVE TO "MOTOR CYCLING"

ALL followers of the sport have heard of Swedish star Bill Nilsson's potent scrambler, which employs an oversize A.J.S. 7R engine. Now this country can boast a similar machine. It is the property of prominent Horsham dealer Harold Lines, who is himself a very well-known ex-moto-cross rider.

Harold bought a second-hand 7R of 1948 vintage about 18 months ago. Since then, his Crawley branch manager, Alan Willsher, has been working intermittently to convert it to a 500 c.c. mount suitable for scrambling. Since the machine was purchased, Brian Leask has joined Harold's staff and he has helped considerably on the project, dealing with most of the cycle parts.

The engine now has dimensions similar to those of some 500 c.c. Nortons—in fact, it employs a piston of that make which has been only slightly altered. The connecting-rod and big-end assembly is of 500 c.c. competition A.J.S. pattern and is carried in a pair of flywheels machined from blanks. The outside of the big-end eye has been shaped, by cutting away surplus metal, to form a circumferential rib. This has reduced its weight appreciably. The main-shafts and main bearings are standard 7R; a balance factor of 0.74 was used on the flywheel assembly.

To accommodate a 100 mm. stroke, the engine is $\frac{15}{16}$-in. taller than the standard 7R and this has made it necessary to cut the timing chaincase halves and weld an extra piece in each. As these parts are made of Elektron, the job was rather ticklish! The $\frac{15}{16}$-in. dimension is dictated by the timing chain, which has $\frac{3}{8}$-in. links, so that one extra link allows the centre points of the sprockets to be $\frac{3}{16}$-in. farther apart; five extra links are used in the chain on the Lines engine. Wellworthys made up a new liner for the original 7R alloy cylinder muff, with a cylinder base flange the requisite $\frac{15}{16}$-in. deeper. New securing studs had also to be made.

(Left) Brian Leask aviates the Ajay.

Except for proprietary fittings, all parts of the Lines special are of A.M.C. origin.

ROAD TEST NO. 103

By the CYCLE Staff

The AJS Typhoon showed remarkable ability to climb the steepest grades without the least tendency to side slip.

AJS TYPHOON SCRAMBLER

Any average competition rider who tries this unique British import, specifically designed for the U.S. market, will agree that its performance is well beyond his most optimistic expectations, especially on the toughest rough going of the Californian scrambles.

The engine of the new AJS Typhoon single has a displacement of well over 30.50 cu. in. In fact, the original 86 mm bore and 85.5 mm stroke of the actual 500 cc engine are increased .050″ and 3/8″ respectively in the test bike of this month.

The performance of this big single will also enable riders to compete against the many 650 cc twins now so popular with dirt riders.

The test bike was furnished to the CYCLE staff by Cooper Motors, of Los Angeles, California, U.S. distributors of Matchless and AJS machines. Incidentally, the Typhoon showed us during the test that she definitely was a "matchless" AJS member of the Plumstead family!

During the test the front fork was the highlight of the suspension department. It scored high in all phases of the punishment that we put it through. Even the most vicious and deliberate attempts to make it bottom proved futile. This teledraulic fork might look too soft when you try it with the machine stationary, but don't cheat yourself! She will disappoint you if you ever expect to feel the characteristic knock when bottoming while riding on the rough. The smoothness of the recoil stroke was also remarkable. The rear shocks can be adjusted to provide a stiff, medium or soft action. Thus they offer a wide and versatile operational range to suit most of the combinations of rider's weight and conditions of track.

Another high point of the test was the braking efficiency, achieved thanks to the generous dimensions of the front and rear brakes and their full width

This is where the grip comes from! Notice the thick, straight type spokes.

AJS and Matchless U.S. distributor Frank Cooper discusses the fine points of the Typhoon with CYCLE tester Castro.

Deeply finned cylinder, large bore carb and sleek gas tank are outstanding features.

deeply finned drums. They remained absolutely fadeproof for the duration of the test and could be operated almost effortlessly.

The Flanders-made wide handlebar accounted for first class steering and road holding qualities, on and off the track.

The Burman gear box provided positive and fast shifting under all conditions. This is a remarkable feat in any motorcycle but it is even more appreciated on a competition mount.

There is a wide variety of gearing available (see specs. chart) but we feel that the standard ratio of the Typhoon will meet the requirements of most scrambles riders. All in all, the ultra discriminating competition rider can find among the great variety in stock, the ratio of his particular choice.

In spite of the straight 9:1 compression ratio of this fabulous single its traffic manners have not been impaired and it has no reason to envy the smooth going of any twin available in the market today. Of course it will not be velvet smooth (if you are too fussy about it!) but will provide you with exciting all-around transportation plus the additional convenience of owning a machine that can be ridden to, on and from the competition events.

Digging through mud and grass is where the Typhoon demonstrates at its very best the surging power of the engine.

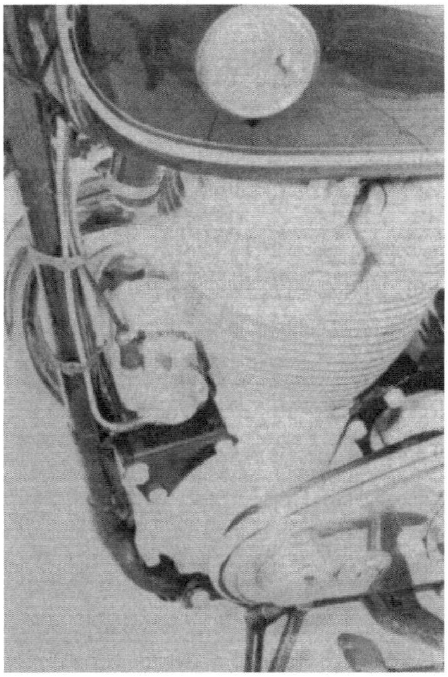
The neatly located Lucas competition type mag is very easy to check.

The acceleration through gears is worth mentioning and the high torque qualities of the all-aluminum alloy engine can be appreciated as soon as you ride it. The AJS Typhoon can be ridden in top gear at 15 mph without chain snatching and in a few seconds you can wind it up to 60-plus without the slightest protest from the engine or transmission!

As the test bike was also ridden in public streets it was fitted with a muffler which didn't seem to impair to a great extent its performance in the dirt.

The clutch lever was very light to operate. The Typhoon has no steering damper, nor does it seem to need any.

The ignition advance control and the air levers are located on either side of the handlebar. Both are easy to reach and operate. On the LH side of the handlebar there is also a valve lifter lever which helps to obtain easier starts.

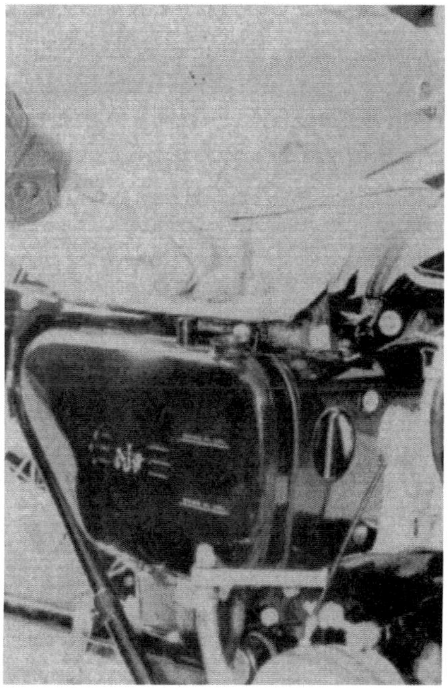
The small dimension of the oil filler cap can be noticed in the above photo.

The rear and front fender are made of aluminum and highly polished Both wheels are fitted with thick and straight spokes. The advantages of the straight pattern spokes over the hook type are obvious. The wheel hubs and the low-

The AJS reliable brakes always provided surefooted stops riding downhill.

er legs of the front forks are also made of aluminum in order to cut the unsprung weight as much as possible.

The oil tank is neatly located on the RH side of the machine and right opposite to it, on the LH side, there is a roomy tool box. It is our impression that in a competition mount like the AJS Typhoon the oil tank filler cap should be a little bigger and easier to remove.

The carb air filter is located next to the oil tank. This tank has a protective outside metal panel so that the rider's leg cannot be affected when the oil gets too hot.

The valve clearance can be checked without disturbing the gas tank. A chromed and business-like filler cap is located on the top of the gas tank.

The folding-type KS lever is conveniently reinforced to stand the stress caused by the high compression engine. The husky and gusseted footpegs are designed and built to cope with the toughest beat ever imagined by a competition rider. The short and positive gear shift lever is equipped with a gear indicator.

A sturdy and thick skid plate protects the underneath of engine and transmission. On the LH side of the machine there is a handy jiffy stand.

The roomy and comfortable twin seat provides for a more than adequate number of riding positions.

To save weight the primary chain cover is made of pressed steel instead of aluminum alloy as the one fitted to the AJS or Matchless twins. The chain lash can be checked through a small port located at the lower section of the chain cover.

In spite of the fact that nowadays the pressed steel chain case is used very seldom by motorcycle manufacturers and has been replaced by aluminum alloy in practically all makes of motorcycles, it is worth mentioning that the Typhoon's primary chain case remained absolutely oiltight during the various stages of the test.

A hypercritical tester might have complained that the exhaust pipe and muffler should be a little more upswept; that the engine was a trifle hard to start when cold and that no adjustment appeared to be provided for the footpegs which—to be honest—didn't need it because they fit the average rider without any problem.

But beyond these minor points of detail design, nothing adverse could be said, and they count little compared with the overall excellence of the layout, handling and performance of this "big lunger."

In brief, the AJS Typhoon provides all the performance required of a competition single 500-plus, without the noise, intractability and extravagance once accepted as the inevitable price for riding a scrambler. It is this blend of punch and charm which is the special appeal of the model.

Excellent handling on sandy surfaces, even with the rear brake fully applied was a nice feature.

After test minor clutch adjustment was required.

AJS TYPHOON

Specifications

ENGINE: Single OHV; alloy cylinder head and barrel; 9:1 compression ratio 587.8 cc.; 47 HP at 6750 RPM; Amal 389/12 Monobloc carburetor; Lucas competition-type magneto with manual advance; spark plug located at the top of the combustion chamber.

TRANSMISSION: Burman four-speed with foot operated shift lever. Multiplate clutch in oil bath. Standard ratios: 1st: 15.48:1—2nd: 10.26:1—3rd: 7.83:1 and 4th: 5.80:1.

Optional ratios:

First Gear	Second Gear	Third Gear	Fourth Gear
18.39 to 1	12.19 to 1	9.30 to 1	6.89 to 1
17.30 to 1	11.47 to 1	8.74 to 1	6.48 to 1
16.34 to 1	10.83 to 1	8.26 to 1	6.12 to 1
14.71 to 1	9.75 to 1	7.43 to 1	5.51 to 1
14.01 to 1	9.29 to 1	7.08 to 1	5.25 to 1
13.37 to 1	8.86 to 1	6.76 to 1	5.01 to 1

GAS MILEAGE: 50 mpg. (overall avg.)

CAPACITIES: Gas tank, 2 gals.; oil tank, 4 pts.; gear box, 1 pt.; primary chain case, 12 ozs.

WEIGHT: 302 lbs.

TIRE SIZE: 4.00-19 rear, 3.50-19 front; both knobby type.

GROUND CLEARANCE: 8½"

ROAD IMPRESSIONS OF NEW MODELS

347 c.c. Model 16 A.J.S.

By DAVID DIXON

FAMOUS SINGLE PUT THROUGH ITS PACES IN SOLO FORM

THE plot sounded good to me. The 347 c.c. Model 16 A.J.S. would be available for a couple of weeks in solo form. After that it would be harnessed with a Wessex Sports sidecar for a further period. Most, of course, are familiar with the Model 16: orthodox, proved to the hilt, a roadster with all the flywheel effect and low-speed punch one normally associates with those magnificent trials models on which Gordon Jackson and others do such incredible things on mud and rocks, weekend in and weekend out, the country over.

I felt at home the moment I cocked my leg over the saddle. Everything about the riding position was just right. Some might have found the handlebar a shade long and the tank a trifle wide between the knees. But there was no impression of being on the biggish model the A.J.S. *looks* to be. Once under way in traffic, the machine was as docile and tractable as a horse going to hay. In spite of its top hamper, the A.J.S. could be filtered through traffic gaps in effortless style. Low-speed steering was excellent if a trifle heavy. And a really good steering lock proved most helpful in maneouvring the machine in confined spaces.

Mechanical quietness has for long been a feature of the A.J.S. singles and the model under review was exemplary in that respect. A whine from the primary chaincase was audible, but only just. Apart from that there was merely a gentle rustling from the valve gear. Not a trace of piston slap could be detected whether the engine was hot or cold. Provided that the throttle was used with discretion, exhaust quietness was average for a single. When the throttle was tweaked hard, however, the exhaust took on a stentorian note which though pleasant, perhaps, to those who appreciate that sort of thing was not so much fun for pedestrians.

When delivered, the A.J.S. had covered only 280 miles. Even allowing for that, the performance was sprightly if not super-sporting. Good engine tractability and sweet transmission encouraged normal upward gear changes at 15, 20 and 30 m.p.h. in traffic. In the initial stages a comfortable cruising speed was around the 50 m.p.h. mark. Fully run-in, the engine would push the job along at 65 m.p.h. plus all day long without, as it were, turning a hair. Throughout there was virtually no vibration.

The clutch was light to operate, smooth in its take-up of the drive and impervious to abuse. A caress of its lever was all that was necessary when changing gear. The gear box, too, came in for high praise. Pedal movement is moderately light and moderately brief and the length of the pedal is such that upward and downward changes could be made merely by pivoting the right foot on its rest. For the sheer delight of it, the gear box was used rather more than was necessary; it can therefore be assumed that my overall fuel-consumption figure of 77 m.p.g. could reasonably be improved upon.

I am something of a fuss-pot where handling is concerned and in that respect the A.J.S. was good but not perfect. I mean nothing derogatory: what I am trying to get across is that the Model 16 is a touring rather than a super-sporting mount and it has, for one thing, a $3\frac{1}{4}$ gallon tank. The top hamper is such that the machine has to be put round corners, and a conscious effort has to be made when it is being cranked swiftly from lock to lock. The A.J.S. was rock steady over most types of road surface and straight-head steering was above reproach.

Oddly enough—for the A.J.S. front fork is normally soft in action—the front suspension proved to be on the hard side, the result, it was felt, of tightish oil seals and bushes. The rear springing was adequately soft. From high or low speeds, the braking was first class and both tyres could be made to squeal without the wheels being locked. I would have felt happier had a rear stop light been fitted.

Those deep, stylishly valanced mudguards were truly effective. I know—I rode through a couple of downpours! But as the rear guard extends well down behind the wheel and does not have a detachable tail piece, wheel removal was a difficult, exasperating

business. Accessibility for other routine maintenance jobs, on the other hand, was excellent and the tool kit was commendably comprehensive. Small points I liked were the knurled finger adjusters for both brakes, the slotted cover in the cast-aluminium chaincase giving access to the clutch thrust-rod adjustment, a sensibly large fuel reserve and the stout rubber retaining strap that effectively eliminates chafing of the battery. I failed to see the point of fitting a 120 m.p.h. speedometer: too many figures clutter the dial.

How would the machine perform with the sidecar? The outfit was taken to the Isle of Man for a two-week period. With the Wessex fitted, what a beautifully balanced outfit it looked! Favourable comment was forthcoming from all who saw it. And the handling was even better, if that were possible, than was the outfit's appearance. Unladen, the lean-out of the machine was 1in and the toe-in of the sidecar wheel was $\frac{3}{4}$in. With the sidecar empty there was a barely noticeable pull to the right; with a load in the chair and the steering damper screwed down by one turn, straight-ahead steering was hands-off. Heavy-duty sidecar springs had been fitted to the front fork.

For the run to Liverpool I left London a little before 7 p.m. and just managed to catch the 1 a.m. boat—and Liverpool was 217 miles away by the route I took. When London was left behind the speedometer registered only 550 miles, so I had to beware of flogging the engine too hard. Yes, I'll admit I was a trifle worried. However, after starting cautiously, I soon found that the outfit would whistle along quite happily at 50 m.p.h. on the 6.48 to 1 sidecar top gear (the solo top ratio is 5.8 to 1). Even with a luggage load which included Vic Willoughby's leathers and my own—and enough stationery to start a business—the chair wheel was inclined to lift on sharpish left-handers. But there seemed to be no limit to the speed at which the outfit could be driven round right-hand bends; overall there was a feeling of absolute tautness.

As the miles totted up and time grew short, I forgot my scruples about new engines. The throttle was tweaked harder, the exhaust assumed an urgent note and the exhaust pipe turned from straw colour to blue. The speed rose from 50 to 60 m.p.h. The landing stage was reached at exactly 1 a.m.! Gang-planks

AND LATER WITH A WESSEX SPORTS SIDECAR ATTACHED

Above left: The Mode 16 is finished in black and chromium and the tank is lined in gold. Above: The author pauses during an out-of-town run on the solo. On the right the A.J.S. and Wessex outfit is in the Isle of Man—by the new Bungalow near the summit of Snaefell

were already being pulled away from the ship's side as I purchased my tickets and hustled the outfit up the ramp and on to the deck. It says much for the handling and alignment of the outfit that after a cup of tea I felt fresh enough to have started all over again. Though there was no brake on the sidecar wheel, the A.J.S. stoppers had proved more than adequate and, even when applied hard, halted the outfit in a perfectly straight line. The overall average speed was 38 m.p.h. and that included stops for re-fuelling and direction finding; petrol consumption worked out at 57 m.p.g. The lighting had proved good enough to allow the use of all the available performance. On the return trip from Liverpool to London a week later, the average was stepped up to 40 m.p.h. and the fuel consumption improved to 58 m.p.g.

Everyone who drove the outfit in the Island was as impressed with its handling and surprised, too, at the way in which that 347 c.c. power unit coped with every situation. The secret of obtaining the best results was, of course, to keep the engine on the boil. A standing start on a 1 in 6 hill in Douglas was made three-up. Even though the all-up weight was in the region of 40 stones, the outfit pulled away smoothly.

A lap of the Mountain course with me in the sidecar proved stimulating. The construction of the sidecar is interesting. A stressed-steel middle section forms a backbone to which glass-reinforced plastic nose and tail sections are bolted. Four steel plates bolted to the middle section provide pick-up points for the connections to the machine. No fewer than seven straight tubes from the sidecar pick up to five points on the machine, making an extremely rigid, cross-braced structure. Wheel suspension is of the plunger variety: the stub axle is carried by a light-alloy bracket which slides on two vertical guides; movement is controlled by an Armstrong spring-and-hydraulic unit.

Rather a lot of engine and transmission noise was transmitted via the connections and bulkhead; doubtless it could be overcome by better insulation. The seating position was quite comfortable, though care had to be taken to keep one's calves away from the steel body former. There was ample leg and elbow room. A footrest is provided but it was too far forward for all

The power unit remained remarkably clean externally. The contact breaker is reached by removing the cover plate on the end of the timing chest

but tall passengers. Draught from the left side was traced to air deflected upward from the integral wheel fairing. Ease of entry and exit was aided by the hinged screen and a cast light-alloy step. Luggage-carrying capacity in the boot behind the seat was admirable: there seemed to be no end to the amount of gear that could be stowed.

All in all, in both solo and sidecar forms, the Model 16 proved itself a thoroughly sound proposition for the enthusiast who requires robust and reliable transport for everyday use.

Below: The battery is housed in a compartment on the left-hand side and retained by a rubber band—a feature that ensures against chafing of the case. The bottom picture shows the rocker clearance being adjusted. The job is easily carried out with the tank in position

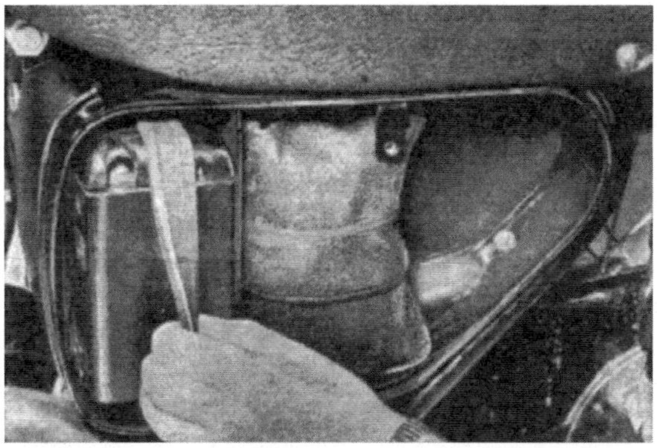

SPECIFICATION

ENGINE: A.J.S. 347 c.c. (69 x 93mm) overhead-valve single; crankshaft supported in two ball bearings on drive side and a bronze bush on timing side; caged roller big-end bearing; light-alloy cylinder head; compression ratio, 7.5 to 1.

CARBURETTOR: Amal Monobloc; air slide operated by handlebar lever.

TRANSMISSION: A.M.C. four-speed gear box; positive-stop foot control. Solo gear ratios: bottom, 15.48 to 1; second, 10.26 to 1; third, 7.83 to 1; top 5.8 to 1; sidecar ratios: 17.3, 11.47, 8.74 and 6.48 to 1. Multi-plate clutch with oil-resistant friction material and incorporating vane-and-rubber shock absorber. Primary chain, ½ x 0.305in in cast-aluminium oil-bath case. Final drive by ⅝ x ⅜in chain; guard over top run.

BRAKES: 7in diameter front and rear.

SUSPENSION: Telescopic front fork with two-way hydraulic damping. Pivoted rear fork controlled by three-position adjustable Girling units incorporating hydraulic damping.

WHEELS and TYRES: 19in-diameter wheels with straight-pull spokes and light-alloy full-width hubs. Dunlop 3.25 x 19in tyres, ribbed front, Universal rear.

ELECTRICAL EQUIPMENT: Lucas RM15 a.c. generator with emergency-start switching. Coil ignition. Lucas 7in-diameter headlamp with pre-focus light unit containing pilot bulb and 30/24-watt main bulb. Lucas 6-volt, 12-amp-hour battery.

FUEL CAPACITY: 3¾gallons.

OIL CAPACITY: 4 pints.

DIMENSIONS: Wheelbase, 56in; ground clearance, 6½in; seat height, 32in.

PRICE: £185 10s; with purchase tax (payable only in Great Britain), £223 15s 3d.

MANUFACTURERS: A.J.S. Motor Cycles, Plumstead Road, Plumstead, London, S.E.18.

WESSEX SIDECAR

CONSTRUCTION: monocoque; stressed sheet-steel middle section, glass-reinforced plastic nose and tail. Hinge-up screen, canvas hood, canvas-and-celluloid all-round screens. Wheel axle carried by light-alloy bracket sliding on two vertical guides; movement controlled by Armstrong spring unit with hydraulic damping. Avon Triple Duty 3.25 x 19in tyre.

DIMENSIONS: Overall length, 7ft 9in; seat back to footrest, 45in; seat to hood, 30in; width at elbows, 35in; seat back, 16½ x 16½in; seat, 16½ x 17in.

PRICE: £93; with purchase tax, £112 3s 7d. Chromium-plated bumper, £2 10s extra.

MANUFACTURERS: Wessex Sidecars Ltd., High Street, Othery, near Bridgwater, Somerset.

The 646 c.c. MATCHLESS Model G12CSR

An 'over-the-ton' machine with superb handling on rough stuff or motorway

Specification

ENGINE

Type	Parallel-twin four-stroke
Bore	72 mm.
Stroke	79.3 mm.
Cubic capacity	646 c.c.
Valves	Overhead (push-rod)
Compression ratio	8.5 : 1
Carburetter	Amal Monobloc, 1⅛-in. bore
Ignition	Lucas K2F magneto manual control
Generator	Lucas E3L 6-v., 60-w.
Makers' claimed output	Not released. Maximum power at 6,200 r.p.m.
Lubrication	Dry sump with two gear pumps
Starting	Folding kickstarter

TRANSMISSION

Separate gearbox with footchange

Ratios	4.8, 5.8, 8.1, 12.2 : 1
Speed at 1,000 r.p.m. in top gear	16 m.p.h.

Speed equivalent to revs at maximum power rating:

Second gear	60 m.p.h.
Third gear	84 m.p.h.
Top gear	104 m.p.h.
Primary drive	Chain
Final drive	Chain
Clutch	Multi-plate in oilbath
Shock absorber	Rubber-vane type in clutch centre

CYCLE PARTS

Frame	Duplex cradle with brazed lugs
Front suspension	"Teledraulic" telescopic forks with coil springs and two-way hydraulic damping
Rear suspension	Swinging fork with three-position Girling hydraulically damped spring units
Wheelbase	55 in.
Tyres	Dunlop: ribbed 3.25 × 19-in. front; studded 3.50 × 19-in. rear
Brakes	7-in. dia. in full-width hubs front and rear; total lining area 23¼ sq. in.
Fuel tank	Welded steel; two taps
Oil tank	5 pints with de-aeration tower
Lamps	30/24-w. head; 3-w. pilot; 18/6-w. stop/tail; 1.8-w. speedometer
Horn	Lucas
Battery	Lucas MLZ9E, 13 a.h.
Speedometer	Smiths 120 m.p.h. with trip
Seating	Dualseat
Stands	Centre and prop
Tool kit	Spanners: 3 double-ended, 1 ring, 1 adjustable, 1 tappet, 1 contact-breaker, 2 box, 1 Girling, 2 tyre levers; 2 Allen keys; screwdriver; clutch-spring tool; tommy bar; grease gun; pliers; circlip tool.
Toolbox	Semi-pannier type on left side
Finish	Black frame and forks, Cardinal red tank and sundries

OTHER EQUIPMENT

Siamesed pipes; polished light-alloy mudguards; pillion rests; competition rear number plate; detachable chromium tank panels.

PRICES

Machine	£280 (inc. £47 17s. 6d. P.T.)
Extras	None
Total as tested	£280
Tax	£4 10s. p.a.; £1 13s. for four months
Makers	Matchless Motor Cycles (Associated Motor Cycles, Ltd.), Plumstead Road, London, S.E.18

'Motor Cycling' Test Data

Conditions. *Weather: Cold, dry (Barometer 29.95 in. Hg. Thermometer 40°F.). Wind: North, 10 m.p.h. Surface (braking and acceleration): Dry asphalt. Rider: 11½ stone, wearing one-piece racing leathers, boots and safety helmet, normally seated (except for "Best certified M.I.R.A. maximum" and flying lap). Fuel: "Super" grade (101½ research method octane rating).*

Venue: *Motor Industry Research Association Station, Lindley.*

Speed at end of standing 1,000 yd.:	
East	96 m.p.h.
West	80 m.p.h.
Best certified M.I.R.A. maximum (rider prone)	108.5 m.p.h.
Flying lap speed (rider prone)	106.2 m.p.h.
Braking from 30 m.p.h. (all brakes)	10 yd.
Fuel consumption:	
At constant 30 m.p.h.	96 m.p.g.
50 m.p.h.	66 m.p.g.
70 m.p.h.	53 m.p.g.
500-mile overall figure	51 m.p.g.

Speedometer

30 m.p.h. indicated = 28.2 m.p.h. true
40 m.p.h. indicated = 37.8 m.p.h. true
50 m.p.h. indicated = 46.2 m.p.h. true
60 m.p.h. indicated = 55.4 m.p.h. true
70 m.p.h. indicated = 65.0 m.p.h. true
80 m.p.h. indicated = 73.8 m.p.h. true
90 m.p.h. indicated = 83.5 m.p.h. true
100 m.p.h. indicated = 92.3 m.p.h. true
110 m.p.h. indicated = 100.8 m.p.h. true
120 m.p.h. indicated = 108.0 m.p.h. true

Mileage Recorder .. .. Accurate

Electrical Equipment

Top gear speed at which generator output balances:

Minimum obligatory lights	20 m.p.h.
Headlamp main beam	26 m.p.h.

Weights and Capacities

Certified kerbside weight (with oil and 1 gal. fuel)	430 lb.
Weight distribution (rider normally seated):	
Front wheel	40%
Rear wheel	60%
Tank capacity (metered):	
Total	4.13 gal.
Reserve: right side	2 pints
left side	1½ pints

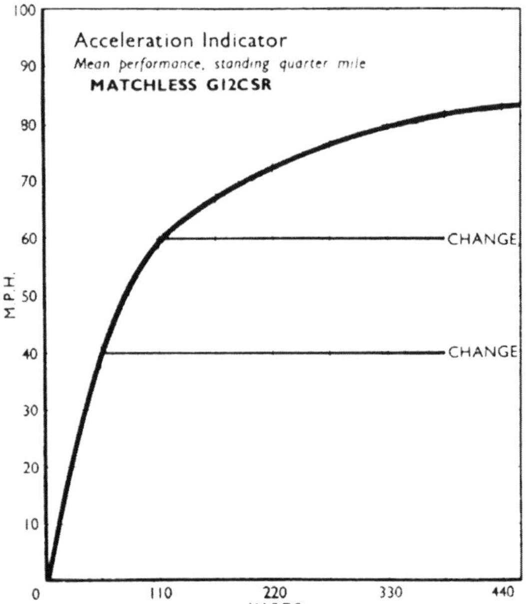

LARGEST, fastest and most powerful machine in the current Matchless range, the 650 c.c. G12CSR is essentially designed for sport. This its specification—tuned motor, siamesed pipes, good ground clearance and semi-competition equipment—shows at a glance.

But it is also at home under normal road conditions. Here, in fact, is a dual-purpose mount.

For fast riding over rough terrain—the kind of country one would cover in following this week's "Scottish," for example—there could be few machines more suitable. And the CSR is equally happy back on the metalled highway, when the rider can thrill to its bend-swinging properties, its road-holding, its slick gearbox and ample power.

The centre-stand tucks up in a manner few roadster components can equal (nobody would ever ground a CSR stand!) whilst the dual-seat is intended primarily as a saddle of generous length for one only. *Vive le sport!*

That's the spirit in which the machine as a whole has to be judged.

First, the engine. Here is the classic design of vertical twin (plus the refinement of centre main-bearing crankshaft), developed to a high degree—and tuned also. Just when many mounts are beginning to "tail off" in top gear, the power of the CSR continues progressively. This happens at about 80 m.p.h. At 75 things are quiet; 5 m.p.h. later everything gets into step, and the Matchless is scorching off to the horizon.

"The CSR really excelled when being thrown from side to side through fast bends. . ."

(Left) The siamesed exhaust pipes are jointed, with a pinch-bolt fitting, on the centre-line of the machine. The dynamo is mounted on the front of the crankcase.

Here is the type of performance for which the customer has bought *this* mount—top-end go. But we found no objection to using the gearbox and revving the CSR very freely indeed, for there was no serious vibration. Some could be felt, but not enough to dictate the choice of cruising speed or to produce any signs of stress in sheet-metal components. A carburetter bellmouth shook loose; that was all.

At the bottom of the scale, the engine was docile, responsive and thrustful. Opening up at 30 m.p.h. in top was easy and useful. Consequently, the traffic manners of the CSR were rated as good. But for overtaking at main-road speeds it was usual to use the box freely.

Generally, the control lever for the ignition was left fully advanced. It was retarded mainly for starting and when slogging along, clutch right home, in bottom gear under adverse conditions. This drill applied when 100-octane fuel was in the tank. If premium (95 to 96 octane) was used, then so was the control lever, frequently.

Fuel consumption averaged 51 m.p.g. overall, a normal return for the kind of riding which this sportster invited. Oil consumption ran at some 500 miles per pint and no leaks of any kind, anywhere, were evident after our 2,200 miles.

The engine was beautifully quiet mechanically. Exhaust noise was—well noisy. To the enthusiast it had a distinctly thrilling quality, like a giant tearing calico. Use of the throttle in built-up areas had to be adjusted accordingly.

Maximum speed recorded at M.I.R.A. was 108¼ m.p.h. There is little doubt that this could be exceeded by the man obsessed with the desire to do the "ton-ten," or some such fascinating figure. The simple expedient would be to cog up. At 108 the Matchless had "run on" past its peak, and could therefore manage a higher fourth ratio. But this would be to its detriment in other respects—acceleration, for instance. Bottom gear is already highish (and fit for clocking 40 in short time as a regular practice).

Actually, in relation to each other, the existing ratios are well chosen. Second was particularly useful for brisk overtaking and for scorching up awkward hills with twisting,

Continued on page 82

The 646 c.c. MATCHLESS Model G12CSR

Characteristic features of this manufacturer's parallel twins are the centre-mainbearing crankshaft and twin gear-type oil pumps mounted on an outrigger plate and driven from the ends of the camshafts. At the base of the "Teledraulic" forks, shown part-sectioned, can be seen the tapered ram which gives progressive hydraulic damping on heavy compression. The four-speed gearbox is of the same basic design as that fitted to the factory's racers.

MOTOR
with
SCOOTER WEEKLY
CYCLING
COPYRIGHT

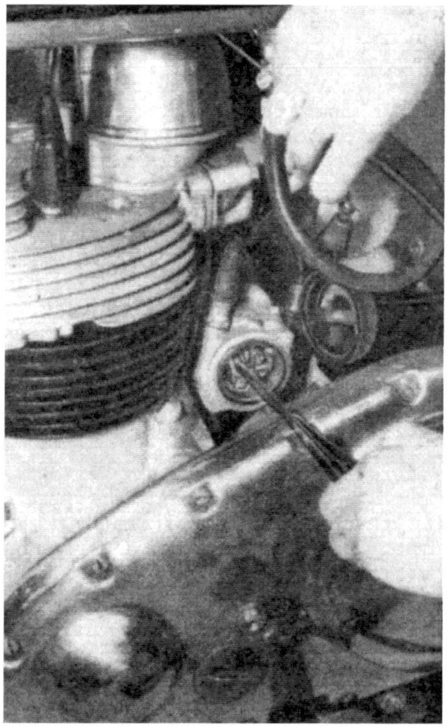

(Above) Topping-up the oil in the primary chaincase. (Top right) Adjusting the contact-breaker points of the separate magneto. (Right) A built-in tommy bar assists withdrawal of the q.d. rear wheel spindle. Unusual is the near-vertical speedometer drive pick-up.

MATCHLESS G12CSR ROAD TEST

" nadgery " bits. The gearchange action was beyond criticism. This box—Matchless' own—is undoubtedly one of the best on the road.

Clutch behaviour, most of the time, was as pleasant. Initially, there was a bout of slipping but matters were rectified when the level of the oil in the chaincase was lowered to a point at which the single-row chain just caressed the surface.

A motor as quick as this one demands good roadholding. A 1961 Matchless, with its duplex frame, has it. The CSR really excelled when being thrown from side to side through fast bends, when traffic conditions permitted, and one consequently well understood its selection by T.T. travelling marshals for Island use.

Both front and rear suspension systems were on the firm side, as befits a motorcycle designed for this usage. If the machine was flung at high speed over long-pitch undulations, the ride remained completely steady and there was absolute freedom from prancing. At low speeds the machine exhibited all the delightful feet-up stability for which this factory's products are renowned in the trials sphere. Handlebar lock is generous.

On the rough, the navigation was up to scrambles standards. At the M.I.R.A. track (which was once a Bomber Command aerodrome) the tester hit a 10-in.-high conical runway-light cover of cast concrete. At the time he was riding one-handed, looking over his shoulder. It is to the credit of the steering qualities of this Matchless—and to nothing else—that not only did he remain aboard, but the machine could be kept on course, although the steering had been deflected on to almost full lock. Convincing. . . .

The resilient dual seat was comfortable, but offered too high a riding position for the tester's particular stature.

The Matchless recorded a perfectly satisfactory braking figure from 30 m.p.h. It was also capable of pulling up from really high speeds in complete safety, provided the rider made maximum use of both units.

It was thought that the 7-in. diameter front unit (lining width, $\frac{7}{8}$ in.) was rated fairly close to its limit but the conveniently placed adjuster permitted easy setting while on the move.

The rear brake always did everything that was asked of it, and every time the limiting factor was tyre adhesion. Neither brake squealed. The front unit was not entirely proof against water.

The CSR is fitted with Lucas magneto ignition and 6-v. D.C. dynamo lighting. There was never any difficulty with starting. The dynamo kept the battery well up and its output was automatically regulated.

Lighting intensity was adequate, but the horn's usefulness was somewhat limited.

No machinery is ever perfect—motorcycle or otherwise—and the fact that only a few minor criticisms could be made of the CSR serves to emphasize the sterling qualities of this Matchless as a particularly fine sports model—a connoisseur's mount which will appeal most strongly to the enthusiast seeking a motorcycle of decidedly individual personality.

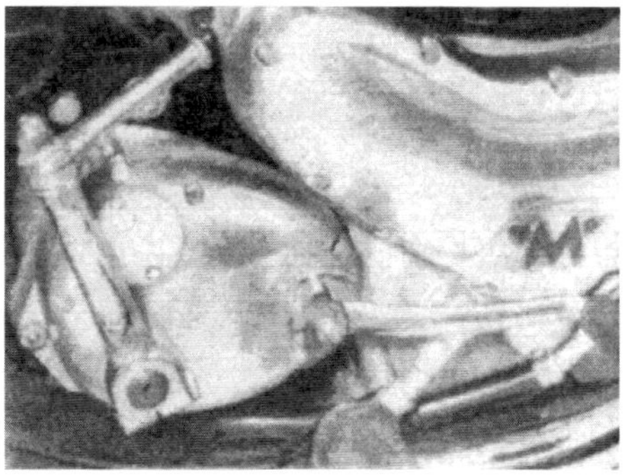

Folding kickstarter, gearchange lever and gear indicator.

646 c.c. Matchless G12 de Luxe

Fast but Tractable Super-touring Parallel Twin with a Liking for the Open Road

DOCILITY and high performance make a rare combination. Top-gear tractability you can have, speed you can have; but both? Yes, both! In the latest 646 c.c. Matchless G12 de Luxe you can. A bulldog with all the get-up-and-go of a greyhound, this is a twin to which everything comes alike, whether it be a gentle potter around the lanes at 30 m.p.h. or a blast along M1 at 80 m.p.h. on half throttle.

From a near-walking pace in bottom gear, juicy, beefy power is on tap right through the range—and that is a characteristic which diehards say vanished with the big vee-twins of old. Above all, the charm of the Matchless lies in the way such urge is developed. There is commendably little fuss, and speed builds up smoothly and with deceptive rapidity.

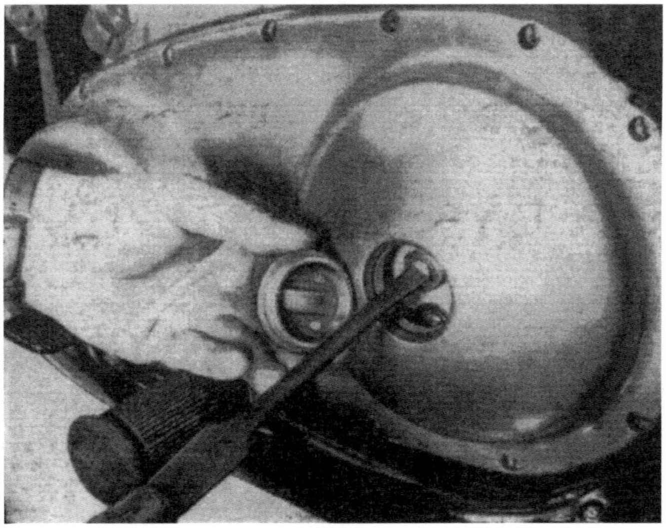

Ready access to the clutch thrust rod adjusting screw and lock-nut is obtainable after removing a screwed cap in the primary chaincase

With such ample power in hand high point-to-point averages could readily, and safely, be maintained; a mere whiff more of gas and the model would sail past slower-moving traffic, so that full advantage could be taken of gaps in the vehicle stream.

Response to the throttle was instantaneous. From 25 m.p.h. in second, a tweak of the grip was sufficient to send the speedometer needle scurrying round to the 50 mark; then up to third gear, and 70 would arrive in a satisfyingly short time before top was engaged. On the other hand, if there was no urgency about the ride, top could be held to well down below 30 m.p.h.—and the engine would pull away smoothly, unobtrusively, with not a hint of transmission roughness.

Long main-road gradients meant not a thing. Steeper hills were a joy to tackle. Second gear sufficed for the really severe stuff, while a standing start on the 1 in 3 test hill at the M.I.R.A. proving ground was easily accomplished in bottom.

ROAD TESTS

Though the engine was exceptionally smooth at speeds up to 60 m.p.h. (or 63 m.p.h., if the speedometer reading was accepted) beyond that some vibration could be felt through the handlebar and footrests—though it was not serious enough to deter one from holding the eighties for mile after mile; for some unaccountable reason the tremor seemed to disappear when accelerating hard. By allowing the model sufficient time to gather speed, a genuine 100 m.p.h. was reached on M1 on several occasions with the rider adopting a slight crouch to reduce windage.

The engine seemed perfectly willing to take harsh treatment indefinitely, with the only external evidence of hard driving a slight seepage from the oil-tank filler cap. If the tank was topped up much above the minimum level, excess oil blown out through the crankcase breather left a film on the rear of the machine. (However, in 750 miles, only a pint of oil was required to restore the level in the tank.)

From the twin megaphone-type silencers came a deep-throated growl which, though pleasing to an enthusiast's ears, was perhaps a shade too loud for these days.

As to be expected with a coil-ignition model, starting was "first time, every time" provided that, when cold, the carburettor was flooded lightly and the air lever closed; within a few seconds the lever could be fully opened, when the engine would settle down to a slow and reliable tickover. One minor crib concerned the very light action of the ignition switch of the test model. Mounted on the right of the headlamp shell, it was possible to brush against the switch, unknowingly turn it on, and so cause the battery to run flat.

So that the full performance of the G12 may be enjoyed, some care should be taken in tailoring the riding position to suit, and for that reason the very wide range of adjustment of the footrests, rear-brake pedal and gear lever was appreciated.

The controls are well laid out and convenient to operate. Horn push and dip switch are within easy reach of the left thumb (though a more far-reaching horn blast would be an advantage on such a roadburner as this). Handlebar grips are of plastic, and it was felt that a bolder patterning might have been more comfortable, in addition to affording greater friction.

Drive-side aspect of the well-proportioned, six-fifty twin power unit

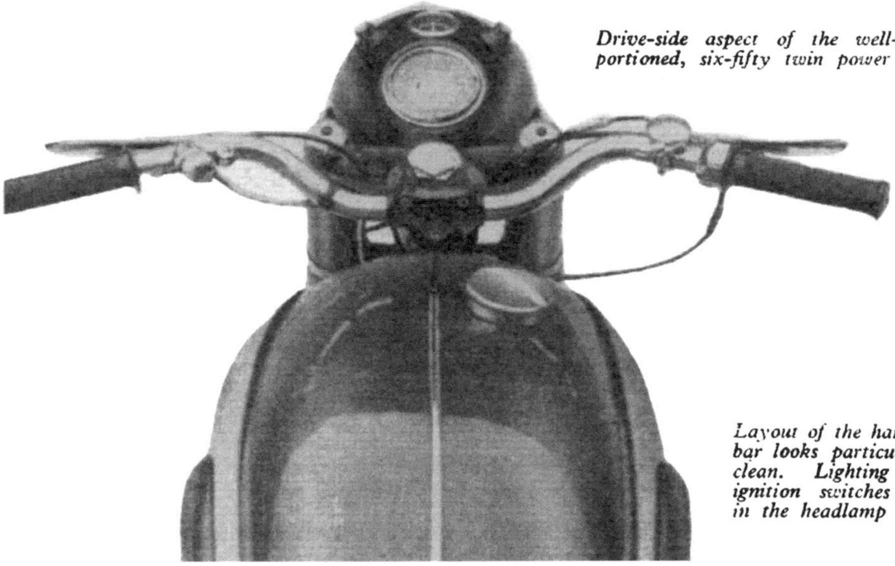

Layout of the handlebar looks particularly clean. Lighting and ignition switches are in the headlamp shell

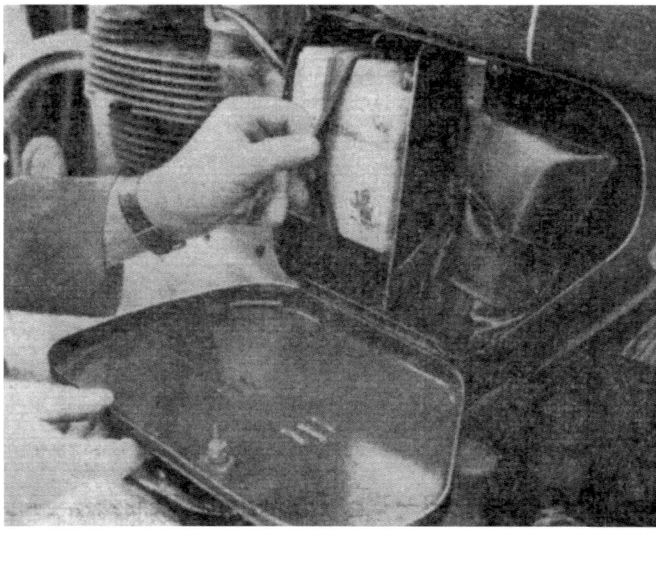

A roomy box on the left of the model, with hinged lid and coin-slot Dzus fastener, contains the battery and tool roll

The dual-seat is generously dimensioned and after a non-stop two-up trip of over 200 miles, neither the rider nor female passenger felt any trace of stiffness or fatigue—and when the passenger volunteers a remark that her half of the seat is comfortable, that can be counted as high praise indeed!

A man-size mount, the Matchless tips the scales at 420 lb and so is a little too bulky to be thrown around as though it were a two-fifty. Nevertheless, once a line through a corner was chosen, it would adhere to that line without deviation and was particularly stable if the power was fed in through the corner. Low-speed balance was good and it was easy to weave the model through slow-moving traffic.

Straight-ahead steering was impressively positive, though at speeds below 35 m.p.h. a trace of front-wheel flutter could be felt on very smooth surfaces. Cornering on unusually bumpy surfaces produced slight (although never troublesome) yawing, but in general the suspension was well up to the job of ironing-out road shocks without pitching.

Handsome fast tourer from a famous stable, the Matchless G12 de Luxe has a duplex cradle frame

Like the engine, the clutch did not object to abuse, and engaged the drive smoothly with a pleasantly light handlebar-lever action. When the engine was allowed to idle slowly, bottom gear engagement was noiseless. Movement of the gear lever was rather heavier than is usual with an A.M.C. gear box; this was noticeable on the upward change from second to third. The ratios matched the engine characteristics perfectly.

Smooth, powerful, and reasonably light to operate, the brakes were capable of making the tyres squeal without locking the wheels. The rear brake drum was unaffected by water, but a downpour lessened the effectiveness of the front brake; however, the full power returned after a few minutes of running with the brake lightly applied.

Rather more intensity of light from the main beam of the headlamp would have permitted higher cruising speeds to be maintained at night. The cut-off on dip was adequate and prevented oncoming drivers from being dazzled.

The only maintenance required during the test period was one adjustment of the rear chain after 600 miles. Accessibility to the power unit and cycle parts is first class and normal routine tasks could be carried out quickly and simply. The Matchless has one of the most comprehensive tool kits to be found on any machine.

That's the Matchless G12 de Luxe, a really robust twin with well-bred manners but all the punch of a prizefighter. The beef is pedigree, and handsome enough to win all the rosettes that may be going.

SPECIFICATION

ENGINE: Matchless 646 c.c. (72 x 79.3mm) overhead-valve parallel twin. Crankshaft supported in one plain and two roller bearings; plain big-end bearings. Aluminium-alloy cylinder heads. Compression ratio, 7.5 to 1. Dry sump lubrication; oil-tank capacity, four pints

CARBURETTOR: Amal Monobloc with air slide operated by lever on handlebar.

IGNITION and LIGHTING: Coil ignition. Lucas RM15 alternator driven by crankshaft, charging 6-volt, 13 amp-hour battery through rectifier; emergency start switching. Lucas 7in-diameter pre-focus light unit with 30/24-watt main bulb and integral pilot light.

TRANSMISSION: A.M.C. four-speed gear box with positive-stop foot control. Gear ratios: bottom, 12.23 to 1; second, 8.15 to 1; third 5.85 to 1; top, 4.79 to 1. Multi-plate clutch with bonded friction faces. Primary chain, ½ x 0.305in in cast-aluminium oil-bath case. Final drive by ⅝ x ⅜in chain with guard over top run. Engine r.p.m. at 30 m.p.h. in top gear, 1,825.

FUEL CAPACITY: 4¼ gallons.

TYRES: Dunlop; front, 3.25 x 19in ribbed; rear, 3.50 x 19in Universal.

BRAKES: 7½in-diameter front, 6½in rear; finger adjusters.

SUSPENSION: A.M.C. Teledraulic front fork with two-way hydraulic damping. Pivoted rear fork controlled by adjustable Girling spring units with hydraulic damping.

WHEELBASE: 55½in unladen. Ground clearance, 6in unladen.

SEAT: A.M.C. dual-seat; unladen height, 31in.

WEIGHT: 420 lb fully equipped with full oil tank and approximately one gallon of petrol.

PRICE: £221 6s 11d; with purchase tax (in Great Britain only), £267

ROAD TAX: £4 10s a year; £1 13s for four months.

MAKERS: Matchless Motor Cycles, Plumstead Road, London, S.E.18.

PERFORMANCE DATA

(Obtained at the Motor Industry Research Association's proving ground Lindley Leicestershire.)

MEAN MAXIMUM SPEED: Bottom: *50 m.p.h.
Second: *75 m.p.h.
Third: 88 m.p.h.
Top: 95 m.p.h.
* Valve float occurring.

HIGHEST ONE-WAY SPEED: 97 m.p.h. (conditions: light three-quarter breeze; rider wearing two-piece suit and overboots).

MEAN ACCELERATION:

	10-30 m.p.h.	20-40 m.p.h.	30-50 m.p.h.
Bottom	2.6 sec	2.8 sec	3.4 sec
Second	3.8 sec	3.6 sec	3.6 sec
Third	—	5.0 sec	5.6 sec
Top	—	7.2 sec	6.6 sec

Mean speed at end of quarter-mile from rest: 82 m.p.h.
Mean time to cover standing quarter-mile: 16.4 sec.

PETROL CONSUMPTION: At 30 m.p.h., 64 m.p.g.; at 40 m.p.h., 62 m.p.g.; at 50 m.p.h. 59 m.p.g.; at 60 m.p.h., 51 m.p.g.

BRAKING: From 30 m.p.h. to rest, 34ft (surface, dry tarmac).

TURNING CIRCLE: 15ft 6in.

MINIMUM NON-SNATCH SPEED: 18 m.p.h. in top gear.

WEIGHT PER C.C.: 0.65 lb.

The 347 c.c. o.h.v. A.J.S. Model 16

Rugged, reliable and economical transport for two

Specification

ENGINE

Type .. .. Single-cylinder four-stroke
Bore .. .. 69 mm.
Stroke .. .. 93 mm.
Cubic capacity .. .. 347 c.c.
Valves .. .. .. Overhead (push-rod)
Compression ratio .. .. .. 7.5 : 1
Carburetter .. Amal "Monobloc," 1 1/16 in. bore, handlebar air control
Ignition Battery-powered coil with automatic control
Generator Lucas RM15 6v. alternator
Makers' claimed output .. No. released. Maximum power occurs at 5,800 r.p.m.
Lubrication .. .. Plunger pump
Starting .. .. Kickstarter

TRANSMISSION

Ratios .. .. .. 5.8, 7.1, 9.9, 14.9 : 1
Speed at 1 000 r.p.m. in top gear .. 13 m.p.h.
Speed equivalent to revs. at maximum power rating:
Second gear .. .. .. 45 m.p.h.
Third gear .. .. 63 m.p.h.
Top gear .. .. 77 m.p.h.
Primary drive Single-row chain in oilbath
Secondary .. Single-row partially enclosed chain
Clutch .. .. Multi-plate in oilbath
Shock-absorber Rubber-vane type in clutch centre

CYCLE PARTS

Frame .. Duplex cradle with brazed lugs
Front suspension "Teledraulic" telescopic forks with coil springs and two-way hydraulic damping
Rear suspension Swinging fork with three-position Girling hydraulically damped spring units
Tyres Dunlop 3.25×19 in.; ribbed front, studded rear
Brakes 7-in. dia. front and rear; total lining area, 23½ sq. in.
Fuel tank Welded steel; two taps; rubber knee grips
Oil tank .. .. .. 4 pt. capacity
Lamps .. 30/24-w. head, 3-w. pilot, 18/6-w. stop/tail, 1.8-w. speedometer
Horn .. .. Lucas
Battery .. Lucas MLZ9E 6-v. 13 a.h.
Speedometer Smiths 120 m.p.h. with trip
Seating .. .. Two-level dual seat
Stands .. .. .. Centre and prop
Toolkit Spanners: 3 double-ended, 1 ring, 1 adjustable, 1 tappet, 1 contact-breaker, 2 box, 1 Girling; 2 tyre levers; 2 Allen keys; screwdriver; clutch-spring tool; tommy bar; grease gun; pliers; circlip tool
Toolbox Semi-pannier type on left side
Finish .. Black stove enamel with gold lining

OTHER EQUIPMENT

Pillion rests; tyre pump; comprehensive instruction book.

PRICES

Machine £240 9s. 4d. (inc. £44 9s. 4d. P.T.)
Extras .. .. None
Total as tested .. .. £240 9s. 4d.
Tax £4 10s. p.a.; £1 13s. for four months
Makers A.J.S. Motor Cycles (Associated Motor Cycles, Ltd.), Plumstead Road, London, S.E.18.

'Motor Cycling' Test Data

Conditions. *Weather: Dry, blustery. (Barometer 29.45 in. Hg. Thermometer 60°F.) Wind: West, 10 m.p.h. with brief calm spells. Surface (braking and acceleration: Dry asphalt. Rider: 11½ stone, wearing two-piece riding suit, boots and safety helmet, normally seated except where stated. Fuel: Premium grade (96 research method octane rating).*

Venue: *Motor Industry Research Association Station Lindley.*

Speed at end of standing 1,000 yd.:
East .. .. 74 m.p.h.
West .. .. 54 m.p.h.
Mean .. .. 64 m.p.h.
Best one-way speed (rider prone, tail wind) .. .. 79.8 m.p.h.
Mean, two-way speed (rider prone, calm air) .. .. 76.2 m.p.h.
Flying lap speed Not taken, conditions unsuitable
Braking from 30 m.p.h. (all brakes) 10 yd.
Fuel consumption:
At constant 30 m.p.h. .. 112 m.p.g.
50 m.p.h. .. 96 m.p.g.
500-mile overall figure .. 86 m.p.g.

Speedometer

30 m.p.h. indicated=26.6 m.p.h. true
40 m.p.h. indicated=35.4 m.p.h. true
50 m.p.h. indicated=45.4 m.p.h. true
60 m.p.h. indicated=56.2 m.p.h. true
70 m.p.h. indicated=62.8 m.p.h. true
80 m.p.h. indicated=73.2 m.p.h. true

Mileage Recorder .. Over-reading 2%

Electrical Equipment

Top gear speed at which generator output balances:
Ignition only All engine speeds above tickover
Minimum obligatory lights 27 m.p.h.
Full lights .. .. .. 27 m.p.h.

Weights and Capacities

Certified kerbside weight (with oil and 1 gal. fuel) .. .. 403 lb.
Weight distribution, rider normally seated:
Front wheel 41%
Rear wheel 59%
Tank capacity (metered):
Total .. .. .. 4.15 gal.
Reserve (left side) .. 1¾ pints

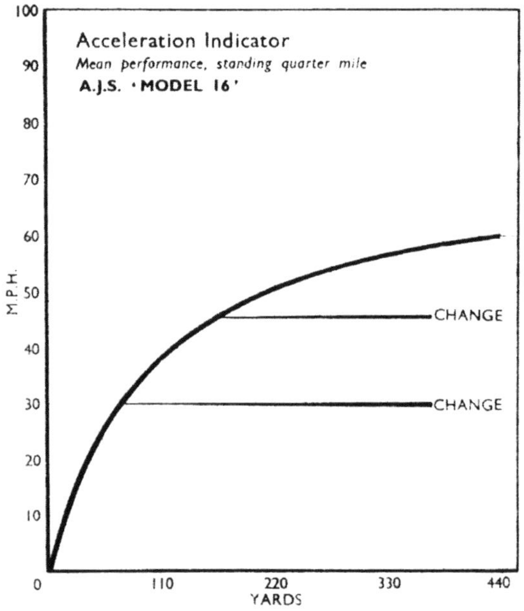

SOLIDLY and soundly made to an established design by a world-renowned British manufacturer, the A.J.S. Model 16 350 c.c. o.h.v. is all that the everyday rider seeks. For the man who needs reliable transport to and from work each day, plus the means of taking himself and his wife for long, safe runs at week-ends, this "Ajay" is a good answer. It is a machine without frills or fuss, as easy on the pocket as it is on the eye.

The test example came to us run in. Immediately, therefore, it was put to work. No glamour attaches to daily duty runs in and out of London, but they provide most exacting tests of low-speed handling and safety. Manageable, unobtrusive and sure-footed, the Model 16 passed these tests with very high marks.

Consider a typical morning's run to *Motor Cycling's* office, in the heart of London, from the tester's home on the fringe of the Surrey Green Belt, some 15 miles out.

After spending a night in the open (*not* typical, this, but a deliberate part of the routine) the Model 16 would be ready to start without hesitation. A light flood, full choke, ease over compression with the exhaust valve lifted, ignition on at the handy headlamp switch, and the next gentle depression of the kick-starter brought certain results, followed by regular fluff-free running with the handlebar choke lever restoring full air straight away.

Of all the attributes of this endearing machine, the facile starting, hot or cold, was perhaps the most memorable.

Bottom gear slipped home in a scrunch-free way, the clutch took up the drive sweetly, and without even the barest trace of spitting-back or other cold-running temperament, the A.J.S. was on its way. An experimental dab on each brake separately revealed that both units were fully proof against ingress of overnight rain (and they remained this way over wet roads).

After running up gently through the gears to 30 m.p.h. one found that at this speed there was a complete and utter smoothness. Not the barest trace of vibration—only well-subdued exhaust pulses to mark one's progress.

Into the 40 limit, with the motor warm by then, and brisk top-gear acceleration poured in; there was not the slightest need to change down to bring about the rapid 10 m.p.h. speed increase. Along the Kingston by-pass—a derestricted road except at week-ends—a tweak of the throttle brought a happy cruising speed of anything between 45 and 65 m.p.h., depending on traffic.

Here is one of the "Ajay's" strongest points—effortless open-road cruising at any speed up to the ceiling that the rider cares to dictate. It was as happy when by-lane pottering at 25 as when really hurrying at, if need be, continued full throttle in the 70s—and even, on downhill stretches, into the lower 80s.

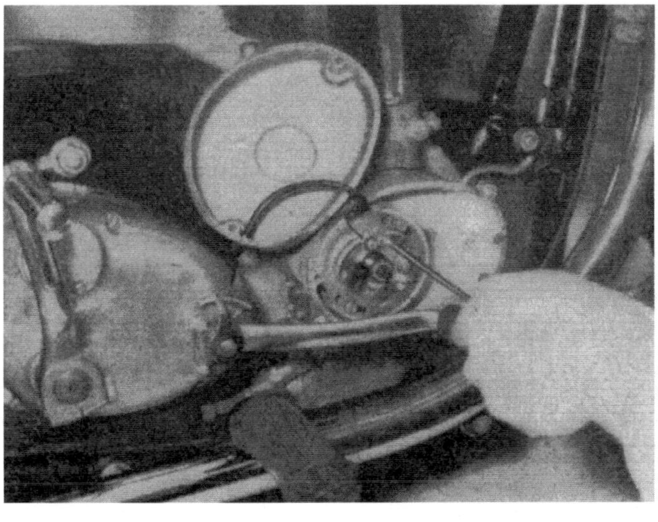

Accessibility is a strong point of this "classical" o.h.v. single, with its separate gearbox. Here the contact-breaker points are being adjusted after swinging back the two-screw cover.

Above some 45 m.p.h., there set in a trace of the internal combustion engine at work. Of what vibration there was, the handlebars manifested the most; the tank, dual seat and footrests none at all at any speed. This factor helped to make 60 m.p.h. cruising the untiring business that it was.

Through the roundabouts of the by-pass the excellent roadholding became most apparent. And when the weather clamped down during the middle of August, the tester developed a high regard for the handling and safe navigation in the wet. Not only was roadholding good and cornering accurate, but the machine as a whole was Safe with a capital S.

On the protruding expansion strips

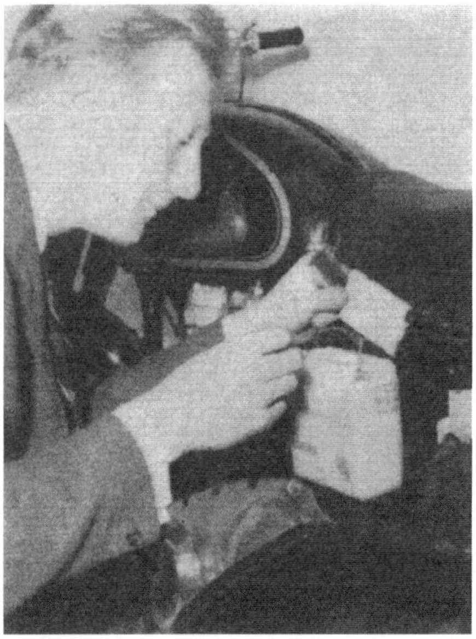

(Left) The battery shares a pannier-type compartment with the tool-roll. (Right) Oiling the front brake cable where it passes through the fork-mounted adjuster.

between the concrete slabs of A3, which can punish the rider in an irritating and fatiguing way, the comfort imparted by the excellently sprung Teledraulic forks at the front and the Girling-controlled rear end was of a high order. And over really rough roads these characteristics were maintained.

There, the two-level, well-upholstered dual seat played an important part. Worthy of note is its great length and general roominess. But the front portion of the seat is too high for most people; it is only a true six-footer who can put a foot down flat on the ground—and then only one foot, not both at the same time.

So far as riding position is concerned, the handlebars could be narrower and set slightly farther to the rear; though the bars can be swung in the clamp, anything other than a forward setting results in the right bar striking the filler cap on full lock.

The 4¼-gal. fuel tank was deemed to be overwide across the knee-grips; however, a smaller tank is available in the A.J.S. range of equipment for those who might regard this feature as a personal handicap. The tank is mounted on two rubber buffers at the front; at the rear it is pulled down on to a foam pad by a rubber strap.

To continue on our typical journey: the end of the by-pass and its roundabout demanded hard braking and then fairly vivid low-speed cranking over. The brakes were truly first class, fully up to the job of arresting a "350," and produced only slight dipping at the front. An abundance of ground clearance on both sides gave the rider complete freedom in choice of "line."

It would be a daring pilot who could touch down the centre stand. Two-up, the Girlings reset to cope with the added load, the silencer was free from grounding on adversely cambered right-handers.

Deep in London's hurly-burly, rarely using top gear and frequently trickling along in bottom or second, the "Ajay" gave a good account of itself. Feet-up traffic-threading was a delight, stability being of top quality, with the brakes ever able to bite the drums hard for those sudden low-speed stops. Generous lock permitted the most to be made of openings in stationary traffic.

At lengthy traffic baulks, the single-cylinder long-stroke o.h.v. engine settled down to the slow plonk-over for which this type is famed. At such times, the ammeter showed a 2-amp. discharge to the ignition coil, but the loss seemed to be made good higher up the scale, for the battery was always at a high level of charge.

The lights were found to be adequate for 50 to 55 m.p.h. cruising, though the dipped beam had a rather poor pattern with a great deal of stray light where it was most definitely *not* wanted. The horn, out of sight and sound, shrouded by other components, was feeble. The speedometer seemed overflattering (as the tests at M.I.R.A. later confirmed).

So much could be learned from the daily humdrum journey. When a summer evening promised a pleasant jaunt in the countryside, the Model 16 provided transport *par excellence* for enjoying the Surrey scene.

Up the hills of the North Downs, without giving a thought to their severity, it took the tester and his wife with complete unconcern. Down the far side towards the Kentish Weald it glided, with the powerful brakes pulling it up precisely to order. And after a brief or lengthy halt by stream or wayside inn, it was always ready for work again after a firm prod on the kickstarter—and only one at that!

Under these general, everyday conditions, so typical for many town dwellers, a welcome 86 m.p.g. was returned. When really cracking along M1, the m.p.g. was 78. This economy in fuel, and zero consumption of oil, contributed towards the cheapness of running. The Model 16, as even our limited experience confirmed, is known to be light on the pocket, both for running costs and for the equally important, though less easily assessed, maintenance/repair factor. It is a motorcycle with stamina, the current development of a tried and sound design which has given yeoman service over the years—as its justly famed Matchless brother, the G3L, did during and after the last war.

(Left) There is a thick heat insulating block between the Amal "Monobloc" carburetter and the stub. (Below) Straightforward handlebar layout.

348c.c. Matchless G5 Matador

EXCELLENT 1962 LIGHTWEIGHT ROADSTER COMBINING ECONOMY WITH FIRST-CLASS HANDLING, POWERFUL BRAKES AND LUSTY PERFORMANCE

ENDOWED with the new title of Matador for 1962, the G5 Matchless three-fifty offers a skilful blend of the most attractive qualities of both larger and smaller capacities. As against the larger it scores on lower initial cost, cheaper insurance and reduced fuel consumption. Compared with smaller models, it has the edge on performance—especially power in the medium-speed range—is not much heavier and is generally more suited to sustained hard work. It is a good example of a type of machine which shows signs of regaining its former popularity now that five-hundreds—although not the bigger twins—no longer attract so many enthusiasts.

Although the Matador has an affinity with the G2 Monitor two-fifty and is, therefore, lighter than earlier Matchless three-fifties, it has a robust, comforting "big-bike" feel. The riding position is uncramped and allows a relaxed, easy posture.

Clutch and front-brake levers and the handlebar are fully adjustable. The twist-grip was extremely light in operation—just as it should be for sensitive control of the throttle. Clutch operation was slightly—but only slightly—heavy, but the clutch took up the drive without a trace of judder or slip. It withstood without protest a series of standing-start getaways when the performance figures were being obtained and in a long mileage required no adjustment.

During the test the Matador was ridden under a variety of weather conditions over many different sorts of surface. At all times the steering and general stability were well above average. At first there seemed a trace of heaviness in the steering at very low speeds but it went unnoticed as one became more familiar with the machine which could, in fact, be ridden feet-up without difficulty at a walking pace.

At the other end of the scale, the Matador could be laid into main-road curves and bends at really high speeds yet it would hold its line without waver. Swift changes of direction could be achieved effortlessly, thanks to the lightish weight and excellent all-round handling characteristics.

Of particular interest to those with limited garaging space is the fact that the weight disposition and wide steering lock made manhandling an easier task than might be expected.

The front fork, which is of the well-proved Teledraulic pattern, was sufficiently firm in action to prevent pitching on undulating surfaces but not so stiff as to give a harsh ride. Controlling the rear pivoted fork are Girling units with three-position adjustment. On the test machine they were set in the softest position for one-up use. The hydraulic dampers were thoroughly effective. No bottoming was experienced.

Despite its strong family resemblance to the two-fifty, the G5 is not simply a larger-bore version of the smaller model. The bore is 2mm larger but the stroke is over 20mm longer. As might

Sleek, businesslike, light: the Matador is finished in tartan red and black

The contact-breaker assembly is located behind a plate on the right-hand engine casting. Remove two screws and the entire assembly is readily get-at-able

ROAD TESTS OF NEW MODELS

be expected the comparatively long stroke endows the engine with a wide spread of beefy power which would not disgrace a five-hundred. Its excellent pulling characteristics were noticeable as soon as the clutch was engaged at low r.p.m. Even when restarting on a 1-in-6 gradient with the engine running at only a fastish tick-over the clutch could be let home quickly and the Matador accelerated smartly at a tweak of the throttle.

Given average road conditions, the G5 would hustle along indefinitely at between 65 and 70 m.p.h. as its cruising speed without the engine showing signs of tiring. High-frequency vibration could be felt through the handlebar and footrests at speeds above about 60 m.p.h. in top gear and at corresponding engine revolutions in the indirect gears. The effect on the hands would, it was thought, be reduced almost completely by handlebar grips of a more resilient type than the smooth plastic ones fitted as standard.

During normal open-road riding, there was no appreciable advantage in buzzing the engine in the lower gears. Changes from top to third were unnecessary until the speed fell below about 40 m.p.h., yet point-to-point averages of between 45 and 50 m.p.h. were commonplace without recourse to spirited riding. Easy gradients and normal-strength headwinds had no marked effect on top-gear performance; a 1-in-12 main-road hill could be climbed at 60 m.p.h. in top.

Exhaust silencing was adequate if the G5 was ridden sensibly. Without fuss or noise the miles could be gobbled up most pleasantly. Only when unnecessarily wide throttle openings were used for hard and prolonged acceleration did the note become loud by current standards.

Throughout the test the Matchless could be started easily whether the engine was cold or hot. An exhaust-valve lifter enabled the piston to be eased past compression after which a single swinging

Specification

ENGINE: A.M.C. 348 c.c. (72 x 85.5mm) overhead-valve single with light-alloy cylinder head and cast-iron barrel. Crankshaft supported in a ball and a roller bearing on the drive side and a plain bearing on the timing side; roller big-end bearing. Compression ratio 7.5 to 1. Oil reservoir bolted to the right side of the crankcase; capacity 2¼ pints.

CARBURETTOR: Amal Monobloc; air slide operated by handlebar lever.

IGNITION and LIGHTING: Coil ignition. Wipac 54-watt alternating-current generator mounted on drive-side crankshaft charging Exide 12-amp-hour battery through rectifier. Wipac 6in-diameter headlamp with pre-focus light unit and 30/24-watt main bulb.

TRANSMISSION: A.M.C. four-speed gear box clamped to rear of engine. Gear ratios: bottom, 18.66 to 1; second, 11.82 to 1; third, 8.3 to 1; top, 6.39 to 1. Multi-plate clutch with bonded friction facings. Primary chain, ⅜in duplex in cast aluminium case. Rear chain, ½ x 0.305in in pressed-steel case. Engine r.p.m. at 30 m.p.h. in top gear, 2,600.

FUEL CAPACITY: 3¼ gallons.

TYRES: Dunlop: both 3.25 x 18in; front ribbed tread, rear studded.

BRAKES: Both 6in-diameter x 1in wide.

SUSPENSION: A.M.C. telescopic front fork with hydraulic damping. Pivoted rear fork controlled by Girling spring-and-hydraulic units with three-position adjustment for load.

WHEELBASE: 54in unladen. Ground clearance, 6in unladen.

SEAT: A.M.C. dual-seat; unladen height, 30½in.

WEIGHT: 352 lb fully equipped, with full oil tank and approximately one gallon of petrol.

PRICE: £180 10s; with purchase tax (in Great Britain only), £220 4s 2d.

ROAD TAX: £4 10s a year; £1 13s for four months.

MAKERS: Matchless Motor Cycles, Plumstead Road, London, S.E.18.

DESCRIPTION: *The Motor Cycle*, 14 September 1961.

PERFORMANCE DATA

MEAN MAXIMUM SPEED: Bottom: *29 m.p.h.
Second: *47 m.p.h.
Third: *66 m.p.h.
Top: 75 m.p.h.
*Valve float occurring.

HIGHEST ONE-WAY SPEED: 76 m.p.h. (conditions; light three-quarter wind; rider wearing two-piece suit and overboots).

MEAN ACCELERATION:

	10-30 m.p.h.	20-40 m.p.h.	30-50 m.p.h.
Second	4.2 sec	5.0 sec	—
Third	—	6.2 sec	7.2 sec
Top	—	7.4 sec	8.8 sec

Mean speed at end of quarter-mile from rest: 67 m.p.h.
Mean time to cover standing quarter-mile: 19.6 sec.

PETROL CONSUMPTION: At 30 m.p.h., 104 m.p.g.; at 40 m.p.h., 80 m.p.g.; at 50 m.p.h., 72 m.p.g.; at 60 m.p.h., 64 m.p.g.

BRAKING: From 30 m.p.h. to rest 30ft (surface, dry tarmac).

TURNING CIRCLE: 13ft 9in.

MINIMUM NON-SNATCH SPEED: 18 m.p.h. in top gear.

WEIGHT PER C.C.: 1.0 lb.

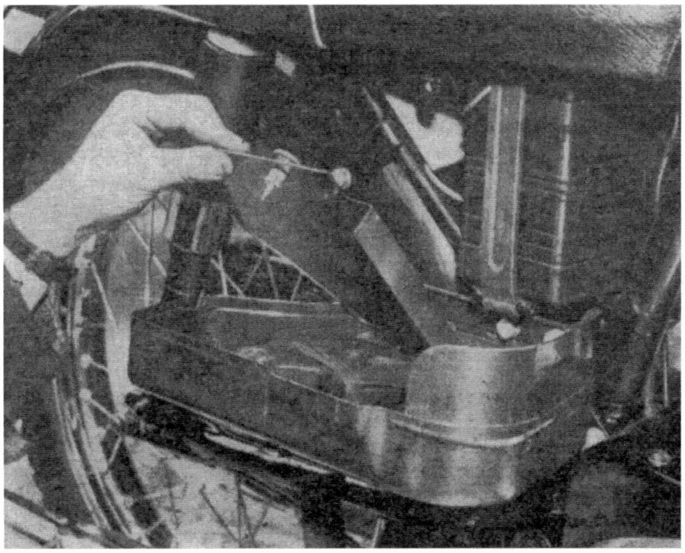

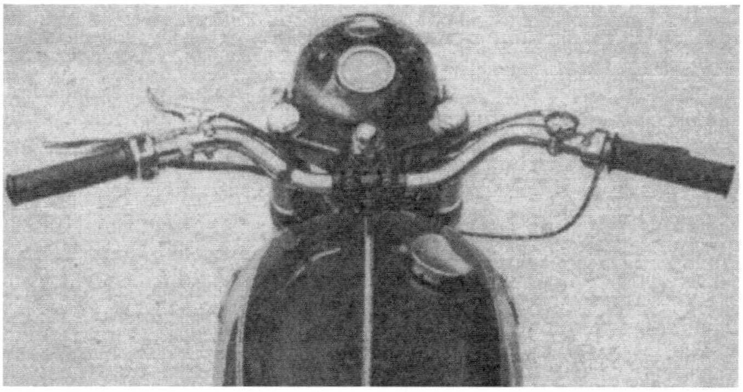

Left: Tools are carried in a waterproof compartment incorporated in the battery cover. The battery is thus readily accessible for topping up

Above: The Matador is fitted with an exhaust-valve lifter, the trigger for which is mounted on the left of the handlebar. Controls are adjustable

kick sufficed on every occasion. The engine started equally well whether the ignition switch was in the normal or emergency position. Idling was slow and reliable once the engine was warm. Mechanical quietness was markedly above average; even when the engine was cold there was no piston slap.

Gear pedal movement was light and gear engagement positive. Nothing unorthodox in the way of control manipulation was required to ensure fast, clean upward and downward changes. The clutch freed perfectly; hence when the machine was stationary with the engine idling bottom gear could be engaged noiselessly and there was never any difficulty in selecting neutral. At 18.66 to 1 bottom gear is rather low but has its compensations for traffic threading without clutch slipping.

Both of 6in-diameter, the brakes were well up to the task of matching the performance. Light and positive in action and convenient to use, they provided first-class braking in every way and, as the performance data show, were capable of giving a stopping distance of 30ft from 30 m.p.h.

An excellent main beam and a satisfactory dipped beam from the Wipac headlamp made 60-plus m.p.h. after-dark cruising possible in safety. Mounted in a single drum on the left side of the handlebar the dip-switch and horn button are conveniently placed for thumbing.

A more elaborate speedometer would be justified. That fitted is of the non-trip type and the total-mileage recorder reads only to the nearest mile. Speed readings were approximately 5 per cent fast almost throughout the range.

Neither the primary nor the rear chain required adjustment during 1,000 miles covered on test. The rear chain has full enclosure as standard and is lubricated by oil mist from the crankcase breather. Tappet adjustment can be carried out after removing a plate on the right side of the rocker box. Also readily accessible is the contact-breaker assembly which is located behind a detachable plate on the right side of the engine.

Undoing a hinged cover on the right beneath the dual-seat gives access to the battery which can then be removed for topping-up. The tools are housed in a compartment on the inner face of this cover.

Finished in tartan red and black, and with its smooth, cranny-free engine-unit castings, the Matador is of attractive appearance. It offers lively, lusty performance, excellent brakes, steering and road holding. It is an excellent all-round proposition.

Left: The opposite side of the battery to that shown above is enclosed by a matching cover concealing the horn and ignition coil

Right: Clutch adjustment could hardly be more simple: the adjuster screw is reached after undoing a plug in the primary chaincase

The 248 c.c. MATCHLESS G2S 'MONITOR SPORTS'

A handsome roadster with an appetite for sustained fast cruising

Specification

ENGINE

Type	Single-cylinder four-stroke
Bore	70 mm.
Stroke	65 mm.
Cubic capacity	248 c.c.
Valves	Overhead (push-rod)
Compression ratio	7.8 : 1
Carburetter	Amal "Monobloc," 1 1/16-in. choke, handlebar air control
Ignition	Wipac battery-powered coil with EMG circuit
Generator	Wipac 6-v. 54-w. A.C. with full-wave rectification
Makers' claimed output	Not given; peak power occurs at 6,800 r.p.m.
Lubrication	Dry sump with oil reservoir integral with crankcase castings
Starting	Kickstarter (folding pedal)

TRANSMISSION

Four-speed footchange gearbox clamped to engine.

Ratios	6.9, 9.0, 12.8, 20.1 : 1
Speed at 1,000 r.p.m. in top gear	10 m.p.h.
Speed equivalent to revs. at maximum power rating:	
Second gear	38 m.p.h.
Third gear	54 m.p.h.
Top gear	71 m.p.h.
Primary drive	Single-row chain
Final drive	Chain
Clutch	Multi-plate in oilbath
Shock-absorber	Rubber-vane type in clutch centre

CYCLE PARTS

Frame	Cradle type supporting engine-gearbox as a unit
Front suspension	"Teledraulic" telescopic forks with hydraulic control
Rear suspension	Swinging fork; two Girling oil-damped spring units with clevis lower ends; adjustable for load
Tyres	Dunlop 3.25×17 in.; ribbed front, studded rear
Brakes	6-in. dia. front and rear. Total lining area, 23½ sq. in.
Fuel tank	Pressed steel with rubber knee grips and twin taps
Oil reservoir	Compartment below and integral with crankcase
Lamps	30/24-w. head, 3-w. pilot, twin 3-w. tail with separate 18-w. stop
Battery	Exide 6-v. 13 a.h.
Speedometer	Smiths 100 m.p.h., non-trip type
Seating	Two-level dual seat
Stand	Centre
Toolkit	Spanners: 5 open-ended, 1 box, 1 adjustable, 1 "C." 2 tyre levers. Allen key. Screwdriver. Feeler-gauge set
Toolbox	Box within battery compartment under seat
Standard finish	Black with red and white tank; chromium-plated mudguards and usual cycle parts

OTHER EQUIPMENT

Dropped handlebars; pillion rests.

PRICES

Machine	£212 5s. 7d. (inc. £38 5s. 7d. P.T.)
Extras	None
Total as tested	£212 5s. 7d.
Tax	£2 5s. p.a.
Makers	Associated Motor Cycles, Ltd., Plumstead Road, London, S.E.18

'Motor Cycling' Test Data

Conditions. *Weather: Dry, mild, blustery (Barometer 29.50 in.Hg. Thermometer 52°F.). Wind: W., 6–10 m.p.h. Surface (braking and acceleration): dry asphalt. Rider: 11½ stone wearing two-piece suit, safety helmet and boots, normally seated except where stated. Fuel: Premium grade (96 research method octane rating).*

Venue: *Motor Industry Research Association Station, Lindley.*

Speed at end of standing 1,000 yd.:	
East	72.3 m.p.h.
West	57.1 m.p.h.
Mean	63.8 m.p.h.
Speed at end of standing 1,400 yd. (rider prone):	
East	74.1 m.p.h.
West	66.9 m.p.h.
Mean	70.3 m.p.h.
Best one-way speed (rider prone)	74.1 m.p.h.
Braking from 30 m.p.h. (all brakes):	13 yd.

Fuel consumption

At constant 30 m.p.h.	124 m.p.g.
50 m.p.h.	86 m.p.g.
500-mile overall figure	82 m.p.g.

Speedometer

30 m.p.h. indicated = 27.3 m.p.h. true
40 m.p.h. indicated = 37.4 m.p.h. true
50 m.p.h. indicated = 48.7 m.p.h. true
60 m.p.h. indicated = 57.7 m.p.h. true
70 m.p.h. indicated = 67.1 m.p.h. true

Mileage Recorder .. .. Accurate

Electrical Equipment

Top gear speed at which generator output balances:

Ignition only	All speeds above min. non-snatch
Minimum obligatory lights	
Headlamp main beam	

Weights and Capacities

Certified kerbside weight (with oil and 1 gal. fuel)	322 lb.
Weight distribution (rider normally seated):	
Front wheel	43%
Rear wheel	57%
Tank capacity (metered):	
Total	3.30 gal.
Reserve	3 pints

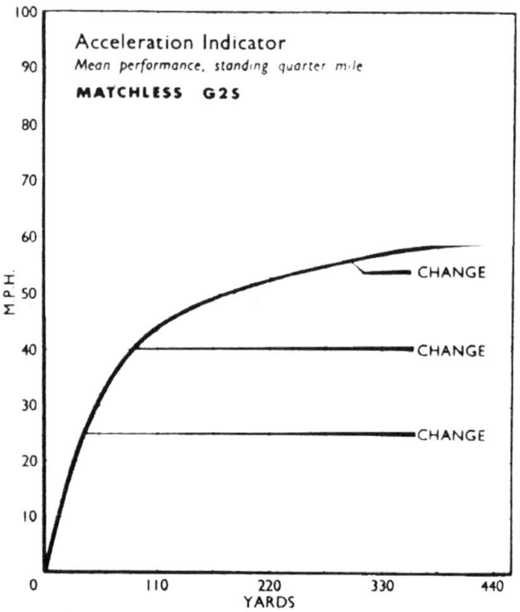

The G2S "Monitor Sports" has the speed and stamina for motorway work. Here it is negotiating the roundabout at the Watford end of M1.

(Below) Exceptionally clean in appearance is the engine-gearbox assembly. The "skewed" vertical fins on the cylinder head deflect air towards the plug.

PLEASING in appearance and performance, the Matchless G2S "Monitor Sports" (like its A.J.S. stablemate, the "Sapphire Sports") is a welcome contribution to the important 250 c.c. sports sphere by Associated Motor Cycles. In the largest capacity class legally open to a solo learner, its single-cylinder engine—which is said to develop 19 to 20 b.h.p.—gives it a potential that will satisfy him long after he has passed his test.

The model which forms the subject of *this* test was only 23 miles old when we took delivery of it. During a careful running-in drill its motor noticeably freed as the miles mounted, and it had over 1,000 miles on the clock before being taken to M.I.R.A. for an electronic assessment of its capabilities.

There, its maximum speed, taken as the mean of representative two-way runs after sufficient distance for build-up, proved to top 70 m.p.h. with the rider well down on the tank and making good use of the dropped handlebars. The best speed of the day was 74 m.p.h., which well illustrates the potential of this technically remarkable motorcycle.

Considerably over-square, the engine of the "Monitor Sports" is désaxé, the centre-line of its cylinder lying slightly ahead of its crankshaft. Besides this, its light-alloy cylinder head is skewed horizontally, with rockers and ports at an angle of 20 degrees to the fore-and-aft axis of the machine—an arrangement which increases the efficiency of valve gear operation, improves cooling, and gives easier access to the carburetter.

One characteristic of over-square engines is that they have to be kept on the boil for the best results. Yet though the Matchless could be revved hard and long without showing signs of fatigue, it also had a good torque lower down the scale. In fact, the engine was both flexible and extremely willing—given the slightest encouragement it would instantly strain at the leash.

Once the unit had warmed up, carburation was usually clean but, when cold, generous use of the choke was needed. There was no hesitation at low engine speeds, as can occur with "hot" valve timings, and the "Monitor" zipped straight up the scale without faltering. Use of such a surge of revs and power in, say, second gear, up to the 6,800 r.p.m. peak showed no serious vibration. Mechanically, the unit was quiet, though the gearbox whined under power in second and particularly in third; bottom was silent. The sharp exhaust bark might, however, be muted with advantage.

Little, if any, tickling was normally needed to start easily from cold, full choke being sufficient unless the temperature was almost down to freezing point. When hot, the engine was just too simple to start; it never kicked back, and hot idling was certain and pleasantly slow. But the Amal carburetter had a distressing tendency to drift from its setting and so cause the engine to race even when the slide was completely closed.

With the machine under way, clutch action was delightful and the lever pressure quite light. When engaged from rest, the clutch

Its level has to be maintained a specific distance below the return hole. This we found difficult to judge, and a dipstick would certainly be advantageous.

For handling, the " Monitor Sports " scored high marks. Steady and free from pitching or back wheel wavering, it kept bang on line and, apart from the rather limited ground clearance, navigation was well up to rapid bendswinging standards. Low-speed traffic threading was also good.

On one occasion, our tester hit a series of bumps, when travelling at high speed. Even so, the Matchless maintained course with equanimity and without the slightest trace of a wobble.

Stability under hard braking was noticeably good, too. True, the front 6-in. unit on the test mount seemed below par and had to be applied hard (the back was normal), but braking effort was controllable and smooth. Both anchors were waterproof and usable on wet roads without fear of wheel locking. In addition, the G2S was remarkably stable in the rain and inspired bags of confidence under these trying conditions.

had a tendency to grab even after factory attention. Gearchanging was good, particularly at speed when one could slip through the cogs very slickly; and, given nothing more than normal matching of engine and road speeds, the changes were absolutely clean.

Gear ratios however, were not so ideal. Although top is slightly low for all-out flat-on-the-tank speed, it is right for the fastest possible progress with the rider normally seated—and we prefer it this way. But third is a long way down and bottom is exceptionally low. In fact, the tester frequently pulled away on a level road in second gear rather than make unnecessary use of bottom, which seemed only for uphill starting and very slow traffic. It appeared that all three intermediate ratios might be higher with advantage.

Removal of the left-hand under-seat panel exposes the coil and horn. The intake of the downdraught carburetter also lies behind the panel.

Inspection plates on the offside give ready access to the contact-breaker assembly and the gearbox end of the clutch cable.

Simple Adjustment

The drum-shaped gearbox fits neatly into a concave recess at the back of the engine and has only to be rotated sightly to adjust the primary chain. The adjuster bolt heads are easily accessible after a q.d. pressing has been removed. The new kickstarter is not so sensible. Its longer crank saves the toes being stubbed, as happened with earlier versions, but now the top of the foot tends to scrape underneath the footrest. It is surprising that the effort to rectify one fault should have produced another.

Perhaps the riding position could also be better tailored for the market at which the " Monitor Sports " is aimed. For the footrests are rather low and scraped when spirited cornering was attempted—in fact greater ground clearance in general would be an advantage. Also the handlebars are hardly a good compromise—though they might suit some statures—and their range of adjustment is limited. In fact, it seems as though a touring position has been adapted without giving the rider the option of returning to a more upright attitude. However, we understand that modifications to the handlebar shape are under consideration.

The large, comfortable dual seat was greatly appreciated by all passengers and certainly has advantages for the ordinary rider. But, being a two-level design, it seemed hardly the right wear for a sports machine, although it is resilient without being soggy.

On the other hand, a slimmer tank might enhance the " Monitor's " sporty appearance, but since the distance between the rider's knees is determined primarily by the width of the gearbox, anything narrower than the one fitted would be less comfortable. As well as having a useful capacity, the tank provides a three-pint reserve. Its right tap was used for the main supply as the left one, when on, came close to the carburetter tickler so interfering with its ready operation.

Though the " Monitor Sports " asks to be driven hard, and was generally galloped at the slightest opportunity, fuel consumption was not heavy, being 82 m.p.g. over a long-period check. Oil consumption, too, was light. It appeared to be roughly one pint per 1,000 miles, but an oil change intervened before the figure could be confirmed. The oil is put in through an accessible orifice at the front of the right-hand crankcase half.

Lighting intensity provided by the Wipac electrical equipment was excellent and the generator showed good charging at all times, poking out a couple of amps. regardless of services switched on even at the non-snatch speed—about 20 m.p.h. in top. The headlight's main beam pattern was exceptionally good, having a long-range pencil probing ahead with a wide pool nearer to the machine—ideal for illuminating manhole covers. Dipped, the beam was also good but emitted rather a lot of stray light. The emergency starting circuit worked well.

Though we have criticized a few detail points—not all of which would be found by every rider—the Matchless G2S " Monitor Sports " is a good machine. In fact, the minor snags we discovered detract in only the slightest degree from what is a highly desirable British sports " 250 " with abundant Go, good manners and economy of operation. And, in its shining chrome guards with red and white tank, it is a graceful mount of which any new owner could justifiably be very proud.

498 c.c. A.J.S. Model 18 Statesman

DISCRETION, dignity and an ability to cope with hard work without showing signs of stress—these are some of the essential attributes of a statesman. Thus when A.J.S. decided to give their machines names in addition to terse designations it was fitting that they should call the 498 c.c. Model 18 the Statesman.

A robust, good-looking roadster single with an ancestry reaching back to well before the last war it can claim, in the motor-cycle sense, to embody all these qualities. The main source of the Statesman's appeal is that it offers a pleasing compromise between the modern and traditional in matters of styling.

Riding the A.J.S. the first few miles reminded one, at a time when vertical twins are fashionable, of the particular charm of the lusty, tractable single. With a maximum in the eighties, the Model 18 is certainly no sluggard. In fact the machine was capable of sustaining 70-75 m.p.h. indefinitely.

In that range the engine was working well within its limit. Hills or headwinds on main roads could be tackled comfortably, the engine pulling hard in top gear. If the occasion demanded that the speed be maintained a quick change into third was the drill. This ratio (6.13 to 1) proved invaluable when a high average was required especially on hilly, winding roads. Speedy overtaking could be achieved in third—which provided good acceleration between 30 and 70 m.p.h.

At the lower end of the scale the engine would slog like a side-valve. The combination of good flywheel effect, a moderate (7.3 to 1) compression ratio and an efficient auto-advance control made a major contribution to this impression. Indeed, advocates of manual ignition control would find a very hard case to answer. There was no pinking during the test, even when deliberate attempts were made to provoke it. Premium-grade petrol was used throughout.

Possessing all the mechanical quietness which was the hallmark of its predecessors, the Statesman also has an effectively subdued exhaust. Although an air filter is not fitted to the carburettor as standard, induction roar was not obtrusive. This generally low noise level, therefore, plus the fact that there was ample power at low revolutions, enabled the machine to be ridden discreetly in the most congested of urban areas.

Carburation was clean throughout the range. The engine picked up unhesitatingly as soon as the throttle was opened and, once warm, would idle with a steady, reliable beat.

The speedometer read approximately 5 per cent fast, and the instrument would have been easier to read, especially in conditions of poor visibility, had the needle been backed by a black back-

The contact breaker is driven off the inlet camwheel and is readily accessible. Two screws secure the light-alloy cover

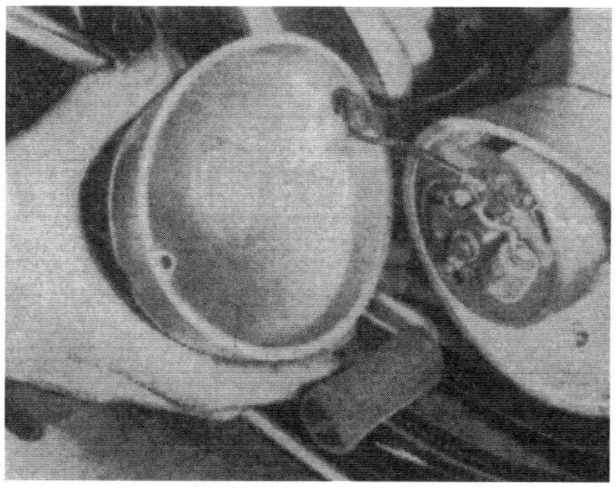

Robust roadster single, combining good power output and tractability: first-class handling and braking

Orthodox handlebar layout is employed. The switch on the left of the headlamp controls the lights while that on the right is for ignition

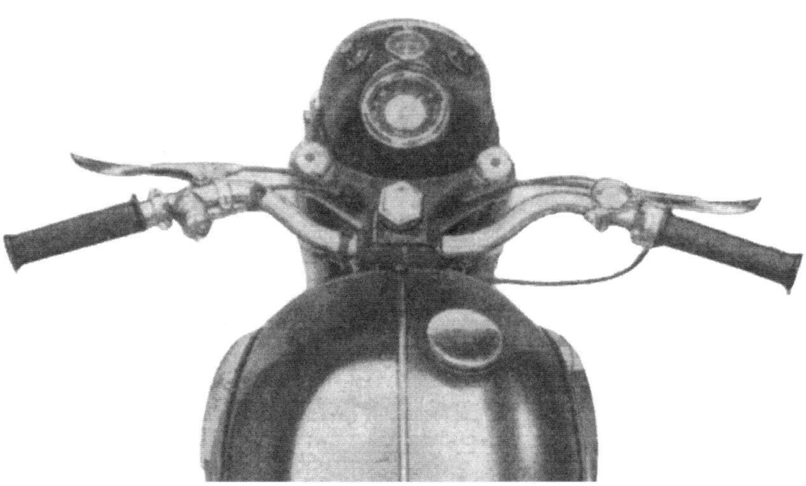

This shot emphasizes the trim, workmanlike appearance of the Statesman. The duplex frame has lugs for sidecar attachment

ground (about two-thirds of the needle rotates over a light grey surface). While the performance figures were being obtained the main mileage record jumped 1,000 miles. A trip mileage recorder is fitted, and could easily be reset by means of the knob protruding from the base of the headlamp.

Handling inspired confidence at all times. The machine could be ridden to a standstill in heavy traffic feet-up without undue effort. Steering lock was felt to be rather limited for a model of this nature. The lack of it made man-handling in and out of limited garaging space, a twice-daily task for many ride-to-work enthusiasts, more awkward than was thought desirable. Centre and prop stands are provided. Both were easily operated and gave safe support.

On the open road the Statesman could be ridden to the limit of its performance without betraying any tendency to deviate from the chosen line. Zestful cornering was encouraged by the way in which the model could, with little effort, be banked from side to side through bends.

Both front and rear springing are hydraulically damped and possessed just that right degree of firmness—an effective compromise between the stiff action needed for stability at high speeds and a soft one for comfort at lower speeds. When the machine was ridden one-up, the three-position Girling units were used in their softest setting. By means of the C-spanner provided in the tool kit they proved quickly adjustable. The Model 18 displayed no tendency to pitching on undulating surfaces and was ridden on unmade, potholed roads without the suspension ever bottoming.

As the miles on a journey mount up so the suitability, or otherwise, of the riding position makes itself known. In this respect the Statesman earned high marks. If any criticism is due, it is that the footrests could with benefit be positioned an inch or so farther to the rear, to relieve the rider's arms from some strain when cruising for long periods in the seventies.

Controls fell readily to hand or foot as the case was. The dual-seat proved very comfortable and generous enough dimensionally to accommodate a passenger without cramping. Fitted as standard, the passenger's footrests had rubbers of the same size as those for the rider and resulted in greater comfort than the usual type. Plastic handlebar grips are fitted. Owing to the nature of the material, and insufficient depth of pattern, the amount of friction they offered, especially during cold or wet conditions, was considered inadequate.

Both front and rear brakes were light in action and could be delicately controlled. During performance tests successive stops from maximum speed

Adjustment of the clutch thrust rod is simply effected. Removal of a threaded cap exposes the adjuster in the centre of the clutch pressure plate

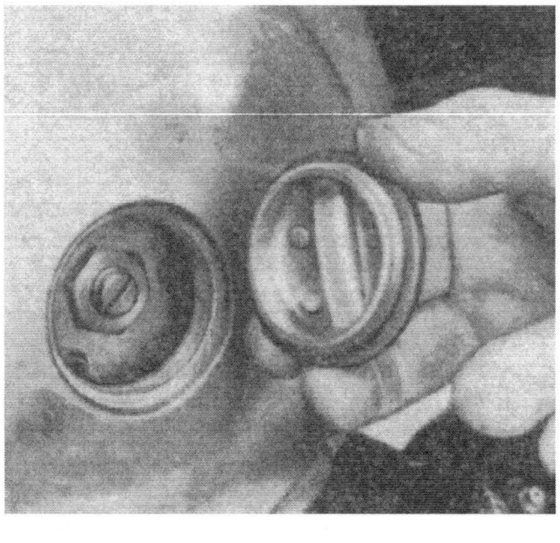

For 1962 a silver-coloured metal motif embellishes the fuel tank of the Statesman

were made without any loss in efficiency or of adjustment.

Starting a big single when overnight temperatures are around freezing point calls for a certain amount of technique. On the Model 18 the drill was to flood the carburettor slightly, close the handlebar-mounted air lever and, with the exhaust-valve lifter in operation, give a long, swinging kick, releasing the lifter when the kick-starter pedal was at the bottom of its travel. In practice the method advised by the makers of easing the engine over compression, releasing the valve lifter and then kicking did not spin the engine sufficiently to start it in severe conditions. It was quite effective enough, though, as soon as the engine was warm.

Clutch operation was progressive, but on the heavy side. Hard usage during the test caused no loss of adjustment. Gear changes in both directions were both positive and silent. Bottom-gear engagement with the engine idling caused, at worst, no more than a slight click. Neutral could be selected without difficulty.

Lighting was adequate for cruising in the sixties after dark. The dipped beam could be readily selected by an easily accessible switch on the handlebar. The headlamp can be adjusted after slackening the two pivot bolts. The charging rate at 30 m.p.h. in top gear was sufficient for the generator to balance the full lighting load.

Mounted under the seat, the horn would be more effective if it had a louder and perhaps higher-pitched note. Complete experimental discharging of the battery proved that the emergency-start system was up to its task and that the generator could restore the battery to a useable state of charge after a few miles running.

Located on the left just below the dual-seat, the roomy tool box could be opened with a coin. The tool kit provided is reasonably comprehensive and contains enough tools for most routine maintenance jobs. A most informative instruction book is also provided. Opening the tool box exposes the battery for inspection. It is held in position by a tensioned rubber strap which has the advantage that, besides being easy to undo, it will not be affected should it become exposed to electrolyte spray. The battery level can be checked with the battery in position.

Removal of two screws allows a light-alloy cover to be detached from the timing chest thus revealing the contact breaker (which is driven by the inlet camwheel).

The Statesman is finished in black relieved by touches of chromium plating. The absence of the familiar A.J.S. gold lining on the fuel tank will be regretted by many enthusiasts. But for those who prefer colour, the model is available in blue with white mudguards.

To sum up, then, the Statesman is a good-looking, well-mannered mount, equally suitable for the daily round or for long week-end runs.

Specification

ENGINE: A.J.S. 498 c.c. (82.5 x 93mm) overhead-valve single. Crankshaft supported in plain and ball bearings; roller big-end bearing. Light-alloy cylinder head; compression ratio 7.3 to 1. Dry-sump lubrication; oil-tank capacity, 4 pints.

CARBURETTOR: Amal Monobloc; air slide operated by handlebar lever.

IGNITION and LIGHTING: Lucas coil ignition with auto-advance. Lucas RM15 alternator with rotor mounted on drive-side crankshaft, charging Lucas 6-volt, 11-amp-hour battery through rectifier. Lucas 7in-diameter headlamp with pre-focus light unit.

TRANSMISSION: A.M.C. four-speed foot-change gear box. Gear ratios: bottom, 12.86 to 1; second, 8.53 to 1; third, 6.13 to 1; top, 5.02 to 1. Multi-plate clutch with bonded friction facings running in oil. Primary chain, ½ x 0.305in in light-alloy oil-bath case. Rear chain, ⅝ x ⅜in with guard over top run. Engine r.p.m. at 30 m.p.h. in top gear, 2,000.

FUEL CAPACITY: 4½ gallons.

TYRES: Dunlop: 3.25 x 19in ribbed front; 3.50 x 19in Universal rear.

BRAKES: 7in-diameter front and rear; finger adjusters.

SUSPENSION: A.M.C. Teledraulic telescopic front fork with hydraulic damping. Pivoted rear fork controlled by Girling spring-and-hydraulic units with three-position adjustment for load.

WHEELBASE: 55¼in unladen. Ground clearance, 5½in unladen.

SEAT: A.J.S. dual-seat; unladen height 31in.

WEIGHT: 403 lb fully equipped, with full oil tank and approximately one gallon of petrol.

PRICE: £208 10s; with purchase tax (in Great Britain only), £254 7s 5d Extras: Blue finish with white mudguards, £3 10s (plus 15s 5d p.t.); quickly-detachable rear wheel, £2 2s (p.t., 9s 3d).

ROAD TAX: £4 10s a year; £1 13s for four months.

MAKERS: A.J.S. Motor Cycles, Plumstead Road, London, S.E.18.

DESCRIPTION: *The Motor Cycle*, 14 September, 1961.

PERFORMANCE DATA

(Obtained at the Motor Industry Research Association's proving ground at Lindley, Leicestershire.)

MEAN MAXIMUM SPEED: Bottom: 38 m.p.h.*
Second: 57 m.p.h.*
Third: 77 m.p.h.
Top: 82 m.p.h.
*Valve float occuring.

HIGHEST ONE-WAY SPEED: 87 m.p.h. (conditions: strong following wind; rider wearing two-piece suit and overboots).

MEAN ACCELERATION:

	10–30 m.p.h.	20–40 m.p.h.	30–50 m.p.h.
Bottom:	3.6 sec.	—	—
Second	4.8 sec	4.4 sec	4.6 sec
Third:	—	7 sec	5.8 sec
Top:	—	9.2 sec	8.6 sec

Mean speed at end of quarter-mile from rest: 68 m.p.h.
Mean time to cover standing quarter-mile: 18.4 sec.

PETROL CONSUMPTION: At 30 m.p.h., 84 m.p.g.; at 40 m.p.h., 72 m.p.g.; at 50 m.p.h., 64 m.p.g.; at 60 m.p.h., 52 m.p.g.

BRAKING: From 30 m.p.h. to rest, 29ft 6in (surface, dry tarmac).

TURNING CIRCLE: 16ft.

MINIMUM NON-SNATCH SPEED: 20 m.p.h. in top gear.

WEIGHT PER C.C.: 0.81 lb.

The 646 c.c. MATCHLESS G12 'MAJESTIC'

1962 ROAD TESTS

All the Best in British 'Big Bike' Tradition

Specification

ENGINE

Type	Parallel-twin four-stroke
Bore	72 mm.
Stroke	79.3 mm.
Cubic capacity	646 c.c.
Valves	Overhead (push-rod)
Compression ratio	7.5 : 1
Carburetter	Amal Monobloc, 1⅛-in. bore
Ignition	Lucas battery-powered coil, automatic control
Generator	Lucas 6-v. 60-w. RM15 alternator
Makers' claimed output	Not released
Maximum power at 6,200 r.p.m.	
Lubrication	Dry sump with two gear pumps
Starting	Kickstarter

TRANSMISSION

Separate gearbox with footchange

Ratios	4.8, 5.8, 8.1, 12.2 : 1
Speed at 1,000 r.p.m. in top gear	16 m.p.h.
Speed equivalent to revs. at maximum power rating:	
Second gear	60 m.p.h.
Third gear	84 m.p.h.
Top gear	104 m.p.h.
Primary drive	Chain
Final drive	Chain
Clutch	Multi-plate in oilbath
Shock absorber	Rubber-vane type in clutch centre

CYCLE PARTS

Frame	Duplex cradle with brazed lugs
Front suspension	"Teledraulic" telescopic forks with coil springs and two-way hydraulic damping
Rear suspension	Swinging fork with three-position Girling hydraulically damped spring units
Wheelbase	55 in.
Tyres	Dunlop: ribbed 3.25×19-in. front; studded 3.50×19-in. rear
Brakes	7-in. dia. in full-width hubs front and rear; total lining area, 23½ sq. in.
Fuel tank	Welded steel. Rubber mounted. Two taps
Oil tank	5 pints with anti-froth device
Lamps	30/24-w. head; 3-w. pilot; 18/6-w. stop/tail; 1.8-w. speedometer
Horn	Lucas
Battery	Lucas MLZ9E, 11 a.h.
Speedometer	Smiths 120 m.p.h. with trip
Seating	Two-level dualseat
Stands	Centre and prop
Tool kit	Spanners: 3 double-ended, 1 ring, 1 tappet, 1 contact-breaker, 2 box, 1 Girling. 2 tyre levers; 2 Allen keys; screwdriver; clutch-spring tool; tommy bar; grease gun; pliers; circlip tool
Toolbox	Semi-pannier type on left side
Finish	Red frame and tank, white guards, grey seat

OTHER EQUIPMENT

Twin exhaust pipes and silencers; pillion rests; twin rear lifting handles.

PRICES

Machine	£262 6s. (inc. £47 6s. 0d. P.T.)
Extras	Red and white colour scheme, £4 5s. 5d. (inc. 15s. 5d. P.T.)
Total as tested	£266 11s. 5d.
Tax	£4 10s. p.a.; £1 13s. for four months
Makers	Matchless Motor Cycles (Associated Motor Cycles, Ltd.), Plumstead Road, London, S.E.18

'Motor Cycling' Test Data

Conditions. *Weather: Perfect, bright and cold (Barometer 29.20 in. Hg. Thermometer 36°F.). Wind: Nil. Surface (braking and acceleration): Dry asphalt. Rider: 11½ stone, wearing two-piece suit, safety helmet and boots, normally seated except where stated. Fuel: Super grade (101½ research method octane rating).*

Venue: *Motor Industry Research Association Station, Lindley.*

Speed at end of standing 1,000 yd.:	
East	90 m.p.h.
West	79 m.p.h.
Mean	84 m.p.h.
Speed at end of standing 1,400 yd. (rider prone):	
East	100.3 m.p.h.
West	90.9 m.p.h.
Mean	95.4 m.p.h.
Braking from 30 m.p.h. (all brakes)	10 yd.

Fuel consumption

At constant 30 m.p.h.	108 m.p.g.
50 m.p.h.	72 m.p.g.
500-mile overall figure	58 m.p.g.

Speedometer

30 m.p.h. indicated=28.9 m.p.h. true
40 m.p.h. indicated=38.9 m.p.h. true
50 m.p.h. indicated=48.2 m.p.h. true
60 m.p.h. indicated=57.4 m.p.h. true
70 m.p.h. indicated=67.0 m.p.h. true
80 m.p.h. indicated=76.2 m.p.h. true
90 m.p.h. indicated=84.3 m.p.h. true
100 m.p.h. indicated=93.3 m.p.h. true

Mileage recorder .. Overreading 1%

Electrical Equipment

Top gear speed at which generator output balances:	
Ignition only	20 m.p.h.
Minimum obligatory lights	31 m.p.h.
Headlamp main beam	30 m.p.h.

Weights and Capacities

Certified kerbside weight with oil and 1 gal. fuel)	441 lb.
Weight distribution (rider normally seated):	
Front wheel	40%
Rear wheel	60%
Tank capacity (metered):	
Total	4.13 gal.
Reserve	3 pints

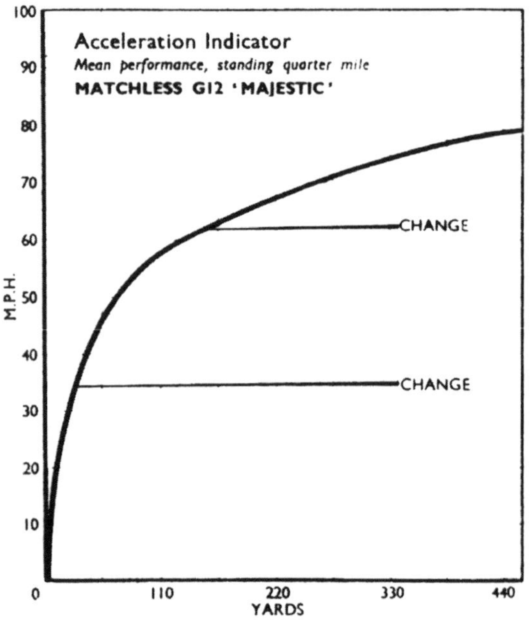

VERSATILITY is the keynote of the Matchless G12 " Majestic "—650 c.c.-worth of useful and satisfying performance. This Plumstead product is as suitable for daily commuting over short and long distances as it is for high-speed galloping with a passenger at week-ends. It shines equally on the speedy let's-get-to-the-seaside run as it does on the leisurely by-lane Sunday tour—a real maid-of-all-work.

It is powered by a lusty, oiltight o.h.v. twin-cylinder engine built along the established lines for which the British motorcycle industry is renowned. But it is now alone amongst this country's four-stroke products in having a three-bearing crankshaft. It is also one of the few remaining marques on which it is possible to dismantle engine or gearbox separately without breaking down the other.

The G12 always did precisely and exactly what the tester asked it to do. The gentlest jaunt at 28 to 30 m.p.h. in top was free from snatch or roughness and sufficiently far up the engine scale for worth-while acceleration to be obtained if the throttle was opened smoothly. Indeed, even an indiscreet handful brought no protest. The point is stressed to illustrate the engine's excellent spread of power and to emphasize the effectiveness of the rubber-vane cush-drive in the clutch centre.

When flitting from a 30 to a 40 m.p.h. speed limit one just didn't bother to slip momentarily into third. Complete tractability extended down to tickover, which was of plonkity-plonk slowness with infallible reliability. There was clean carburation when coming off the throttle stop and a willing, fuss-free response well suited to nudging oneself gently through traffic without attracting notice or creating offence.

The low-speed engine and transmission characteristics were so good that 10 out of 10 marks were awarded on the tester's check card. The same score was also given for starting from both the hot and the cold conditions. When icily cold (literally!), the engine burst into life readily; the mixture was weak enough to demand generous use of the choke for moving away. On a cold January morning, with black ice on the road, full air could be restored within three-quarters of a mile. The twin usually liked a flood for a cold kickstart. When hot, it asked for no more preparation than the turning on of the headlamp-mounted coil-ignition switch.

The presence of the third (centre) main bearing—together with plain big-ends—did not create cold-cranking difficulty, though its slight extra drag was detectable.

Middle-speed manners also earned high marks. Here was 650 c.c. performance without temperament. When well under way—at 80 m.p.h., for example—the " Majestic's " motor was a willing worker and would climb straight up the remainder of the rev band with alacrity.

Cruising in the 70s or 80s was almost playing, and even faster mile-swallowing made no demands the power unit could not meet. In fact, rider-toleration of wind pressure was probably the prime limiting factor. Vibration was present above r.p.m. corresponding to some 75-ish in top. It could be felt in all gears and was mainly manifest through the bars and the rests. The rubber-mounted tank and the seat were free from it. It did not exceed a tolerable level but was more marked than in several contemporary twins, some of which are virtually free from vibration.

" Majestic " progress, even on slippery, wet roads.

Hard driving produced not a vestige of an engine oil-leak. Weeping from the primary chaincase—a Matchless mannerism of yesteryear—was utterly non-existent, though the twin inspection caps of this latest-style cast-alloy assembly could be sited better. Engine oil consumption was moderate.

The hard rider ought to relish the acceleration available through the gears. This was well worth having. It was accompanied by a fairly hearty medium-tone exhaust note with the calico-ripping quality beloved of the enthusiast. Mechanically, the unit was quiet; what noise there was came from the valve gear and from the clutch plates floating in the grooves in the drum, when lifted.

Gear-changing during full-throttle thrashing was excellent. The pedal moved lightly and positively and gave not the slightest trouble. Downward changes were just as easy as upward ones. If the rider was using rather less than full performance—such as when commuting to and from work—the cogs moved in an equally satisfying manner. Then, a typical change-up pattern was 22, 45 and 65 on the Smiths speedometer.

Downward shifting needed mild blipping with no particular effort at road/engine speed synchronization. Neutral was readily found from either second or bottom, when at rest as well as when coasting to a standstill. The adjustable footchange pedal could be well placed in relation to the adjustable footrest. The kickstarter crank did not interfere with the leg when putting the foot out; but no rubber was fitted so the foot sometimes slipped off at full depression in wet weather.

Hard braking to a full stop, in a short distance, was well within the G12's capabilities from traffic speeds. The front anchor was delightful to use and needed no more than a loving caress to bring instant, controllable response. Dipping at the front was slight. The rear brake, the footpiece of which was properly placed, did all that was asked of it. Neither brake became waterlogged in the wet, even when the machine had been standing outside. Neither grabbed in the damp and no fade was experienced—though, in view of the relatively small diameter and area of the front stopper, it must have been working well up towards its rated maximum.

The keyword, versatility, also applies to the model's handling. The " Majestic " was equally at home in traffic and on the open road. Finding one's almost pedestrian way

Left : Close-up of the sturdy front wheel mounting showing also the fork leg drain plug and items of the brake assembly. Right: The "Majestic" engine features separate barrels with light alloy heads and exhaust pipes which are pushed in without locating rings. The engine steady plate from the cylinder head to the frame and the cast aluminium chain case are other points worthy of note.

to the head of a stationary queue without losing balance revealed good weight distribution, highlighting a quality of importance to the man faced with the need to negotiate city traffic, fight his way out of a crowded factory gate, or thread a cautious passage along a narrow track to a remote farm or house.

On byway and highway, the duplex frame did its share to promote safe navigation. Cornering at steep angles and on generous lock at, say, 15 m.p.h., was good—though the weight could be felt a little. At really high speeds, such as when making time on a long run, the bendswinging properties of the Matchless "Majestic" were a great asset. The machine was very sure-footed, in the wet and the dry, and possessed better road-holding than many of its contemporaries.

In fact, when cornering on wet roads the handling was completely free from vices. There was never any fear of the back end suddenly breaking away and the approach of mild drifting could be controlled and used—a fact which speaks well for the good characteristics of frame and suspension.

Braking on a wringing-wet highway, granite-chipped or shining black, was of the straight-ahead variety and most stable.

Comfort-wise, the springing absorbed low-speed road shocks well; at high velocities it managed to retain the necessary firmness to hold line and to refrain from giving the rider that awful jelly-like feeling of a too softly sprung, under-damped mount. The Girling rear units are adjustable for passenger load and use was made of this advantageous feature even though alteration is a toolkit task.

The dualseat, of the slight split-level type, is wide and resilient. It is also long enough to commend it to those who have experienced the cramped accommodation of many other so-called dual seats.

The riding position was on the high side, perhaps too high for those with short legs. But, then, this machine is big in stature in every respect.

No he-man tactics were needed to get the "Majestic" on to its 1962 centre stand. This is a good device and required no more than firm foot pressure on the special tread bar and a flip up from the right hand grasping the edge of the pressed-steel basepan to the seat. The tread bar's pressure pad is so small that it can be uncomfortable when operated in soft-soled shoes. Support when on the stand was excellent.

The fairly accessible prop stand demanded careful use, however. It would not hold the mount on a soft surface, or much out of the vertical. In short, it demanded a hard, flat surface. Also, it tended to spring up as soon as weight was relieved.

Both stands tucked up out of the way and were fully proof against grounding during vivid cornering and when thrusting fast along unmetalled roads. In fact, the abundant ground clearance, generous lock, and wide handlebars would commend themselves to many faced with daily journeys of the sort undertaken by, say, the forestry worker.

The bars, however, seemed a little too wide. On the left one the plastic grip worked loose several times and its collar prevented the hand being slid inboard to reach the dipswitch horn button easily. The layout also produced an awkward reach to the clutch lever.

Electrically, many high marks were given to the G12. The RM15 is one of Lucas' more powerful alternators. It kept the battery fully charged during the short days of December and January and provided sufficient light at night. The block-lens light-unit gave rather better beams than those we have previously experienced. The main had good range and spread with a subsidiary pool of light thrown close up to the mount; the dipped beam gave equally effective illumination. Both filaments produced a vivid white light indicative of plenty of voltage. The horn was an advance on earlier patterns, though towards the end of the test it went out of action, as did an ignition condenser, most accessibly located in the distributor.

Our test machine, a 1962 model, was not fitted with the new-season battery nor a detachable key to the ignition switch. These items, though announced, are not yet featured on production models. But it did have the massive and attractive chromium-plated Matchless medallion now emblazoned on the tank for the coming year. It is a zinc-base die-casting.

The "Majestic's" colour scheme of red frame and tank, with white guards, is pretty to look at and proved surprisingly serviceable. At first, we expected that it would show every dirt smear. But the guards have sleek lines and offer few crevices to trap road filth. Consequently, they could easily be swabbed down with a wet rag.

Indeed, as well as being eminently versatile, the G12 is an eyeable and thoroughly practical mount. It is a Matchless of which to be proud—every bit as majestic as its name.

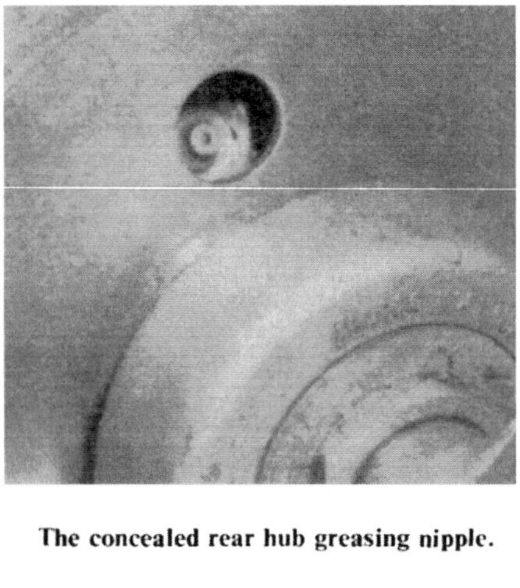

The concealed rear hub greasing nipple.

Continued from page 13

suit the very different dimensions of the vertical single-cylinder engine.

The large-capacity—five-gallon — petrol tank is rubber mounted, pannier-wise, on the top tube; it has deep sides and towards the rear these are recessed to provide a comfortable knee-grip. The one-gallon oil tank extends well round the rear mudguard, thus presenting a considerable surface to the cooling airstream.

Both wheels—front 3 ins. by 21 ins., rear 3.25 ins. by 20 ins.—are of straight-spoke construction and have high-tensile steel rims unless light alloy rims, which can be supplied at an extra charge, are specified. The hubs are of magnesium alloy and have non-adjustable journals.

The 8¼-in. brakes are exactly as fitted to the twin. They are of the double-leading-shoe type and are reputed to be extremely efficient. When I was in the A.J.S. racing department the other day one of the twins was being overhauled and I was able to see how well the brakes have stood up to a season's racing. The brake plates and shoes are, of course, of light alloy, the whole assembly, for its size, being remarkably light.

Bearing in mind the very short space of time which elapsed in the case of the twin, between its leaving the drawing board stage and its appearance in the I.o.M., I am confident that quite a few brand-new 348 c.c. o.h.c. A.J.S.s will be crossing to the Island towards the end of May. In fact, having seen the number of components already in course of production in the works, I am convinced that not a few of these models will be in the hands of private owners by the time the first 1948 race commences.

Their advent will undoubtedly be welcomed by the all-too-few British manufacturers who already produce racing motorcycles.

Continued from page 21

right angle that rising to a post position can be done quickly and with little effort.

The handlebars, although not of the extreme high type, are a perfect match both in rise and sweep back. This is not a factory bar, but is optional according to rider's choice, in the California area.

Another aid to riding control is the narrow, trim-looking gas tank which allows the knees to be kept close in when picking a path through a difficult section of ground, and seems to make for more sensitive control.

Down in the engine room Ajay differs in several ways from many of the singles. Few cylinders have the massive finning area that has been dedicated to cooling. Head hold-down bolts run clear through the cylinder to the cases for added top end rigidity. Racing type hairpin valves, enclosed in those two large domes atop the barrel, are claimed to have more snap than conventional coil valve springs and are less affected by heat because of their more remote position. An especially heavy formation of finning can be noticed about the exhaust port area.

We were also interested to learn that a new type of steel flywheel is now being used that allows a much tighter fitting of the crank pin; thereby reducing chances of flywheel slippage. A large flanged bronze bushing on the timing side of the main bearing now stands ready to back up the main shaft in case of failure of the shock absorber spring on the engine sprocket. Continued running after breakage of the shock absorber spring has in the past occasionally resulted in end play which allowed the main shaft to pound through the opposite side of the case. This condition can no longer be a threat.

Fortunately for those who will be unable to obtain true Competition Scrambles jobs in '52 (yes, production is limited on the Competition model) there is little basic difference between it and the standard single cylinder model. The stock job can be converted by replacing its 19 inch front wheel with a 21 inch hoop, switching fenders, fuel tank, exhaust pipe and cams. The standard 500 has hairpin valve springs this year, and with the addition of the longer duration competition cams, only the alloy cylinder remains as a major difference.

For the lad who just can't seem to get the hang of backwoods riding, or the elder who's about to toss in the towel because the runs are getting too tough, the Ajay is a shot in the arm. It may be worth $865* to your ego. To those who are already Scrambles-mounted I say, "So long—wait for me at the next check."

**Plus tax and license at Los Angeles, Calif.*

PERFORMANCE SUMMARY

Maximum in Low	32 mph
Maximum in Second	51 mph
Maximum in Third	65.45 mph
Maximum in High	81.81 mph

BRAKING

From 25 to stopped, rear brake only	43'4"
From 25 to stopped, front brake only	36'5"
From 25 to stopped, both brakes	24'

ACCELERATION

*1/10 mile drag (9.2 sec.) 39.1 mph avg.
8.51 mph/sec. (avg. velocity change rate)
*1/4 mile drag (17.3 sec.) 52.0 mph avg.
6.01 mph/sec. (avg. velocity change rate)
*Low, second and third gears used

SLOW RUNNING

High gear without snatch	14 mph

TURNING CIRCLE

Minimum Diameter	12'6"

MILEAGE

Under competition circumstances	44 mpg

Continued from page 67

body in which are mounted the speedometer, ammeter and the separate ignition and lighting switches. Adjustment for the grip angle of the handlebar is permitted by hemispherical ends on the short risers on the bar; in these ends are studs which pass through slotted holes in dimples in the yoke. Clutch and front-brake controls have welded-on pivot blocks and the levers are hooded. On the left side of the handlebar is a combined dip switch and horn button of the latest Wipac ring design.

Both mudguards are very shapely and of deep helmet section. The rear guard has no visible stays and for the close-fitting front guard there is only a tubular hairpin stay at the lower end. To add to the clean appearance, it is intended that the front registration numbers should be painted on or attached to the valances of the mudguard. Wheel size is 17in, admirably suited to the proportions of the machine, and the rims are chromium plated; tyre section is 3.25in front and rear. Brake diameter is 6in and hubs are of full-width pattern. The rear-brake pedal pivots on a sleeve on the pillion footrest bolt and has a concealed return spring.

In standard form the machine has a deep-section guard over the upper run of the rear chain but, as mentioned, a chain-case is available at extra charge. The case is divided horizontally and embraces the fork arm to which the upper and lower halves are bolted; the edges of the halves overlap to prevent the ingress of road dirt and water. Forward end of the case is shielded by a sheet-metal section bolted to the back of the primary chaincase.

Battery, horn, coil and rectifier are concealed under the dual-seat by sheet-metal cases. The tool kit is stowed within the other case

Nominal wheelbase is 53in and the weight is about 325 lb. Finish of frame and mudguards is black stove enamel. Capacity of the petrol tank is 2¾ gallons and it is finished in Mediterranean blue with gold lining on the A.J.S. and in Cardinal red with silver lining on the Matchless. The appropriate tank colour is repeated in a flash on each side of the engine castings which also carry a small replica of the tank badge.

It is anticipated that production will start in May but initially the output will be exported. Foreign prices are being announced at the Geneva Show and home-market prices as soon as possible thereafter; in both cases they are expected to be highly competitive. Manufacturers are Associated Motor Cycles, Ltd., Plumstead Road, London, S.E.18.

THROUGHOUT the world, it is customary for manufacturers to build single- or twin-cylinder engines to power the machines they sell for general road work, and to go on to more cylinders for their racing motorcycles. Fours, sixes, and even one V-8 has been turned out for Grand Prix racing – but they have almost never been offered to the general public.

In England, however, this order of things is completely reversed; they turn out twins by the carload for touring motorcycles, and have even built a few fours for mass-sales, but when they go racing, they inevitably seem to revert to the single-cylinder layout. One reason for this may be that their multi's have almost always been a disappointment, while they are the acknowledged past-masters of the single-cylinder art. No one gets more power, torque, smoothness or reliability out of the big "banger" than the British. In fact, if the builders of those spectacular multi's had the English designer's infinite capacity for attention to detail, their creations would be even faster than they are.

One builder of racing multi's, Honda, seems to have acquired this knack over the past couple of seasons of racing and the results speak for themselves. However, the lessons are more easily learned with a single, there being fewer parts to confuse the issue, and the best "text-book" available is the English racing engine, of which the Matchless G-50, presented here, is a superb example.

As a design, the G-50 is almost an exact copy of AMC's AJS 7R, which was introduced in 1948. The 7R, which is still very much in the picture, has the same stroke, 3.07 inches, and a bore of 2.972 inches, giving a displacement of 21.35 cubic inches, or 348 cubic centimeters. For the G-50, a new cylinder and piston were fitted (and some minor revisions were made to the cylinder head and crankcase) to give a bore of 3.54 in. and a displacement of 30.25 cu. in., or 496cc.

In assessing the 7R/G-50 design, it should be understood that it was designed for production from the first, and while the original design specifications included those features that give a high power output, it was considered equally important that a high degree of reliability be retained. Moreover, it was essential that the design be simple enough so that the private owner, for whom it was intended, would be able to maintain it without elaborate facilities or undue expense. A tall order? Yes indeed, but as the years have demonstrated, they did very well.

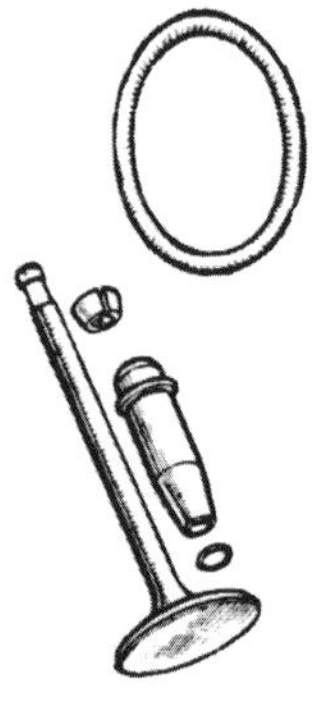

BY GORDON H. JENNINGS

TECHNICAL ANALYSIS:

MATCHLESS G-50 ENGINE

The power plant for the G-50 and G-50 CSR, an excellent example of the fine British art of single-cylinder racing engine design.

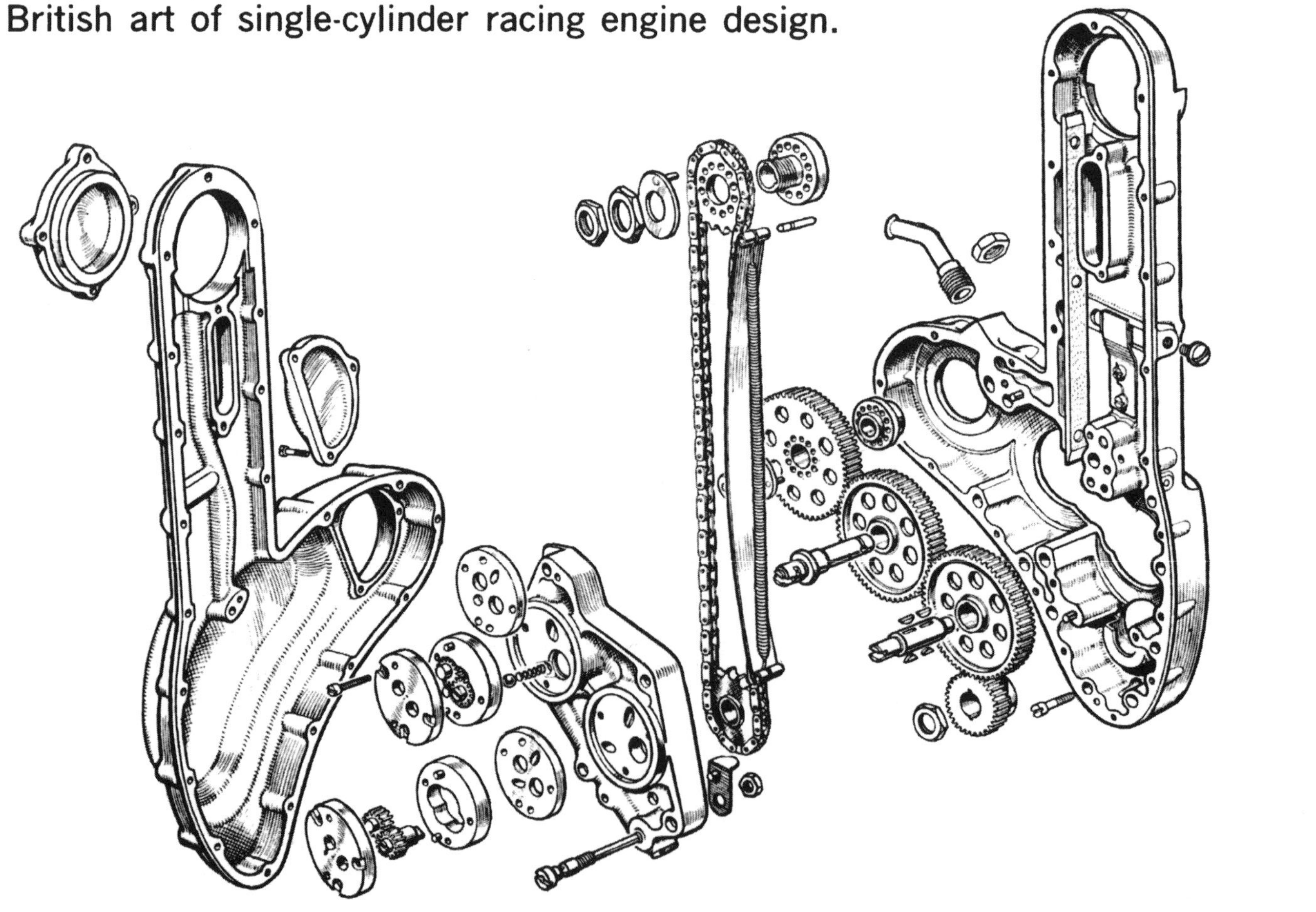

Initially, it was appreciated that no matter how cleverly the upper end of the engine was designed, it would all go for nothing if the crankshaft could not carry the load. Accordingly, every effort has been made to provide strength in that area. The flywheels are forged from nickel-chrome molybdenum steel and are joined by a two-piece crankpin. This has a straight-through section, with threaded ends, and a hardened collar that is the inner bearing race. The through section is drilled to feed oil to an annular groove inside the collar and oil is fed to the rod bearing via a pair of holes in the collar itself. Rigidity is supplied by pulling the flywheels together against the sides of the bearing-race collar by tightening nuts on the threaded ends of the crankpin.

The mainshafts are flanged on their inner ends and are an interference fit. Their flanged end fits into a recess in the center of the flywheel and the space-consuming securing nut is eliminated. The width of all bearings, shafts and the flywheels is held to a minimum to reduce the span between main bearings and insure maximum rigidity. The lengths to which the designers have gone may be judged by the odd crankpin fastening nuts used. These have an extended cylindrical section that fits into recesses in the outer sides of the flywheels and the hexagonal portion, to which the wrench is fitted. stands free. After tightening, this hexagonal head is sawed away, leaving only enough of the narrowed end of the nut to hold the flywheel. This system has the advantage of saving space and has been completely successful – although it is necessary to break away the nut in pieces and fit new ones when the engine's lower end is overhauled.

The crankcase of the G-50 is quite narrow, and is cast

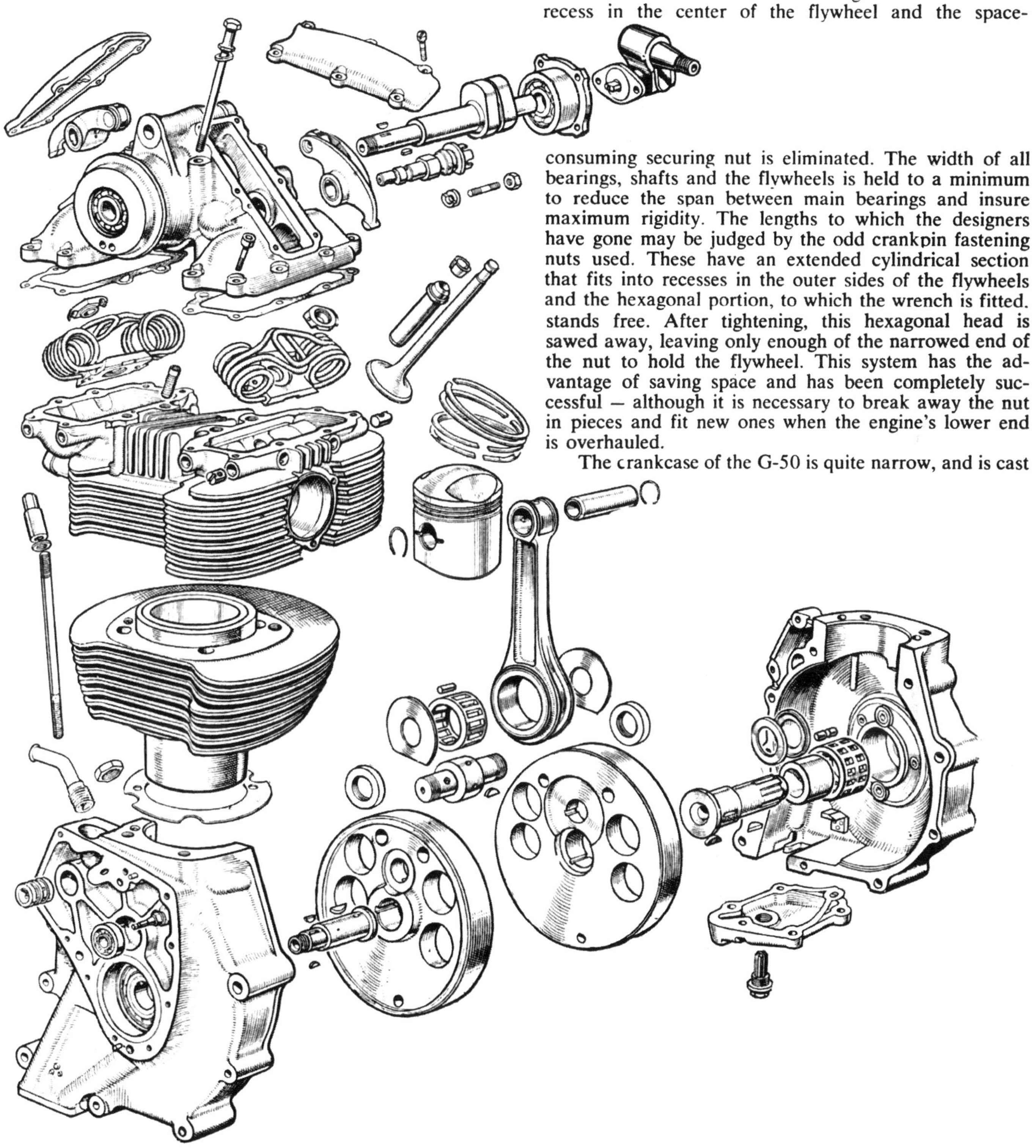

Though essentially straightforward in conception, the G-50 has no lack of interesting detail points. Examples are the separate oil pumps housed in the timing case, the rubbing strip and friction damped tensioner on the chain drive to the camshaft, the roller cam followers and the unusually deep spigot of the barrel into the head. The crankpin has been turned through a right-angle in the drawing, to show the two oil outlet holes. A single row of long rollers is employed for the big end bearing; the outer race of the drive-side main bearing is positively located in the crankcase.

from a magnesium-aluminum alloy. Like most crankcases of this type, it is split on its centerline and forms both crankcase and oil sump for the engine. Unlike the others, though, the casing is not extended to contain the ignition and camshaft drives. These are carried in a separate casting bolted on the "right" side of the engine.

The drive to the single overhead camshaft is an interesting bit of machinery. A set of 2:1 reduction gears take the drive from the crankshaft and from this gearset a single-stage chain drive leads up to the cylinder head.

This is very unusual in a racing engine; gear-train or bevel-gear and tower shaft drives are the conventional methods used. However, due to the reduction gearing, the G-50's cam-drive chain is only running at 1/2 engine speed and the arrangement has been giving good results for approximately 14 years. One reason for this may be that the designers have taken such pains to see that there will be no chain flutter. A spring-loaded, blade-type tensioner runs against the slack side of the chain and this has a friction damper at its free end to stop the vibrations caused by "feedback" from the cams. A rub-strip opposite the tensioner rides against the chain and it, too, serves to dampen-out chain flutter.

One might logically ask why, with the need for all of this tensioning and damping, chain drive was selected at all? There are several reasons. First, an all gear drive was eliminated because it is so difficult to maintain a proper mesh with all of the expanding and contracting of an aluminum alloy engine, and the spur gears used in such a drive tend to be very noisy even when in proper mesh.

Secondly, the tower shaft arrangement was decided against because of the extreme precision with which such drives must be constructed. All that was left was the chain, which is relatively cheap, quiet (if that counts for anything) and insensitive to small changes in on-center distances.

Thus, in the G-50 engine, it is possible to change the compression ratio by simply stuffing spacers of different thicknesses under the cylinder. This would, of course, not be possible with an all-gear drive and the towershaft layout would require a lot of adjusting with the changes in cylinder height.

A final comment on the camshaft drive: by juggling a series of keyways, sprockets and drive-pins, the valve timing may be brought to within 0.5° of the "exact" setting.

Contained with the same casing as the camshaft drive are the oil pumps and the magneto drive. There

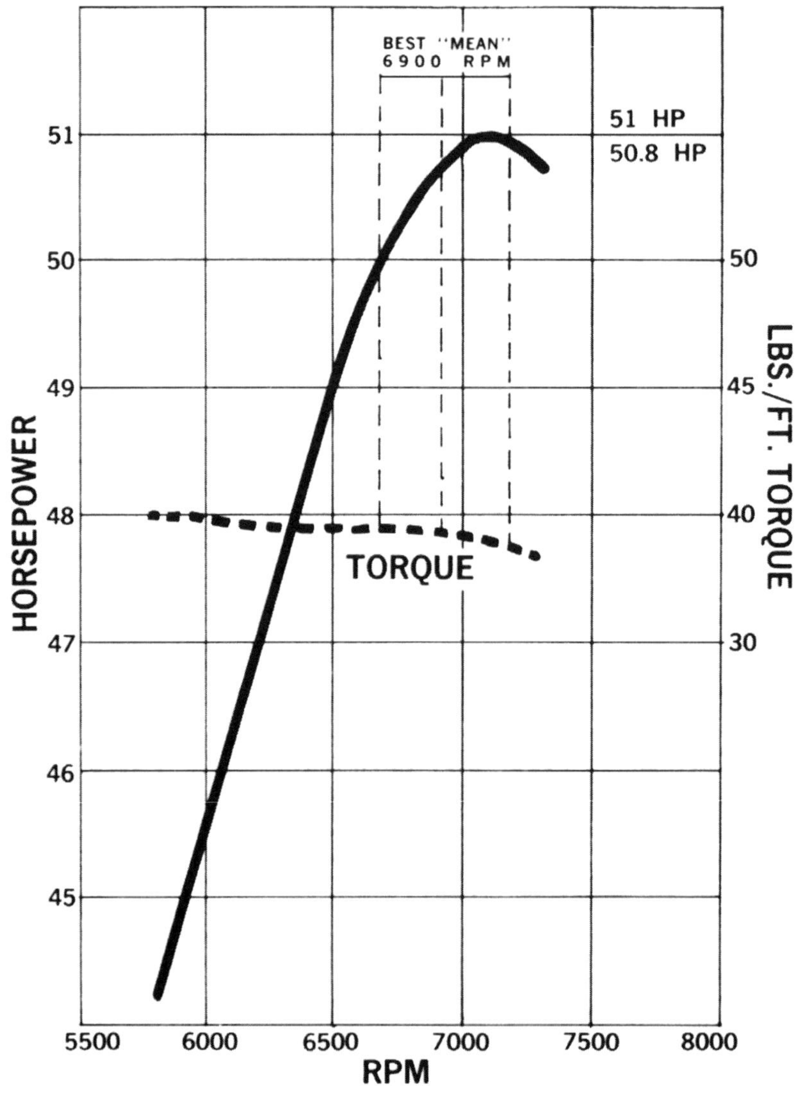

Horsepower and torque curves of the G-50.

is a gear train out to the magneto and it, like the camshaft, can be jiggled one degree at a time. The idler gears that take the drive from the crankshaft to the magneto also serve to drive the oil pumps. Two entirely separate gear-type pumps are used – one for pressure feed to the engine, the other scavenging oil from the sump. It is more common to have both pressure and scavenging pumps combined in a single housing, but this makes the pump longer and, in the G-50, it would have necessitated a lump on the side of the engine. It is, naturally, best if a racing engine can be kept narrow and that is the reason for the two pumps.

In the cylinder head, a single, centrally-positioned camshaft operates both valves. Forged steel rockers carry the motion from the cams to the valves, and there are .75-in. rollers in the cam follower end of each rocker. Early engines had pad-type followers, but these proved to be somewhat sensitive to lubrication interruptions and the change was made. The roller is a trifle heavier, but has been 100% reliable – and that is what matters most at Matchless. A nice, if not unusual, touch was the method provided for valve clearance adjustments: instead of having screw-type adjusters, each rocker pivots on an eccentric bushing, and these bushings can be turned to set the clearance.

The valves are set into the hemispherical combustion chamber with an approximately 75% included angle, with the exhaust valve tilted over a trifle more than the intake. The intake port has a pronounced "downdraft" angle and is bent around to the side at about 20%, this angle provided to give a swirl to the inrushing fuel/air charge.

For the G-50 CSR, the G-50 engine has been detuned somewhat – although not so much that it is exactly gentle. The only difference worth mentioning is the carburetor: a throat size of 1-1/2 in. is specified for the road racing G-50, and this has been reduced to 1-3/8 for the "street" model. The carburetor is, in both cases, a great fierce-looking GP-series Amal, with a remote needle and the new-type, narrow remote float chamber.

In either state of tune, some power at the top end has been sacrificed for good torque and power over a broad range, and we think perhaps they just may have the right idea. Very highly tuned engines require five- or six-speed close-ratio transmissions, and on some courses a rider can spend so much time stirring the gears about that he loses any advantage that the extra power might otherwise have given him.

Mostly, the things that make the G-50 engine a good design are its flexibility, reliability, and the fact that it is easy to keep in good repair. These things are essential for the private owner, and the G-50's successes have been because it supplies these things in large measure.

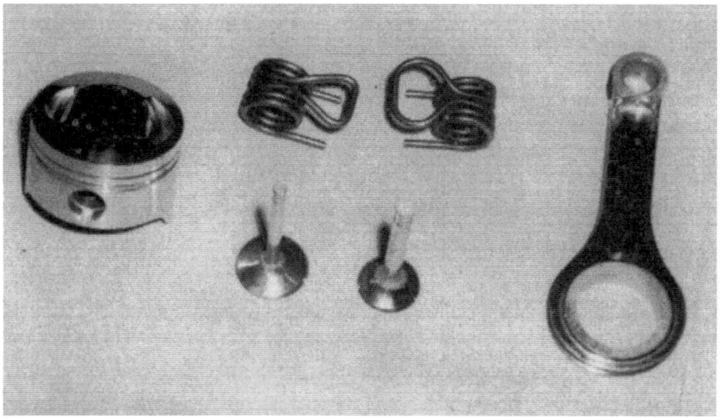

Beautifully finished piston, valves, valve springs and connecting rod reflect immense care to detail.

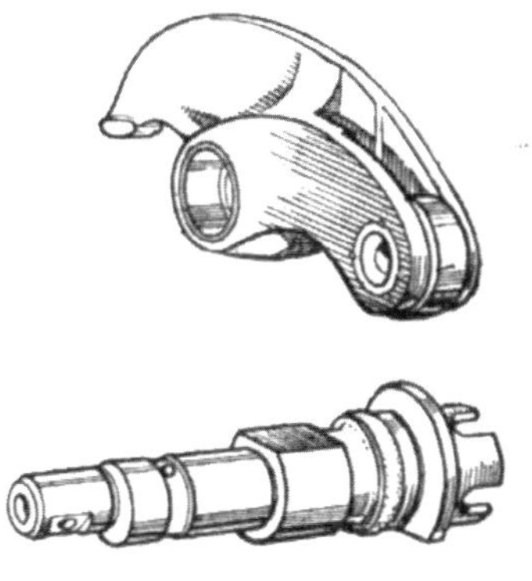

Details of the rocker, which has a roller cam follower and, for clearance adjustment, an eccentrically-mounted spindle.

CHARLES DEANE TRIES OUT THE

SPORTS AJS

By raising the compression ratio of the standard AJS Sceptre or model 16, whichever you like to call it, Associated Motorcycles have produced a sports 350 which is equal to, if not better than, many standard 500 singles.

In fact, if it were not for the smaller cylinder size, one could quite easily mistake it for its bigger brother. It has the same large 4½-gallon capacity fuel tank, same 7 inch diameter front and rear brakes, same duplex frame and in practically every other way is identical to the bigger model.

First impressions always count a great deal when one is considering buying a machine, and after only travelling a few miles on the Sceptre Sports, I felt completely at ease.

The gearchange was positive, the clutch—light in use, and the brakes—marvellous! The figure in our chart will show exactly how efficient they were.

'M.M.' MACHINE TEST REPORT No. 38

Machine A.J.S. *Model* Sceptre *c.c.* 348

Mileage 900 *Price new* £247/13/2d *Examiner* C. E. Deane

Supplied by Associated Motor Cycles, Woolwich.

Maximum Points 10 (compared with machines in same c.c. and price range)

Brakes (front)	9	*Lights*	9
Brakes (rear)	8	*Engine accessibility*	8
Brakes (both)	9	*General performance*	9
Steering at high speed	8	*Overall finish*	9
Steering at low speed	8	*Electrical layout*	8
Gearbox action	8	*TOTAL*	93

Fuel consumption 68 *m.p.g.*

Acceleration 0-50 8 *secs.*

Top speed 84 *m.p.h.*

Remarks Maximum speed - rider prone

BRAKING GRAPH

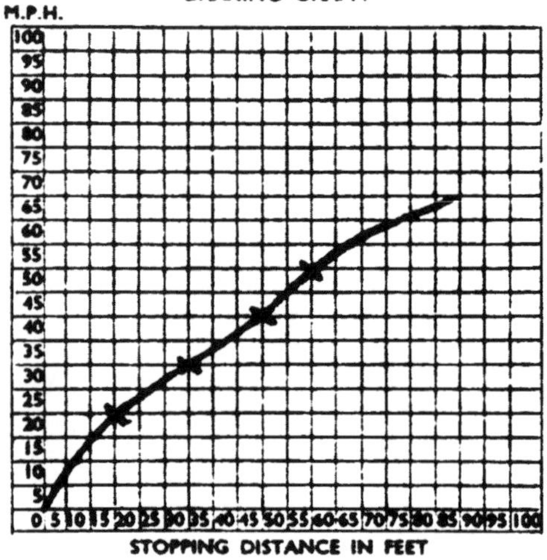

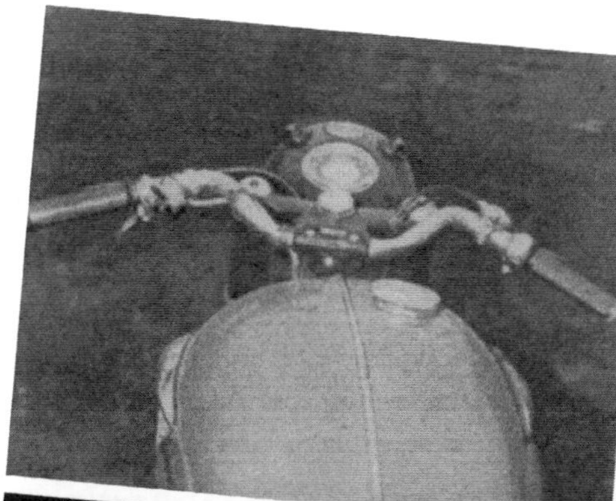

The extremely neat and careful layout of the controls on the Sceptre Sports handlebars. Note the 120 m.p.h. speedo.

Conventional engine and gearbox layout with the pushrods now enclosed in the tubes cast into the cylinder barrel

The rear suspension is fully adjustable and the toolbox is large and contains a fine toolkit. Note brake adjustment

The rear wheel is quickly detachable leaving the sprocket in place. Oil levels are clearly marked on the tank

Note duplex frame to front of engine. This gives much improved handling at speed. Alternator in the chaincase

The chrome mudguard closely hugs the front wheel and gives a very clean look to the unit. Note width of brake

From 30 miles-an-hour, they could stop 382 pounds of machinery, plus a 135-pound rider in 30 feet.

Handling of the Sceptre Sports is also one of its good points. The duplex frame, which is now used on all the large capacity AMC motorcycles, coupled with the adjustable swinging arm rear and teledraulic front suspension, gives hairline steering at all speeds up to maximum. Handling in traffic was also quite reasonable, although being on the small side, I found the weight of the machine rather a handful as the saddle is a little too far from the ground. I could only just stand on tip toes when astride the bike.

Although the Sceptre Sports has an 8.5 to 1 compression ratio, starting was simplicity itself. A valve lifter is fitted to ease the engine over compression, after that it was just a matter of tickling the carburetter, three-quarter closing choke, turn on ignition and one prod on kick-start. Nine times out of ten, the motor would start first time. One minor complaint here, is that anybody could do just that—there is no ignition key or steering lock fitted. Therefore, I had to find some other method of thief-proofing the machine.

The handlebar layout on the Sceptre is also very good. Every single item—dip-switch, horn button, choke control, light switch and ignition switch (on headlamp) fell readily to hand. On no occasion was I caught out and fumbling for one of the controls. This, believe it or not, is quite unusual on sports machines—British manufacturers always manage to tuck vital controls in some out-of-the-way place!

Compared with other single-cylinder machines, the Sceptre Sports has good acceleration, although the silencing could be greatly improved to allow for the larger throttle openings, which brought about a healthy crackle from the exhaust.

Happy Anywhere

When using a road test motorcycle, it receives the same treatment as most other machines on the road. I ride to and from the office every day, to the motorcycle club or local gathering place for motorcyclists one or two evenings a week, and at weekends I cover a few hundred miles touring the countryside.

This I think gives a motorcycle a fair test and with the Sceptre Sports, I found that I could cruise at a steady 70 miles-an-hour two up, providing traffic conditions allowed. Maximum speed in the prone position was just over 80-miles-an-hour.

The gearchange on the Sceptre, as I mentioned earlier, was nice and positive. The gear ratios themselves were also carefully spaced—in first gear it was possible to trickle along at a fast walking pace, with the engine just ticking over, while second gear brought 50 miles-an-hour on to the speedo in just eight seconds. 75 m.p.h. was possible in third gear and 84 m.p.h. in top. Although the engine would drop down to about 28 m.p.h. in top gear, it wasn't really happy below about 35 m.p.h. and third gear was the most sensible gear for cruising about town.

Lighting on any machine is very important, and the Sceptre is fitted with the same lighting equipment as the larger AMC models. It is very efficient and on main beam gave a light powerful enough to enable cruising speeds in the upper fifties. The dipped beam also worked well, and as I mentioned earlier, the dipswitch could be easily operated—with a flick of the left thumb.

The general finish of the machine is excellent, with a nice deep dual-seat for comfortable touring, and chrome mudguards and dropped handlebars to give a sporty appearance. Basically, I think it is a good example of a fine British single-cylinder motorcycle.

HOW LONG DOES IT TAKE?

Top up battery	1½ mins.
Adjust brakes	½ min.
Adjust tappets	4 mins.
Check points	1¼ mins.
Adjust chain	1 min.
Remove and replace rear wheel	10½ mins.

OVER THE TON FROM THE POWERFUL 650 c.c. A.J.S.

THE 31CSR or Hurricane is the largest motorcycle in the A.J.S. range of machines produced by the AMC concern at Plumstead. With cruising speeds in the 80 to 90 miles-an-hour mark and a top speed of about 105 miles-an-hour, it is definitely a motorcycle for the enthusiast. No power output figures are given, but comparing it with sports machines of the same capacity, we would set a figure around 45 brake-horse-power.

The 650 c.c. o.h.v. twin-cylinder engine of the Hurricane seems to have a split personality—if that is possible in motorcycles. About town it burbles along as quietly as any small-capacity run-about machine (see large silencer—pic. 2), yet as soon as one arrives at the open road and gives the machine its head, it becomes a rip-snorting road-burning sports bike.

Although the top speed isn't shattering, one must remember that this machine has been built to cater for the everyday rider who uses his bike to and from work and, on the odd occasion, for high-speed touring. If you require out-and-out speed, a speedkit is available which consists of twin carburetters, special camshaft and high-compression pistons. The speedkit gives better acceleration, plus a top speed of over 110 m.p.h. However, this conversion detracts from the standard Hurricane's about-town riding manners.

Throughout the entire test of over 600 miles, only 1 pint of oil was used. This wasn't really surprising after studying the engine (pics. 3 and 6), for even after thrashing the motor on the M.1, there wasn't a sign of an oil leak from the unit.

Starting the Hurricane was simplicity itself. The first prod filled the cylinders with fuel, the second prod brought forth life from the motor. Ignition was by magneto and an alternator supplied current for the lighting. Both units proved entirely trouble-free throughout the test.

Gear ratios in the four-speed gearbox seemed ideally spaced, and the machine would plod along at walking pace in first gear. However, top gear could never be comfortably used below 35 m.p.h. and for town use the first three gears had to prove adequate. The gearchange was positive on every gear, although a little noisy if one tried to rush through the box.

Good Road Manners

Roadholding is excellent and no noticeable frame whip was encountered on fast bends. No steering damper is fitted and at no time did one prove necessary. Unfortunately, there is one failing concerning the steering and that is the nearness of the handlebars to the petrol tank when on full lock. The staff of M.M. are all at present nursing sore thumbs where they have been crushed between the bars and the tank. The handlebar layout (pic. 8), although somewhat untidy to look at, proved to be well thought out. Everything, including the horn button, is either well positioned or easily adjustable.

The riding position of the Hurricane is more suited to the taller rider, for with a saddle height of 31 inches, short riders have a job to reach the ground when astride the stationary machine.

As mentioned earlier, the electrics proved trouble-free throughout the test and the headlight gave a long penetrating beam which allowed reasonable cruising speeds to be maintained at night. The instruments which were set in the headlamp shell consisted of a 120 m.p.h. speedometer and an ammeter.

Vibration seems to be a failing of the big vertical twins, and although there was a certain lumpiness about this CSR engine, there was no really uncomfortable vibration in the rev. range.

Fuel consumption figures proved quite interesting for a large capacity machine, for when used on slower journeys the Hurricane was covering almost 70 miles to the gallon. Yet, as soon as the tap was opened, they dropped considerably to the 50 m.p.g. mark. However, taken overall, a figure of 58 m.p.g. for a machine of this size is quite fair.

Only two further points regarding the performance of the Hurricane remain. First, acceleration; which was very good and had twin-carbs been fitted it would no doubt have been breathtaking. Opposite to acceleration is deceleration or, in other words, stopping. In this department, the Hurricane proved about average. At speeds up to 50 m.p.h. the brakes, both front and rear (pics. 5 and 7), were good, but above this speed the linings could have been a little harder to be more firm in their action. However, taken all round, the A.J.S. 31CSR Hurricane is a handsome machine (pic. 1). Performance is good, but can be improved with the speedkit. ●

ROAD TESTS OF CURRENT MODELS

348 c.c. Matchless G3 plus Garrard Grand Prix makes a trim, sporting outfit

MERCURY and GRAND PRIX

Matchless three-fifty paired with Garrard open single-seater makes a 60 m.p.h. outfit with excellent handling and first-class fuel economy

IF you have never driven a sporting sidecar outfit, your motor-cycling education is incomplete. Such are the sentiments of hundreds of seasoned enthusiasts.

But wait. Not everyone can take on the high initial cost and heavy drinking habits of a big twin. If this is your problem, then the latest 348 c.c. G3 Matchless Mercury plus the sporty-looking but very comfortable Garrard Grand Prix, might well provide the answer.

With a 60 m.p.h.-plus maximum, this outfit contrives to squeeze no less than 58 miles from every gallon of petrol while cruising at a steady 50 m.p.h. And, what is more, it does this still leaving the firm impression that it is a zestful, sporting machine.

The Matchless was the same machine as tested in solo trim (report: *Motor Cycle*, May 10). Let's recap briefly on the changes embodied in the new model.

Bore and stroke have been altered from 69×93mm to 74×81mm, and, because of the bigger bore, it has been possible to increase the valve sizes.

Carburettor choke is up from

SPECIFICATION

ENGINE: Matchless 348 c.c. (74×81mm) overhead-valve single. Crankshaft supported in plain and ball bearings: roller big-end bearing, light-alloy cylinder head; compression ratio 8.5 to 1. Dry-sump lubrication; oil-tank capacity 4 pints

CARBURETTOR: Amal Monobloc; air slide operated by handlebar lever.

IGNITION and LIGHTING: Lucas coil ignition with auto-advance. Lucas RM15 alternator, with rotor mounted on drive side of crankshaft, charging Lucas 6-volt, 11-amp-hour battery through rectifier. Lucas 7in-diameter headlamp with pre-focus light unit.

TRANSMISSION: A.M.C. four-speed foot-change gear box. Gear ratios: bottom 16.6 to 1; second 11.05 to 1; third 7.91 to 1; top 6.48 to 1. Multi-plate clutch with bonded friction facings running in oil. Primary chain, $\frac{1}{2}$×0.305in in light-alloy oil-bath case. Rear chain, $\frac{5}{8}$×$\frac{3}{8}$in with guard over top run. Engine r.p.m. at 30 m.p.h. in top gear, 2,550.

FUEL CAPACITY: 4¼ gallons.

TYRES: Dunlop 3.25×19in; ribbed front, Universal rear.

BRAKES: 7in-diameter front and rear; finger adjusters.

SUSPENSION: A.M.C. Teledraulic telescopic front fork with hydraulic damping. Pivoted rear fork controlled by Girling spring-and-hydraulic units with three-position adjustment for load.

WHEELBASE: 56in unladen. Ground clearance, 5½in unladen. Seat height, 31in unladen.

WEIGHT: Complete outfit, 567lb fully equipped, with full oil tank and approximately one gallon of petrol.

PRICE: Machine only, £248 4s 5d including British purchase tax.

ROAD TAX: Complete outfit £6 a year; £2 4s for four months.

MAKERS: Matchless Motor Cycles, Plumstead Road, London, S.E.18.

DESCRIPTION: *Motor Cycle*, 14 September, 1961.

SIDECAR

MODEL: Garrard Grand Prix single-seat sports.

BODY and CHASSIS: Garrard integral metal construction with light-alloy panelling. Four attachments to machine. Pivoted fork wheel suspension controlled by Armstrong spring-and-hydraulic unit. Spoked wheel with 3.25×16in Avon Safety Mileage Mark II tyre. Fixed transparent plastic screen. Collapsible hood secured by press-studs. Dimensions: ground clearance, 6in unladen; track, 41in; overall length, 87in; seat back to nose, 45in; seat to hood, 30in; seat cushion and squab combined, 44in long×19in wide; luggage compartment (irregular shape), 36×18×14in

PRICE: Complete, £98 6s including British purchase tax. (This price includes the brake which has been added to the standard specification since this model was submitted for test).

MAKERS: Garrard Sidecars, Ltd., 13, Third Avenue, Bletchley, Bucks.

DESCRIPTION: *Motor Cycle*: 28 September, 1961.

PERFORMANCE DATA

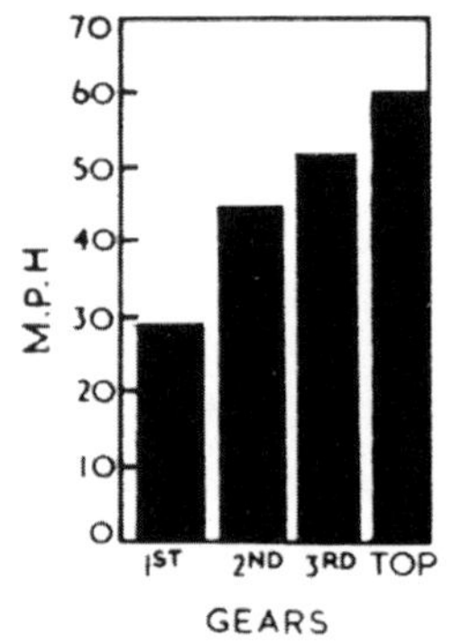

HIGHEST ONE-WAY SPEED: 62 m.p.h. (conditions: light tail wind; rider wearing two-piece suit and over-boots; 12-stone passenger in sidecar.)

Speed in bottom and second gears limited by valve float.

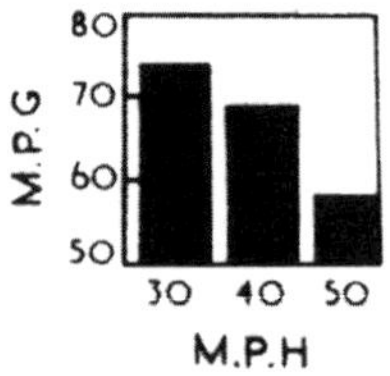

MEAN ACCELERATION:	10-30 m.p.h.	20-40 m.p.h.	30-50 m.p.h.
Second	7.2 sec	8.2 sec	—
Third	—	13.4 sec	13.4 sec
Top	—	17.6 sec	18.4 sec

Mean speed at end of quarter-mile from rest: 52 m.p.h.
Mean time to cover standing quarter-mile: 22.8 sec.

BRAKING: From 30 m.p.h. to rest, 36ft (surface, dry tarmac).

TURNING CIRCLE: Left-hand, 15ft; right-hand, 22ft.

MINIMUM NON-SNATCH SPEED: 15 m.p.h. in top gear.

WEIGHT PER C.C.: 1.64 lb.

1$\frac{1}{16}$in to 1$\frac{1}{8}$in and exhaust pipe diameter up by $\frac{1}{8}$in to 1$\frac{5}{8}$in. The compression ratio has been raised from 7.5 to 8.5 to 1.

Unlike its predecessor, the Mercury engine now has the pushrod tubes cast integrally with the cylinder barrel and head.

More punch in the middle of the range and a generally crisper response from the engine are the main benefits from these changes. With a sidecar attached, this power bonus is even more noticeable. The outfit could be pushed all day with the speedometer needle hovering on the 55 m.p.h. mark. (The instrument read approximately 6 per cent fast throughout the range.)

Despite having had its power pepped up, the G3 remains extremely tractable—a desirable attribute in a machine used for sidecar work. The outfit could be waffled along entirely unobtrusively at 30 m.p.h. in top gear.

If snap acceleration was wanted and a lower gear selected, the exhaust noise could, however, become obtrusive.

Mechanical noise was muted. Only slight hiss came from the carburettor, which is not fitted with an air filter.

Premium-grade petrol was used throughout the test and no pinking was experienced. Driven hard, and with full use made of the gear box, the Matchless used no more than a gallon of petrol every 50 miles. Oil consumption during the 700 miles covered was negligible. Oil-tightness of the engine and gear box was exemplary.

Whether an outfit is used for shopping or touring, the ease with which it can be started can make or mar its appeal. No conditions encountered during the test, including standing in the open in rain overnight, failed to blemish the Mercury's one-kick record.

Upholding the reputation it has for sweet, positive changes, the A.M.C. gear box enabled the best use to be made of the power available. Top (6.48 to 1) and third (7.91 to 1) are ideally spaced and by "playing tunes" on these two ratios more than merely respectable average speeds could be maintained.

Bottom gear could be engaged quietly. At worst there was no more than a slight click. Neutral could be selected without bother with the outfit at rest.

Close-up of the pivoted-fork sidecar wheel springing. The damper can be seen through the spokes

The clutch engaged the drive smoothly and was not distressed by the additional weight of a sidecar, even when half-a-dozen full-throttle starts were undertaken. The clutch had to be slipped when restarts were made on a 1 in 6 gradient. A 1 in 12 main-road hill was surmounted at 40 m.p.h. in third gear.

Driving position on the Matchless is comfortable and the rider's left leg is not hampered by sidecar connections.

Sidecar alignment was just right and the Matchless and Garrard steered hands-off on a level road. Reduced front-fork trail would have lessened the effort required for cornering. However, correct use of the throttle, particularly to make the machine run round the sidecar on left-handers, helped to reduce steering heaviness.

Handling was such that swervery could be zestfully tackled.

On the road, the outfit proved pleasant to handle and the sidecar was comfortable

Above: A full-width door permits easy access to the luggage locker. The seat is supported on rubber straps

When the outfit was being held hard into tight right-handers, some trace of front-wheel hop was evident, though the suspension absorbed all normal road shocks and provided a comfortable ride. A steering damper is fitted and was set so that it was just biting.

Passenger comfort is good. The Grand Prix seat cushion and squab are of hammock type, supported on broad rubber straps.

Weather protection is provided by a hood, erected in a few seconds, and fastened to the double-curvature, transparent-plastic screen by press studs. A lockable luggage compartment behind the seat provides reasonable stowage space.

Good braking enabled full use to be made of the performance. The figure of 36 feet from 30 m.p.h. is extremely creditable when it is considered that the total weight of rider and passenger was 24 stone. Both brakes were light and progressive in operation and did not require adjustment during the test.

Main-beam illumination was such that all the performance could be used after dark. Dipped beam cut-off was satisfactory. The horn note was inadequate.

Accessibility is usually impaired when a sidecar is fitted. On the Mercury, however, the tappets, contact-breaker points and carburettor can all be reached easily from the right hand side of the machine.

Unhappily, the upper-rear sidecar connection fouls the hinged lid of the battery and tool compartment. The lid can be opened partially, permitting the tool roll to be removed. But to service the battery, the connection had to be parted.

The Mercury-Grand Prix outfit makes a good-looking zestful vehicle guaranteed to make journeys a lot of fun while helping to keep running costs down to sensibly economical levels.

Rear view of the outfit, showing to advantage the rakish lines of the Grand Prix sidecar

ROAD TESTS OF NEW MODELS

248 c.c. SUPER SPORTS MATCHLESS

THIS IS NO CRAMPED lightweight. First impression on straddling the Monitor Super Sports is of the "man-size" dimensions of the riding position. The handlebar is virtually flat and slightly downswept at the grips—this was felt to be a good compromise between sporty styling and practicability and results in a forward lean rather than a semi-crouch.

Clutch and front-brake levers are fully adjustable. The twistgrip on the test model was light in operation and permitted delicate throttle control. Clutch operation, too, was light and the clutch took up the drive without judder or slip. It gave no trouble during a series of standing-start getaways when the performance figures were obtained and required no adjustment during the test.

THE Monitor Super Sports was ridden under a variety of weather conditions over many different surfaces. The positive steering and general stability made the rider feel at home whatever the going.

On first acquaintance a trace of heaviness was apparent at low speeds, but it passed unnoticed as familiarity grew. The machine could be ridden feet-up on full lock, though with the hands in the normal position on the grips, the thumbs were trapped between the grip and the petrol tank when full lock was applied.

Compared with the G2S Monitor Sports, the G2CSR has more power, better braking and more robust suspension. A larger carburettor choke size, bigger inlet valve, longer induction tract, high compression ratio, stronger valve springs, steel flywheels and stiffer connecting rod and crankpin: all these are the means by which more power has been found.

As a result, it has been possible to raise the gear ratios. The front brake now has a light-alloy hub with a cast-iron drum shrunk in. A heavyweight front fork is fitted as standard.

On twisty roads the model

The 6in-diameter front brake has light alloy shoeplate with cast-in air scoops

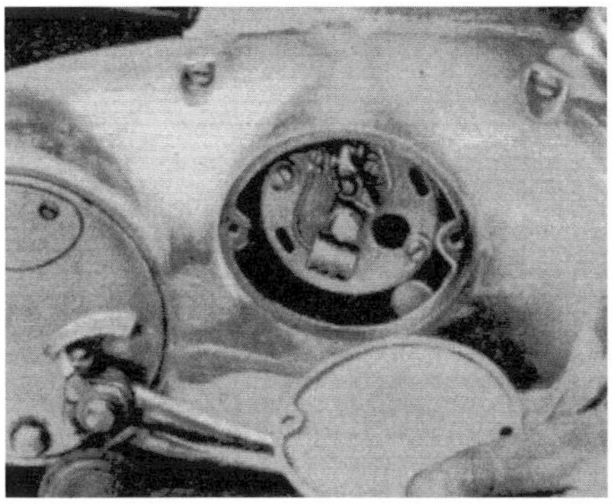

Access to the contact breaker and condenser is simplicity itself

really came into its own. It could be flicked through successive bends without wavering. The same applied when taking main-road curves at the machine's maximum speed.

Contributing to this sort of handling was the renowned Teledraulic front fork; although moderately firm in action it gave a ride free from harshness and prevented pitching on rippled or undulating surfaces.

The pivoted rear fork is controlled by adjustable Girling units. Set in the softest position, these did not bottom and provided excellent damping.

Tractability has not been sacrificed in the search for more power. Restarting on a 1-in-6 gradient called merely for a fastish tickover, and the clutch could be engaged fully after only a yard or so had been covered. The 19 to 1 bottom gear also enables traffic threading to be accomplished without undue use of the clutch.

A 1-in-12 main-road hill was breasted in third, with the speedometer needle on the 60 m.p.h. mark (the instrument read approximately 10 per cent fast throughout the range; no trip recorder is fitted).

Under normal road conditions the Matchless could be cruised at a true 60-65 m.p.h. without the engine showing any signs of tiring and at such speeds only slight vibration could be felt through the handlebar grips.

Attractive Sportster Combining Peppy Performance With Excellent Handling and First-class Braking

Both on narrow, twisting lanes and in fast, main-road cruising, the handling was superb

Sporty styling, yes—but practicability has not been sacrificed in the Monitor Super Sports

SPECIFICATION

ENGINE: *A.M.C. 248 c.c. (70 x 65mm) overhead-valve single-cylinder. Double-row roller big-end bearing; crankshaft supported in a ball and a roller bearing on the drive side and a plain bearing on the timing side. Light-alloy cylinder head; cast-iron cylinder barrel. Compression ratio 8 to 1. Dry-sump lubrication; oil capacity, $2\frac{1}{2}$ pints.*

CARBURETTOR: *Amal Monobloc, $1\frac{1}{8}$in choke; air slide operated by handlebar lever.*

IGNITION and LIGHTING: *Coil ignition. Wipac 54-watt alternating-current generator mounted on drive-side crankshaft charging Exide 12-amp-hour battery through rectifier. Wipac 6in-diameter headlamp with pre-focus light unit and 30/24 main bulb.*

TRANSMISSION: *A.M.C. four-speed gear box clamped to rear of engine. Gear ratios: bottom, 19 to 1; second, 12.04 to 1; third, 8.45 to 1; top, 6.04 to 1. Multi-plate clutch with bonded friction facings. Primary chain $\frac{3}{8}$in duplex in cast aluminium case. Rear chain $\frac{1}{2}$ x 0.305in with metal guard over top run. Engine r.p.m. at 30 m.p.h. in top gear, 2,600.*

FUEL CAPACITY: *$3\frac{1}{4}$ gallons.*

TYRES: *Dunlop: front 3.25 x 17in ribbed; rear, 3.25 x 17 studded.*

BRAKES: *6in-diameter front and rear.*

SUSPENSION: *A.M.C. telescopic front fork with hydraulic damping. Pivoted rear fork controlled by Girling three-position spring-and-hydraulic units.*

WHEELBASE: *53in unladen. Ground clearance, $5\frac{1}{4}$in unladen. Seat height, 30in unladen.*

WEIGHT: *329 lb, fully equipped, including approximately one gallon of fuel.*

PRICE: *£228 10s, including British purchase tax.*

ROAD TAX: *£2 5s a year.*

DESCRIPTION: *Motor Cycle, 10 May 1962.*

MAKERS: *Matchless Motor Cycles, Plumstead Road, London, S.E.18.*

These, by the way, are of the almost-smooth plastic variety.

Full use of the machine's sprightly acceleration was limited in built-up areas by exhaust noise.

Mechanical noise amounted to no more than a subdued tapping from the valve gear. No air filter is fitted but induction noise never obtruded.

Starting was easy whether the engine was hot or cold and it was unnecessary to use the tickler even for the first start of the day. The air lever could be opened almost as soon as the engine fired.

Starting was equally simple with the ignition switch in the normal or emergency position. Idling was slow and regular as soon as the engine was warm.

OILTIGHT

The $2\frac{1}{2}$-pint oil compartment is bolted to the right side of the crankcase. The oil level did not vary during the 500 miles covered on test and the power unit remained oiltight except for a weep from the gear-box inspection plate.

Gear-pedal movement was light and gear engagement positive. No particular technique was required to execute fast, clean upward and downward changes. The clutch freed properly, with the result that neutral could be readily selected with the model at rest and bottom gear could be engaged noiselessly.

Both brakes are of 6in diameter. That at the front has a light-alloy shoe plate with cast-in air scoops. On the test machine the slots had not been cleared of metal to permit air to enter. Perhaps it was partly for this reason that the front brake suffered from fade during heavy applications.

Lighting is adequate for 60 m.p.h. cruising after dark. The horn note is usefully loud. Routine maintenance could be carried out with the tools provided in the tool kit.

Finished in tartan red and black, the Monitor Super Sports can be distinguished by its chromium-plated tank sides, silver enamelling of the cylinder barrel and polishing of the cylinder-head fin edges and rocker covers.

The G2CSR is not temperamental and is a lively, attractive mount. It will undoubtedly appeal to the rider who wants sporting lines plus useful performance.

PERFORMANCE DATA

MEAN MAXIMUM SPEED: *Bottom*: 28 m.p.h.*
Second: 43 m.p.h.*
Third: 63 m.p.h.*
Top: 73 m.p.h.
** Valve float occurring.*

HIGHEST ONE-WAY SPEED: *75 m.p.h. (conditions: slight tail wind; rider wearing two-piece suit and overboots).*

MEAN ACCELERATION:

	10–30 m.p.h.	20–40 m.p.h.	30–50 m.p.h.
SECOND	4.8 sec	5 sec	—
THIRD	—	7.8 sec	7 sec
TOP	—	11.2 sec	10 sec

Mean speed at end of quarter-mile from rest: 67 m.p.h.
Mean time to cover standing quarter-mile: 19.6 sec.

PETROL CONSUMPTION: *At 30 m.p.h., 112 m.p.g.; at 40 m.p.h., 88 m.p.g.; at 50 m.p.h., 72 m.p.g.; at 60 m.p.h., 64 m.p.g.*

BRAKING: *From 30 m.p.h. to rest, 30ft (surface dry tarmac).*

TURNING CIRCLE: *14ft.*

MINIMUM NON-SNATCH SPEED: *20 m.p.h.*

WEIGHT PER C.C.: *1.32 lb.*

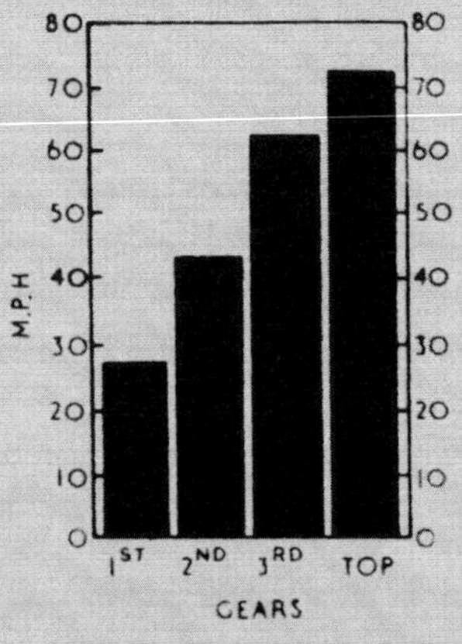

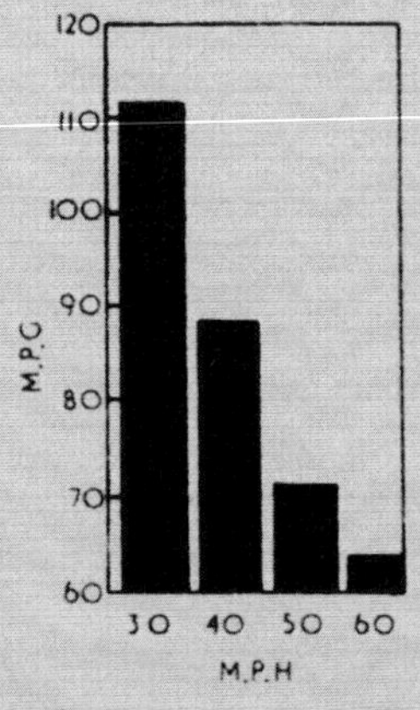

TIRELESS HIGH-SPEED ROAD BURNER

WIDE SPREAD OF BEEFY PUNCH

EXCELLENT ROADHOLDING

646 c.c. A.J.S. Hurricane

ROAD TESTS OF NEW MODELS

Yes, it looks sporting—and the Hurricane's performance more than backs up its appearance. Though flexible, the six-fifty twin engine provides snappy acceleration and excellent high-speed cruising

IN skilful hands, there is no safer vehicle on the road than a large-capacity sporting solo. Ample power makes swift, unobtrusive overtaking possible. High average speeds are effortlessly achieved without taking liberties with machinery or road safety. Just how effortlessly was demonstrated most convincingly by the 646 c.c. A.J.S. Hurricane.

Burbling quietly through town traffic at 30 m.p.h. in top gear or cruising on half throttle at almost 85 m.p.h., the Hurricane was prepared to burn the candle at both ends. A mean timed maximum speed of 99 m.p.h. may seem unexceptional—but a wind gusting to 35 m.p.h. during the MIRA tests did little to help! Under better conditions, a mean speed in excess of 100 m.p.h. could be expected.

Absolute maximum speed counts for little. Of far more importance are

Top left: Removal of the accessible plastic cover exposes the contact breaker. Below: little effort is needed to lay the Hurricane into sweeping bends

acceleration and optimum cruising speed. The sheer, zestful urge which sent the Hurricane's speedometer needle rocketing to the 90 mark (a true 85 m.p.h.) was exhilarating.

In the gears, exceptionally beefy power soon gave an indicated 55 m.p.h. in second and 80 m.p.h. in third. Tweaking the grip wide open in top would send the needle skipping beyond the ton in a commendably short time—and without the rider adopting more than a slight crouch.

Wind pressure, rather than any mechanical limitation, set a top limit of 85 m.p.h. to the comfortable cruising range. Apart from the odd hold-up for road works, this speed was maintained for the 72 miles of M1. Top-gear cruising speed was well maintained in face of main-road climbs and stiff winds.

Flexibility was such that the engine pulled strongly from 30 m.p.h. upward in top gear on full advance—and without pinking, provided premium petrol was used. On full retard, a standing-start was easily made on a 1 in 4 hill. With some coaxing, the model managed to restart on a 1 in 3 gradient.

RUSTLE

For a high-performance unit, the engine was commendably quiet mechanically. At tickover, no more than a light rustle was heard from the valve gear. On small throttle openings, exhaust noise was well muted. Hard acceleration produced a more throaty note and some restraint was desirable in built-up areas. But the note was never harsh enough to be objectionable.

The sporting nature of the unit was evident from the slightly hard feeling of the power delivery. Vibration became noticeable through the handlebar at 75 m.p.h. in top gear and gradually increased, later to be felt through the footrests and seat as well, right up to maximum speed. It did not become severe enough to call for a reduction in cruising speed.

Nothing could have been easier than the starting drill—lightly flood the carburettor, close the air lever, pull the ignition lever back a third and swing on the kick-starter pedal. Invariably the engine burst into song at the first or second attempt. And the air lever could be fully opened within the first mile.

Sufficiently low to allow a walking-pace trickle, bottom gear engaged noiselessly at all times.

Moderately light and delightfully smooth in operation, the clutch withstood a succession of eight full-throttle standing starts—when the quarter-mile acceleration figures were obtained—without sign of abuse. The gear change was superbly light and positive.

When delivered, the model was fitted with a downturned handlebar. This proved uncomfortable. With the steering on full lock, the rider's thumbs were trapped between the handlebar and petrol tank and the enforced riding crouch caused cramp in the shoulders.

Reversing the bar resulted in a more relaxed position yet afforded sufficient forward lean to allow high speeds in comfort. The only snag was that the combined horn button and dip switch, screwed to the handlebar, was then on the right by the twistgrip.

Excessive reach to the front brake and clutch levers was cut by inserting a chain connecting link between each lever and its pivot block.

The latest-pattern petrol tank, with knee recesses at the rear, felt considerably more comfortable to grip than the older tank. But the knees of a tall rider tended to chafe against the trailing edges of the chromium-plated tank embellishments.

The Hurricane has an unusually high dual-seat so that a rider with long legs could only just place both feet firmly on the ground. Softer seat padding would give greater comfort and might obviate a suggestion of cramp at the end of a 300-mile day.

Pleasantly light at low speeds, the steering was reassuringly precise and taut right up the scale. A steering damper is not fitted, nor is there need for one. Because of the highish centre of gravity, some effort was required to heel the model into slow corners—or to switch rapidly from left bank to right.

But little effort was needed to lay the Hurricane into sweeping 80-90 m.p.h. bends. On smooth surfaces it would remain glued on line. Bumpy bends taken at over 70 m.p.h. provoked some mild weaving. On slippery roads the high-cling tyres provided noticeably good grip.

A shade hard at low speeds, front and rear suspension were well matched; they absorbed road irregularities well at high speeds.

Smooth in operation, the brakes were adequately powerful, although the front called for fairly heavy lever pressure. Adjustment was not required throughout the 600-mile test.

The headlamp threw a long, wide beam of sufficient intensity to allow a safe 70 m.p.h. on unlit roads.

OIL-TIGHT

In spite of being driven hard throughout the test, the Hurricane remained completely oil-tight.

Lack of automatic lubrication for the rear chain hastened adjustment at 400 miles—that was the only maintenance required. For other routine jobs that would be called for in time, the tool kit was adequate. The roll-on centre stand was just too easy to use; an excellent prop stand is also provided.

Finished in dark blue and black, with the usual parts chromium - plated and with polished light-alloy mudguards, the Hurricane not only looks a sporty mount—it is thoroughly well made and as rapid as they come: a most attractive road-burner.

SPECIFICATION

ENGINE: *A.J.S. 646 c.c. (72 x 79.3mm) overhead-valve twin. Crankshaft supported in one plain and two roller bearings; plain big-end bearings. Light-alloy cylinder heads; compression ratio 8.5 to 1. Dry-sump lubrication; oil-tank capacity, 4½ pints.*

CARBURETTOR: *Amal Monobloc; air slide operated by handlebar lever.*

IGNITION and LIGHTING: *Lucas magneto, with manually operated advance and retard. Lucas RM15 alternator, with rotor on left-hand end of crankshaft, charging 6-volt 12-amp hour battery through rectifier. Lucas 7in-diameter prefocus light unit with 30/24 watt main bulb.*

TRANSMISSION: *A.M.C. four-speed gear box. Gear ratios: bottom 12.23 to 1; second 8.13 to 1; third 5.83 to 1; top 4.78 to 1. Multi-plate clutch with bonded friction facings running in oil. Primary chain, ½ x 0.305in in light-alloy oilbath case. Rear chain ⅝ x ⅜in with guard over top run. Engine r.p.m. at 30 m.p.h. in top gear, 1,850.*

FUEL CAPACITY: *4 gallons.*

TYRES: *Dunlop: front, 3.25 x 19in GS ribbed; rear, 3.50 x 19in K70 Gold Seal.*

BRAKES: *Both 7in diameter x 1⅛in wide; finger adjusters.*

SUSPENSION: *A.M.C. telescopic front fork with hydraulic damping. Pivoted rear fork controlled by Girling spring-and-hydraulic units with three-position adjustment for load.*

WHEELBASE: *55¼in unladen. Ground clearance, 6¼in unladen. Seat height, 32in unladen.*

WEIGHT: *412 lb fully equipped, with full oil tank and approximately a gallon of petrol.*

PRICE: *£314 8s including British purchase tax.*

ROAD TAX: *£4 10s a year; £1 13s for four months.*

MAKERS: *A.J.S. Motor Cycles, Plumstead Road, London, S.E.18.*

DESCRIPTION: Motor Cycle, 27 September 1962.

PERFORMANCE DATA

(Obtained at the Motor Industry Research Association's proving ground at Lindley, Leicestershire.)

MEAN MAXIMUM SPEED: **Bottom 44 m.p.h.; *second 66 m.p.h.; *third 93 m.p.h.; top 99 m.p.h. *Valve float occurring.*

HIGHEST ONE-WAY SPEED: 104 *m.p.h. (conditions: tail wind gusting to 35 m.p.h.; rider wearing two-piece waxed-cotton suit and overboots).*

MEAN ACCELERATION:

	10-30 m.p.h.	20-40 m.p.h.	30-50 m.p.h.
Bottom	2.6 sec	2.6 sec	—
Second	3.4 sec	3.2 sec	3.6 sec
Third	—	5 sec	4 sec
Top	—	6.8 sec	6.2 sec

Mean speed at end of quarter-mile from rest: 90 *m.p.h.*

Mean time to cover standing quarter-mile: 14.8 *sec.*

PETROL CONSUMPTION: *At 40 m.p.h., 66 m.p.g.; at 50 m.p.h., 64 m.p.g.; at 60 m.p.h., 51 m.p.g.*

BRAKING: *From 30 m.p.h. to rest, 35 ft. (surface, dry tarmac).*

TURNING CIRCLE: *15ft 6in.*

MINIMUM NON-SNATCH SPEED: 19 *m.p.h. in top gear with ignition fully retarded.*

WEIGHT PER C.C.: *0.64 lb.*

Knee recesses of the latest-pattern petrol tank give improved grip

MATCHLESS G-80 CS

OWNER SATISFACTION is the ultimate standard by which all motorcycles must be judged. The best handling, best looking, best performing machine in the world is still a terrible motorcycle if it displays a reluctance to start, or a fondness for time in the repair shops, or a habit of quitting out in the proverbial "middle of nowhere." Conversely, the bike that shows very little that is exciting in its basic specfications can, through sheer willingness, attract a devoted following. Reputation is, in the final analysis, as fine a thing for a motorcycle to have as overhead camshafts, solid-gold handlebars, or anything else one might care to mention. Taken on this basis, the Matchless 500cc single is a jewel beyond price: it is the very soul of reliability, has no handling vices, is reasonably fast, and there are a horde of people passionately fond of the bike who will cheerfully hang from the nearest tree anyone who suggests that they might be just as happy riding something else.

The Matchless G-80 CS that is the subject of this test-report is powered by a further refinement of the 500cc single that has been with us for years, in one form and another, developed from an original design laid down before the Hitler wars. Ancient though its origins may be, however, it is today as modern as any large-displacement single can be. Over the years, the stroke has been shortened and the bore enlarged, to bring it to the present near-square dimensions. The engine has all of its main

castings (cylinder head, barrel, crankcase) made of light aluminum alloys and this has done a lot for both the cooling and the engine's overall weight. Of course, the engine is still no lightweight; those flywheels, which even the torque so beautifully, are about the same size and weight as millstones and they do add considerable heft to the power package.

The cylinder head is heavily finned, and both valves and porting are of sufficient size to permit good breathing at high engine speeds. Hairpin valve-springs are retained as a feature in this engine, and the valvegear chests are unusually wide and low as a consequence. Also retained is the exhaust-pipe mounting system, in which the pipe is simply plugged into a bore at the mouth of the exhaust port. At one time, a great deal more reliance was placed on the ability of the pipe's mounting straps to hold the pipe securely in this bore than was completely justified, and riders found it necessary to add additional support to keep the pipe from creating a fretting corrosion where the pipe joined the cylinder head. Now, this added support has been provided by the factory.

The Matchless engine's lower end is much the same, in general layout, as any other big single, but it does have somewhat larger bearings. On the big-end of the connecting rod, for example, is an exceptionally wide bearing, which consists of triple-row rollers in an aluminum cage. Lubrication for these, and the other, bearings is supplied from a most interesting oil pump. The pump's working element is a double acting piston, which has gear teeth cut in its center. A worm gear on the timing-gear end of the crankshaft rotates this piston, and there is a cam-groove at one end that engages with a pin, and the action of the groove against the pin shuffles the piston back and forth. This provides a two-way pumping action, with more capacity on the scavenging side than in delivery, to insure that the sump does not load with oil. The oil supply is carried in a long, well-baffled tank, which has enough air-space provided to prevent the tank from foaming over under conditions of sustained, high-speed engine operation.

Drive to the transmission passes through a single-row chain to the tried and true Matchless clutch, which will stand more slipping and general abuse than almost any other. The primary chain is adjusted by moving the transmission on its mounting, which is probably a less desirable system than the blade-type automatic tensioners used with unit-constructed engine-transmission units; but it works for Matchless. In any case, there isn't much chain-snatch evident when everything is in adjustment, and one can't ask for much more than that. One feature that helps in producing the drive smoothness is the effective cushion built into the clutch hub.

The Matchless' transmission, although unfashionably separated from the engine and in its own casing, is a particularly nice bit of hardware. The ratios are well spaced, and that is always a delight, but what we found most impressive about the transmission was the smoothness and precision of the shifting action. The gears click right into engagement every time, without forcing, and the transmission could be nudged into neutral from either low or second with ease. And we are told by those who have been riding the "Big-M" in competition for years, and this includes no less than three staff members, that the transmission is all but unbreakable. No amount of stamping and yanking seems to distress it in the least.

A twin-tube, duplex-cradle frame supports all of this machinery, and it too is quite impressive. Very sturdy lugs brace the steering head and most of the other frame-tube junctions, and the makers have not been grudging in their use of masses of iron where the loads are heavy. This particular model, being essentially a competition scrambler (even though fitted with road-gear) had the skid-plates, etc., for that kind of work.

Conforming with what all of the competition riders have been doing anyway, the forks have been extended

on this new model, which accomplishes several things; increasing the fork angle and adding to ground clearance to name but two. These forks, and the swing-arm rear suspension, have a fantastic amount of travel, and are sprung and damped to perfection; this big two-wheeler will bound along over boulders and potholes without a trace of bottoming — or instability and with incredible comfort.

Full-width hubs are featured, per long-time Matchless practice, and this offers the major advantage of straight-in spoke lacing, as we have said many times. The brakes are not outstanding, but they are good; more than adequate for the service the bike will probably see. The front hub is, incidentally, a bit different from that used previously. Although not all of the reasons for the change are apparent, we would guess that the major motivation was added rigidity for the forks.

Another change, and rather unwelcome, is from those fine aluminum-alloy fenders that graced the Matchless for so many years to chromium-plated steel fenders. The new fenders will, no doubt, look better and be more dent and scratch-resistant, but we are not convinced that this change is an improvement. Also, the front fender is quite short, in front and below, and both rider and engine are subject to a steady shower of debris when riding in sand or mud.

Because of the height added by the extended forks, the over-center stand has been eliminated, and one must rely entirely on the "kick" stand — which is not always convenient if service or on-the-spot adjustments must be done. There is some doubt in our minds that the kick stand will be left on many of the G-80 CS scramblers used in competition. On our first trip out through the brush, the stand's return spring was torn away and the stand had to be wired up into place. Even more annoying was the fact that the stand is mounted in such a way that is reduces the ground clearance by approximately an inch — and ground clearance is, you will agree, important for this type of bike.

A mixed blessing were the foot-pegs, which were of the folding type and had a little upturned hook at their ends to prevent one's feet from slipping. This last feature, and the positioning of the pegs, were much appreciated. What we did not like was the way the pegs would stick in the folded position. Several times, when flailing along through low brush, we found that if one's feet are lifted to clear a high bush, the pegs would fold well up out of the way. This leaves the rider in the uncomfortable position of having to grope for the confounded things with his feet and, with luck, push them back down in position. The pegs need a stop, to keep them from folding too far, and should be free to fall down into place.

The seat we definitely did like, and a lot. It is a Bates seat, installed by dealers here, and it is soft enough, big enough, and shaped well enough, to take a lot of the Monday-morning soreness out of week-end riding.

Most of the staff agreed that the handlebars were not up to the high standard set by the seat, and that nearly everyone would prefer something a bit wider.

The Matchless G-80 CS comes with full road equipment, as we have said, but we did not bother to take it out on the road. A short session at the drag strip was all of the pavement riding done with the machine, which accounts for the fact that no speedometer error tabulation is furnished in the data page. To make a full-fledged, competition scrambler out of it, one has only to remove the electrical equipment, which comes away with an ease that tells us the makers had something like this in mind all along, and install a short exhaust extension pipe to replace the muffler. The chain guard should be removed, as it is heavy and makes chain servicing a problem, although it is a good item to have under any but racing conditions. It is usual to fit a special air cleaner for desert scrambling, but this will not be required on the Matchless; the stock air cleaner, which has a large, washable element, will handle the dust problem very well.

As riders, we were most impressed with this big Matchless. It had a lot of power over a phenomenally-wide engine speed range — as long as the rider does not ask it to pull down too far, and the handling is absolutely without fault. Starting, either hot or cold, was easy; perhaps because of the first-rate Lucas racing magneto that supplies the spark. Performance on our test machine was limited by the touring-type gearing, which would have to be changed for serious competition.

Actually, performance, insofar as pure speed is concerned, is probably never going to be outstanding with this bike. It is, for one thing, as heavy as a sackful of anvils, and the engine feels like it is tuned for a wide spread of power instead of maximum power — which makes it a pleasure to ride, but not exactly a tiger at the drag strip.

Be that as it may, we gained during our association with this Matchless a very high regard for the Big-M, and we can understand how they command such owner-loyalty. The G-80 CS is a most impressive all-around performer on clay roads, up sandy washes, over boulder-strewn creek beds and just about every place but on a paved road — and it seems likely that with the right tires it would be pretty good there, too. •

MATCHLESS G-80/CS

SPECIFICATIONS

List Price	$1,095 FOB L.A.
Frame Type	tubular, two-loop
Suspension, front	telescopic fork
Suspension, rear	swing arm
Tire size, front	3.50-19
Tire size, rear	4.00-19
Brake lining area, sq. in.	25.5
Engine type	single cyl., ohv
Bore and stroke	3.39 x 3.36
Displacement, cu. in.	30.4
Displacement, cu. cent.	497
Compression ratio	8.7:1
Bhp @ rpm	not specified
Carburetion	1 3/16″ Amal Monobloc
Ignition	Lucas magneto
Fuel capacity, gal.	2.4
Oil Capacity, pts.	4.5
Oil System	dry sump
Starting system	kick, folding crank

POWER TRANSMISSION

Clutch Type	multi-disc, wet plate
Primary drive	single-row chain
Final drive	single-row chain

Gear ratio, overall:1

4th	5.80
3rd	7.10
2nd	9.90
1st	14.9

DIMENSIONS, IN.

Wheelbase	55.0
Saddle height	33.5
Saddle width	10.0
Foot-peg height	11.7
Ground clearance	6.1 (stand)
Curb weight, lbs.	380

PERFORMANCE

Practical maximum speed (after ½-mile run)	90
Speed in gears @ 6800 rpm	
4th	92
3rd	75
2nd	54
1st	36
Mph per 1000 rpm, top gear	13.6

SPEEDOMETER ERROR

30 mph, actual
50
70 not tabulated (see text)

ACCELERATION

0-30 mph, sec.	2.6
0-40	4.3
0-50	6.1
0-60	8.8
0-70	11.9
0-80	16.9
0-90	25.8
0-100	
Standing ¼ mile	16.2
speed reached	79

ENGINE / ROAD SPEED

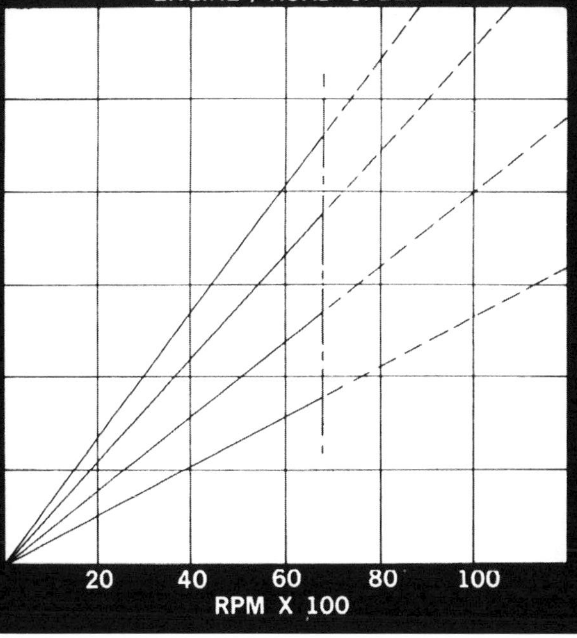

ACCELERATION

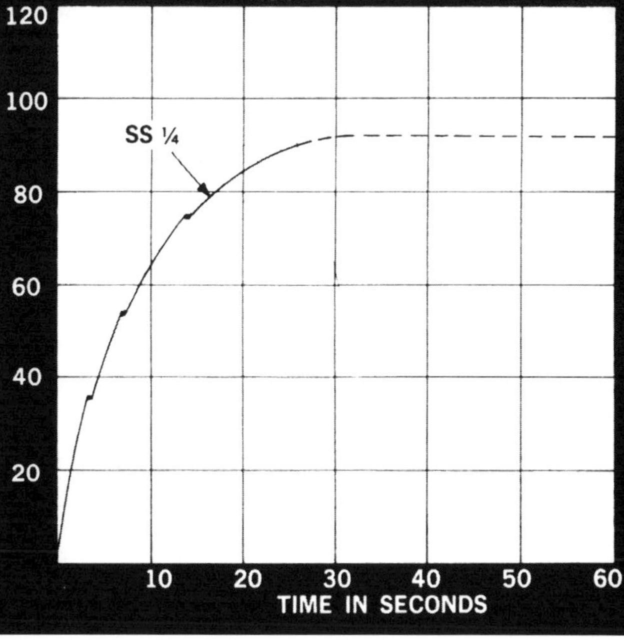

Traditional, attractive styling for a British sporting single: the front fork is the "heavyweight" Teledraulic

ROAD TEST
248cc A.J.S. CSR

THE A.J.S. 250 CSR remains as testimony to long-held, and enduring, articles in the faith of a famous British motorcycle maker: it is a single-cylinder four-stroke, relatively simple . . . "traditional". . . in design and manufacture, with pleasing lines, finish and performance. If its way of doings things—apart from the infinitely superior comfort it affords the rider—is not greatly different from that of its forerunner of, say, 25 years ago that should not be taken as detrimental to the 250 CSR. The fact is that in most respects the "traditional" single had reached a near-peak in development a quarter-century back and aside from such trivia as burying the pushrods behind the cylinder finning and making the piston stroke fashionably short, there was not much that *could* be changed . . . so these comments may be taken as tribute to the precocious skills of the industry in palmier days, rather than as criticism of the ageing face of the 250.

Sociologists and shopkeepers say that teenagers control a great deal of money now . . . and motor cyclists, in the main, are teenagers, passing on to minicars when they are older, with less money and more responsibilities. So any motorcycle manufacturer with a keen eye for sales—not so commonplace, apparently, no matter what you may think!—does well to tailor his machines for teenagers' tastes and in this context the two-fifty is specially important for it is the upper limit in engine size permitted to riders still with L plates. There is much about the 250 CSR to appeal to a chap in leather jacket, jeans and bonedome legally barred from ownership of really fast machinery: all the chrome plating, for instance—on the well-valanced mudguards and the rear-chain guard in addition to the usual parts, and all the polished light alloy, and the narrow, slightly down-turned handlebar . . . The impression is distinctly sporting, the reality less so.

The cylinder, and head in light alloy, are cleanly designed, the pushrods behind the fins, and set at an angle to the wheelbase; the tappet inspection cover is close under the tank but quite accessible. The bore is much bigger than the stroke—69.85mm to 64.84mm—and the axis of the cylinder is offset to reduce piston rock, and slap, when the engine is cold. This model was introduced in May 1962 as a sporting version of the Model 14S, and has continued in the range unchanged. The improvements incorporated in the CSR stem from the Woolwich scrambler and make impressive reading compared with the standard two-fifty . . . smoother inlet passage, bigger inlet valve, a bigger (1⅛in choke) carburettor, higher (8 to 1) compression ratio, scrambler hairpin valve springs with high loading to curtail valve float, steel flywheels, larger crankpin, duplex primary chain: the result is a tough little engine which would, presumably, respond to more ambitious tuning than it currently receives at the factory.

Crankcase and gearbox are separate, a fact difficult to establish from examination of the A.J.S. because the two are enveloped by large, smoothly polished light-alloy cheeks, the righthand one containing the 2½ pint oil tank at the front, the other enclosing the primary chain and provided with two inspection ports for chain and clutch adjustment. The 17in front wheel has a nicely made full-width hub with alloy back plate and shrunk-in cast-iron drum, the shoes being of 6in diameter; the 17in rear wheel has a rather nondescript

Lights are controlled by the lefthand switch in the headlamp, ignition by the righthand one (which has an emergency starting position). The ammeter needle, partially shrouded by a plate in the dial face, can be seen well enough by the rider

Head-on aspect of the A.J.S. is free from knobbly obtrusions : the handlebar droop is comfortable

All controls can be individually positioned, the clutch cable is provided with effective, if somewhat fiddling, adjustment

Girling rear-suspension units have three-position adjustment. Rear brake is effective, if not so impressive as front

hub, full width again but obviously deriving from the James and/or Francis Barnett side of the business . . . but it houses a good brake, with very little sponginess in the control and a reluctance to lock the wheel unless applied really heavily. Front suspension is by Teledraulic fork, with damping on both strokes, rear by long Girlings with three-position adjustment to cope with varying loads or road surfaces.

The finish, as I have remarked, is pleasing—and has quality too as I found after leaving the machine for two or three weeks in a damp, chilly garage . . . there was no dulling of the guards, or of any of the chrome, the only blemish being a slight rusting of front-wheel spindle nuts. After pausing a few minutes to savour this I found my happy mood shortlived as I struggled to start the engine . . . with no success until I changed the FE 250 racing plug for a new FE 80, when the engine fired first kick. Racing-type petrol taps are fitted, the levers stiff to move and so immune to vibration, and all that was necessary for a " cold " start was to push the tickler once or twice, set the air lever on the right side of the handlebar to a " slack " wire, turn the headlamp switch to " I " . . . and kick. After a few seconds, the air lever could be fully opened and then the engine would settle to a slow, reliable tickover, the exhaust note deep but flat, with very little noise from the valves or piston.

Ignition is by coil, with current from a 54-watt Wipac a.c. generator charging the Exide battery through a rectifier : the " E " position for the headlamp switch is for use when the battery is discharged, when the alternator is connected direct to the ignition coil, so providing current independent of the battery.

Forgetting to free the clutch plates on a couple of occasions before starting, I had the salutary shock of leaping forward a foot or two, to an agonized " crunch " from gearbox pinions, when I tried to engage bottom gear; thereafter I had no difficulty in remembering to free the clutch plates with an introductory shove on the kickstarter.

The gearbox was delightful—the gear lever set by the factory to exactly the right spline relative to the footrest and moving lightly and certainly at the gentlest pressure from the foot. It was possible to change gear on the move without clutch accompaniment—and no need for special care or finesse : this exercise probably not to be recommended in general but demonstrating just how biddable was the gearbox on the test machine (and there is no reason to think that " retail " CSR's are inferior).

All the controls, in fact, worked nicely. The clutch, not particularly light, but smooth and with its action unimpaired by plenty of fast starts and some deliberate slipping; both brake controls well placed for handlebar and footrest, and proving powerful and dependable even when hard-used—the front with just a hint of " sponginess " and good enough to be used alone most of the time, the rear " harder " and reluctant to lock the wheel without excessive pressure.

The front brake on the CSR has a tiny air intake, as you may see from the photograph on page 132. On our machine,

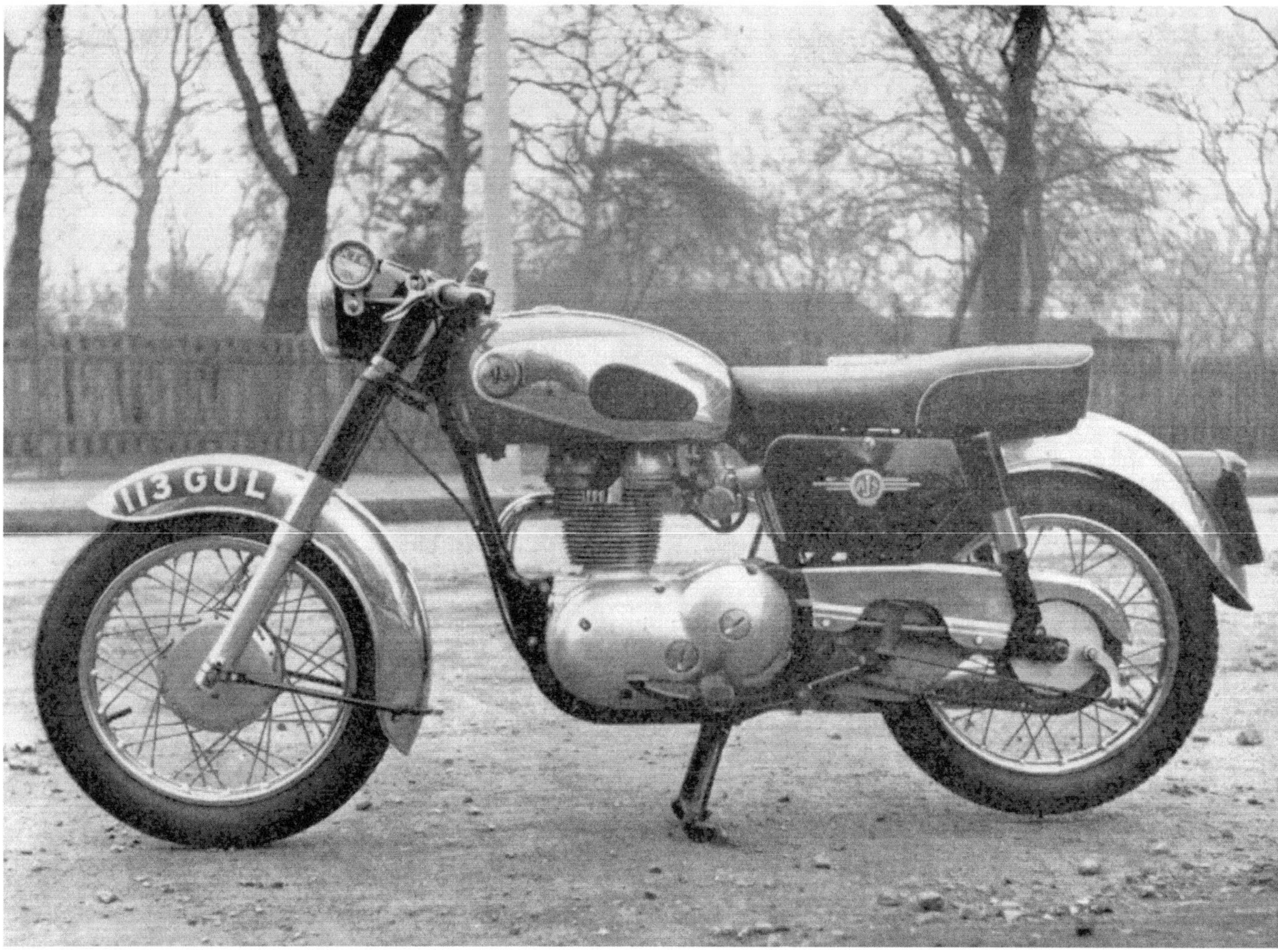

Plenty of chrome plating, of good quality, enhances the CSR's appeal to teenagers. One or two people who rode the machine had criticisms concerning its performance but everybody praised its good looks

and as delivered for roadgoing use, it was inoperative, but a judicious flick with a sharp knife edge, we are told, clears the grille to permit the ingress of cooling air . . . though we could see no sign of the equally important air *exit*.

The CSR riding position seems to show a mild schizophrenia in A.M.C. planning—the handlebar being narrow and down-turned (which means " racer," or at least " sports," to most people), the arrangement of the dual-seat, petrol tank and footrests belonging very definitely to the touring side of the A.M.C. range. What this means is that—in relation to the handlebar—the footrests are too far forward, the seat a little too low, the tank too wide at the rear. This may appear hypercritical to some but the image set by the CSR is super sports—I believe that is the description favoured by factory publicity—and the price, at £234 10s 11d, is high enough to raise expectations of a truly sporting layout. In open-road cruising at about 70 m.p.h., I quickly arrived on the nose of the dual-seat, knees splayed by the tank and carrying rather too much weight on my arms.

Cruising speed, you will have noted, I put at 70 m.p.h.—this as shown on the speedo, and being in fact a reading " optimistic " by about 7 m.p.h. There was some vibration via footrests and handlebar at this speed, with probably the worst vibration period at 55-58 m.p.h. (50 " true "), when the licence disc mounted on the lefthand bolt for the headlamp regularly blurred in sympathy with vibration from the engine.

Mechanical noise at tickover was slight, just a little clacking from the valves, and no amount of flat-out riding caused it to increase. But at lowish speeds, before wind roar rendered criticisms of minor noises academic, the petrol tank, closely shrouding the valves, appeared to magnify engine noise and in some way picked up a general whine from engine pinions and transmission. This was noted, I should say, without the sound-deadening effect of a helmet and neck curtain.

In the first few miles of a first trip on M1 I was conscious that the engine was working quite hard in maintaining a 70-75 m.p.h. (speedo) cruising rate, but as the miles ticked up it became obvious that the CSR could cope with any amount of hard driving; the strain was being felt more acutely by the rider than the motorcycle.

Top speeds in the gears were 48 m.p.h. (43 true) in second and 65 (60) in third, with valve float occurring on both occasions; in top gear, grip against the stop, the A.J.S. would reach a speedometer 82-83 (74) on any slight downgrade. Normal changing-up speeds were : into third at 30 m.p.h., into top at 45—this routine providing pickup to match that from using top revs in the gears. A wide range of (automatic) ignition advance and retard resulted in rather tepid acceleration at low r.p.m., and a " flattish " exhaust note. Always there was the impression that this was a modestly tuned engine working well within its potential and probably—as mentioned earlier—capable of being modified to produce impressive performance.

What *was* really impressive about the CSR as tested was its steering : firm but not heavy, absolutely positive, very responsive. There may be more advanced, more truly contemporary, engine units in other (mainly foreign) two-fifties, but the CSR's steering bears comparison with the best. Once, at night, I was unaware that the road surface was damp—it was not wet enough to provoke tyre hiss, but certainly offered diminished grip for the 3.25in tyres—so I was probably negotiating the Leicestershire bends with less caution than if I had known of the slippery

Front brake measures 6in and is housed in an impressive full-width Motoloy hub; note the air intake, inoperative until cleared by a knife blade. As for the engine—" smooth " would seem to be the word. The kick-starter had to be depressed for the toolbox to be opened fully

tarmac. But the point is—the A.J.S. performed beautifully, with no disturbing lightness at the steering head.

The suspension was comfortable, perhaps a little firm at the front, softer at the rear.

The finish and styling of the CSR are highly satisfactory, in the best traditions of the British sporting machine, and in our experience are unlikely to be sullied by oil leaks; in several hundred miles of motorway driving we found never a smear on the engine.

But if oil kept to its place, inside the engine, petrol insisted on vacating the tank . . . admittedly via the proper, carburettor, path . . . at an impressive rate. Cruising at 60-70 m.p.h., we found it necessary to fill the 3¼ gallon tank at frequent intervals. Exact consumption worked out at 64 m.p.g. At around 50 m.p.h. the CSR's thirst improved to a little over 70 m.p.g. C. A.

And a second opinion . . .

WHEN I first rode the 250 CSR I had come straight off another two-fifty—one of an entirely different type, the T10 Suzuki, a twin two-stroke touring machine. Frankly, I was disappointed in the change. I had expected the " sports " machine to have more snap and better acceleration, but the two-stroke could leave it pretty well any time.

Admittedly the Ajay was faster given a long enough run to work its speed up but on roads where there was not a lot of straight and a few hills it had a job to keep up with the twin.*

The fact that the CSR has an automatic advance and retard mechanism may have an adverse effect on its acceleration—give me a manual control every time. The rather flat exhaust note when one opens up seems to confirm this.

Naturally, compared with a twin the engine did not feel particularly smooth and at low and medium speeds had a lot of " thump," but it did smooth out, and at maximum—in my case 80 on the clock, getting down to it but with rather bulky clothing—there was very little vibration.

What I did like was the riding position, comfortable without being too " sporty," good handling and really excellent brakes; the back stopper in particular was very powerful and smooth in operation. F. P. H.

* Perhaps a slightly higher third gear might have helped but as it was I did not feel happy about " knocking down a cog " as soon as the revs dropped slightly.

A.J.S. 250CSR SPECIFICATION

Engine A.M.C. overhead-valve single cylinder; 65×70mm, 248 c.c. Double-row roller bearing for big-end; ball and roller bearing and plain bearing for crankshaft. Light-alloy cylinder head, cast-iron barrel. Compression ratio, 8 to 1. 1⅛in-choke Amal Monobloc carburettor with handlebar-controlled air slide.

Gearbox A.M.C. four-speed with foot change bolted to rear of engine. Overall ratios: 18.68, 11.32, 8.31, 6.39 to 1. Multi-plate clutch with vane-type shock absorber. Primary drive by duplex chain in cast-aluminium case with oil bath. Chain final drive.

Electrical Equipment Coil ignition, Wipac 54-watt a.c. generator charging 6-volt, 12 amp-hour Exide battery through rectifier. 6in-diameter headlamp with 30/24 watt main bulb.

Wheels and Brakes Chrome-plated steel rims carrying 3.25×17in tyres; 6in diameter brakes in full-width hubs, front with air intake.

Suspension Front Teledraulic fork with two-way damping. Pivoted-fork rear suspension controlled by three-position Girling spring-and-damper units.

Tank Capacities Petrol, 3¼ gallons. Oil, 2½ pints.

Dimensions Wheelbase, 53in. Seat height, 30½in. Weight, 320 lb.

Makers Matchless Motor Cycles, Plumstead Road, London, S.E.18.

Price £234 10s. 11d. including tax.

Performance

1st gear	26 m.p.h.
2nd gear	43 m.p.h.*
3rd gear	61 m.p.h.*
Top gear	72 m.p.h.

*Valve crash occurring.

Braking From 30 m.p.h., using both brakes—32ft.

Fuel Consumption 75 m.p.g. in town and country use; 64 m.p.g. for 65-70 m.p.h. cruising. Premium fuel.

ROAD TESTS OF NEW MODELS

248 cc AJS SAPPHIRE NINETY

LAST SUMMER saw this sports model really hit the limelight with a clear-cut class win in the Thruxton 500-Mile Race. Its success, which remained a talking point long after the meeting, was a reminder that the quarter-litre market is not the exclusive preserve of non-British manufacturers. Following this, in the autumn when the AJS 1965 range was announced, there was news of modifications to give more go to the Sapphire Ninety, as this 248 cc model is now named.

Coil valve springs replace the hairpin variety. Compression ratio is stepped up from 8 to 1 to 9.5 to 1 and gear ratios are closer than before. A larger-capacity silencer is fitted. It is tucked in and raised slightly to keep it out of the way when cornering.

First impressions mean a lot. When you straddle the Sapphire Ninety you are assured that AJS don't regard riders of two-fifties as necessarily on the small side.

Dimensions of the handlebar-seat-footrest set-up are such that whether your pals call you Lofty or Titch you should feel at home.

The handlebar is almost flat with slightly downswept grips and induces a forward lean rather than a crick-in-the-neck crouch. The riding position allows you to tuck in and adopt a relaxed, comfortable stance when hustling.

A criticism of the Matchless equivalent model tested in 1962 was that, on full lock,

Smooth contours give the power unit an attractive appearance and make cleaning easy

Specification and Performance Data

ENGINE: AMC 248 cc (70 × 65mm) overhead-valve single-cylinder. Double-row roller big-end bearing; crankshaft supported in a ball and a roller bearing on the drive side and a plain bearing on the timing side. Light-alloy cylinder head, cast-iron cylinder barrel. Compression ratio 9.5 to 1. Dry-sump lubrication; oil capacity, 2½ pints.

CARBURETTOR: Amal Monobloc, 1⅛in choke; air slide operated by handlebar lever.

ELECTRICAL EQUIPMENT: Coil ignition. Wipac 54-watt, alternating-current generator mounted on drive-side crankshaft charging Exide 11-amp-hour battery through rectifier. Wipac 6in-diameter headlamp with pre-focus light unit and 30/24-watt main bulb.

TRANSMISSION: AMC four-speed gear box clamped to rear of engine. Gear ratios: bottom, 17.97; second, 11.68; third, 8.05; top, 6.51 to 1. Multi-plate clutch with bonded friction facings. Primary chain ⅜in duplex in cast-aluminium case. Rear chain ½ × 0.305in with metal guide over top run. Engine rpm at 30 mph in top gear, 2,800.

FUEL CAPACITY: 3¼ gallons.

TYRES: Dunlop: front 3.25 × 17in ribbed; rear, 3.25 × 17in studded.

BRAKES: 6in-diameter, front and rear; finger adjusters.

SUSPENSION: AMC telescopic front fork with hydraulic damping. Pivoted rear fork controlled by Girling three-position, spring-and-hydraulic units.

DIMENSIONS: Wheelbase, 53in. Ground clearance, 5½in. Seat height, 30in. All unladen.

WEIGHT: 330 lb, fully equipped, including approximately one gallon of fuel.

PRICE: £224 18s 5d, including British purchase tax.

ROAD TAX: £2 5s a year.

MAKERS: Matchless Motor Cycles, Ltd, Plumstead Road, London, SE18.

DESCRIPTION: "Motor Cycle," 22 October 1964.

MEAN MAXIMUM SPEEDS: Bottom*, 31 mph; second*, 48 mph; third*, 70 mph; top, 82 mph.
*Valve float occurring.

HIGHEST ONE-WAY SPEED: 83 mph (conditions: still air, rider wearing two-piece suit).

MEAN ACCELERATION:

	10-30mph	20-40 mph	30-50 mph
Bottom	3.8 sec	—	—
Second	4.6 sec	4.8 sec	—
Third	—	7.8 sec	6.8 sec
Top	—	11.2 sec	9.8 sec

Mean speed at end of quarter-mile from rest: 70 mph.

Mean time to cover standing quarter-mile: 19.4 sec.

PETROL CONSUMPTION: At 30 mph, 112 mpg; at 40 mph, 84 mpg; at 50 mph, 66 mpg; at 60 mph, 58 mpg.

BRAKING: From 30 mph to rest, 30ft (surface, dry tarmac).

TURNING CIRCLE: 14ft.

MINIMUM NON-SNATCH SPEED: 19 mph in top gear.

WEIGHT PER CC: 1.32 lb.

one's thumbs could be trapped against the petrol tank. This does not occur on the Sapphire Ninety.

Almost smooth, plastic handlebar grips are fitted. They do not provide enough friction, especially when they are wet.

Control layout is good. Levers and pedals are correctly positioned for easy use.

During the test the AJS was ridden in a variety of weather conditions and over a variety of surfaces, some of them exceedingly slippery.

No praise is too high for the handling; the rider felt confident of staying in control at all times.

A trace of heaviness could be detected in the steering when traffic threading at walking pace, but this was more than offset by rock-steady adhesion to line at the other end of the scale.

In any case, one soon became accustomed to the heaviness and then it passed unnoticed.

Suspension at both ends was firm and well damped and contributed to predictable handling when flicking the machine quickly through a right-left-right series of bends.

Rippled surfaces on metalled roads or pot holes on unmade tracks failed to catch the suspension off guard.

The power bonus makes itself felt in the middle and upper part of the range. There is an overall impression of increased liveliness, and the unit is remarkably free from vibration.

Bull point is that the added power has not been obtained at the expense of tractability.

The Sapphire would woofle along at 30 mph in top gear on a trace of throttle and without any fuss. At this pace in built-up areas there was just a subdued burble from the exhaust.

Mechanical noise was at a commendably low level, but the exhaust note became obtrusive at wide throttle openings.

The Smiths speedometer is easily read and was no more than 2 mph optimistic throughout the range. There is no trip mileage recorder, but the total mileage is shown to the nearest tenth.

For high-speed, open-road cruising, suitable gear-change speeds were 28, 40 and 60

Top: Handling is exemplary and inspires confidence at all times. Note the new silencer. Above: The Sapphire Ninety contrives to look up-to-date without recourse to unnecessary gimmicks

mph. The 1965, closer spacing of the ratios enhances the appeal of the machine. Happy cruising speed was 60 to 65 mph.

Restarting from a standstill on a 1 in 6 hill was accomplished without difficulty. The clutch could be fully engaged after covering the first yard or so.

Using third gear, main-road gradients were dismissed with nonchalance. This ratio (8·05 to 1) provided brisk acceleration for overtaking whether solo or two-up.

ZIPPY

In fact, acceleration in general, producing as it does a standing-start quarter-mile in under 20 seconds, is decidedly brisk and is more than adequate to hoist the model into the zippy class.

Gear selection is so smooth and easy that it immediately calls to mind the knife-through-butter cliché, although

Removal of the left side cover gives access to the ignition coil and rectifier. The battery is located under the seat

Excellent controllable braking is provided by the British Hub front brake. It maintained its performance during heavy rain

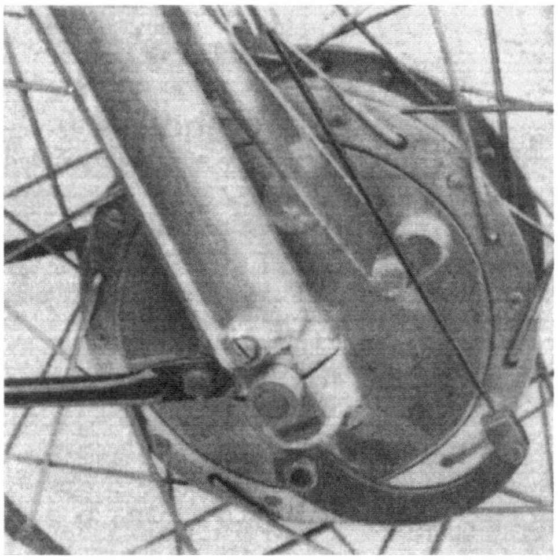

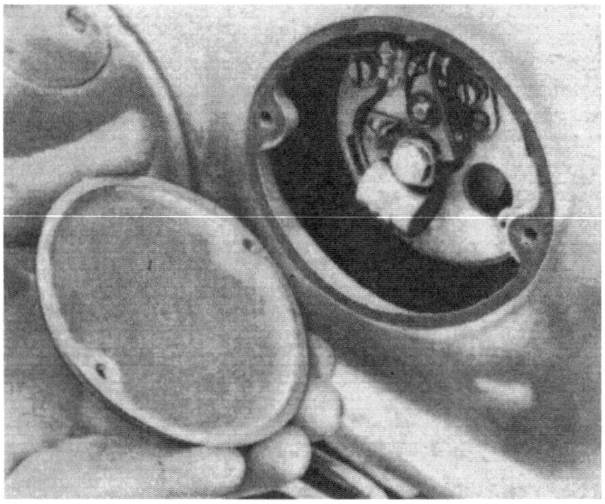

Two screws secure the contact breaker cover on the right of the power unit. Points adjustment is easy

a shade less pedal travel would be preferred.

Ratios should be swapped as fast as the pedal could be operated.

The clutch engaged smoothly in normal usage, but full-power take offs when the standing-start quarter-mile figure was being obtained produced some snatching. There were never any signs of clutch drag. No adjustment was required throughout the period of the test.

Starting proved simple. It was advisable to leave the carburettor tickler alone and use the handlebar-mounted air-slide lever. The engine was sensitive to the air setting until the proper operating temperature had been reached.

NO PINKING

Fuel consumption was light, especially in view of the performance available. The engine performed quite happily on premium-grade petrol and showed no tendency to pink under heavy load.

Oil is contained in a separate sump on the right of the crankcase. Consumption was negligible even during sustained hard riding. Oil tightness was marred by a slight weep from the gear-box end cover.

Performance and handling can be exploited to the full only if matched by the braking. In this respect the AJS earned full marks. Both brakes were light in operation and efficient at all speeds.

Ventilation scoops are provided on the front brake but, on the test model, the slots had not been properly cleared of metal. However, no fading was experienced.

Lighting was adequate for 60-mph cruising after dark. The dipswitch has a wipe action; this is excellent as it ensures that there is no possibility of lag during the change from main to dipped beam or vice-versa.

Action of the lighting and ignition switehes was spongy. A firm action with a positive click when the movement is complete would be more acceptable.

Accessibility for routine maintenance of rocker settings, contact-breaker, battery and other items is good. The tool kit supplied with the machine is adequate.

Thanks to the smooth contours of the power unit and liberal use of chromium-plating cleaning is a straightforward task.

The Sapphire's attractive finish is given the final touch by its blue and chrome and gold-lined petrol tank. Here is a bike which is easy on the eye: easy on the pocket for tax, insurance and fuel; and with its peppy performance, easy to have fun on.

745 cc Matchless G15

ROAD TESTS OF NEW MODELS

TIRELESS, lusty power—that is the keynote of this new number in the Matchless range for 1965. It gives the sort of satisfaction that comes only from plenty of cubic capacity plus a high-performance heritage. Zip when required, yet fair docility for town work.

Heart of the matter is a twin-carburettor Atlas engine in a Matchless duplex frame—a model primarily to please Matchless devotees who want something with even more urge than the established six-fifties in the range. Now they have it—and how! Exceptionally beefy punch from as low as 1,500 rpm, shattering acceleration, effortless 85-to-90 mph cruising and a maximum comfortably in excess of 100 mph.

Peak power, a claimed 49 bhp, is developed at 6,800 rpm, but such is the flexibility of the engine that upward gear changes were seldom notched higher than 5,000 rpm.

For the rider who revels in big, muscular engines, it was sheer delight merely to seek out gradients and flatten them with a tweak of the twistgrip.

Without hesitation, from a pottering 30-or-so mph in top, the model surged forward right up to an indicated 100 mph—a true 95 mph—and there was no need for the rider to adopt more than a half-crouch.

Cruising speed was dictated by comfort under wind pressure rather than by any mechanical limitations. Around 80 mph seemed about right, but this speed fell within a vibration period, felt through the handlebar and seat, between 4,000 and 5,500 rpm. **At** higher speeds the engine became progressively smoother. Mechanical noise was limited to a rustle from the valve gear.

With two carburettors, precise synchronization of the throttles is necessary for even running at low speeds. The Matchless was good in this respect without being perfect. In addition, the big silencers do their work most effectiveley. Hence, it is a satisfactory town model, being reasonably content to pobble along at 30 mph even in its 4.5 to 1 top gear.

Not unnaturally, with such large pistons to move, kick-

The Matchless, big and beefy in every sense, is really at home burning up the open road

Traditional styling enhances the Matchless. Heat sink for the Zene diode is beneath the front of the tank

starting from cold was a man's job, calling for some hefty swinging.

As usual with down-draught carburettors, care was required to avoid over-flooding. The best drill was merely to turn on the petrol, allow a minute or two for the level to settle in the carburettors and close the air lever before swinging on the crank.

The engine responded on the second or third effort; with the engine warm a single swing on crank was enough.

With the carburettors correctly adjusted for slow running—a bit tricky to do, because the pilot air screw of the left carburettor is obscured—the engine tick-tocked reliably at just below 1,000 rpm.

A flexible power unit does not make the choice of gear ratios critical but, nonetheless, those on the Matchless have been carefully plotted. Bottom was low enough to allow effortless re-starting on a 1 in 3 test hill: the others are happily spaced for normal roadster use.

So light it could be operated by one finger, the clutch stood up to abuse—six full-throttle standing starts when obtaining performance figures made no impression on its efficiency. Movement of the gear-change pedal was remarkably light and short in travel and gear engagement was completely positive and noiseless at all times.

A model capable of 90 mph cruising requires some thought on riding position and to give slight forward lean a 26in-wide, flat, almost straight, handlebar is fitted. The footrests remain as on the standard six-fifties; without question, locating them 3in rearward would provide a more comfortable position.

Wide and well shaped, the dual-seat gives good support but would be improved by deeper padding. Only 10in wide across the rear, the petrol tank provides comfortable knee-grip.

The use of a 3.50in-section studded front tyre did not, as expected, impair steering—indeed, steering was considerably lighter than on many models of half the capacity.

SURE FOOTED

For unhurried touring, the roadholding was steady and precise, but high speed on bumpy bends caused slight head-shaking and rear-end twitters. Stability on wet and greasy surfaces was sure-footed at all times.

Sharp left-hand cornering caused grounding of the centre-stand extension, but the model could be leaned way over on right-handers without fear of touching. Only on slowish corners could the effect of top hamper be felt, then only slightly.

Front and rear suspensions are designed more for high

A rubber balance pipe joining the carburettor manifolds assists even tick-over

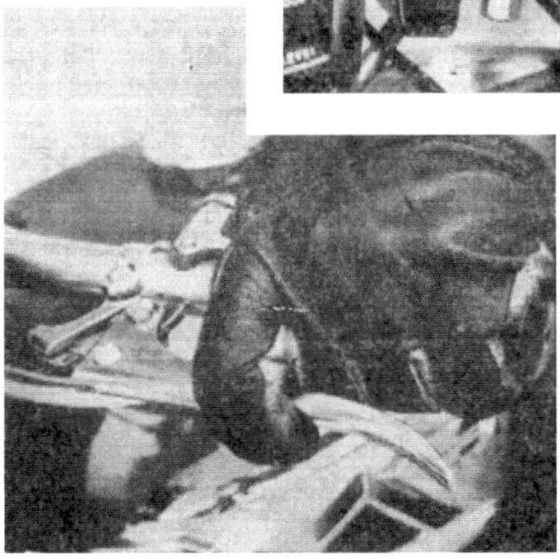

So light is clutch-spring pressure that one-finger operation is possible

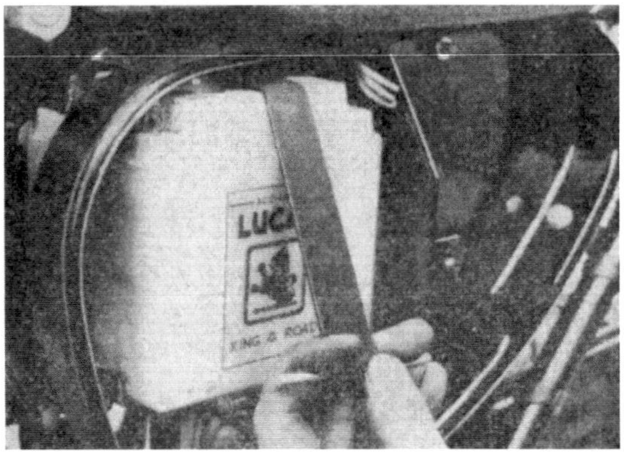

Below: A rubber strap retains both six-volt batteries; the bar is located in two ears on the carrier

than low speeds. They do the job well when the G15 is used as a roadburner, but softer damping—at the rear, at any rate—would be better for town work.

Both brakes were extremely powerful and smooth, were equally effective from high or low speeds, and required adjustment only once after the initial bedding-down. They remained dry in wet weather.

Headlamp of the 12-volt electrical system gives a deep, well-spread main beam, allowing daylight cruising speeds on unlit roads, and the dipped beam was free from dazzle. The horn was totally inadequate, emitting only a weak bleep that was inaudible to car and lorry drivers at speeds over 30 mph.

Rear view shows the roll-on centre stand; just forward is the prop-stand extension. Note also the well-shaped seat, lifting handle and deeply-valanced mudguards

Massive power plant, showing the revmeter drive; air filter for the carburettors is compact and neat

Effortless in use, the roll-on centre stand held the model steady on most surfaces. For casual parking, the prop stand was excellent; as the float chambers are on the left, there was no tendency for the carburettors to flood if the petrol was left turned on.

Routine maintenance was straightforward; the exhaust rocker clearances could be checked without removing the tank but to reach the inlets or work on the carburettors, the tank had to come off, but it is a fairly simple and quick job: A spanner to adjust the rockers was not included in the tool kit.

Both batteries are retained by a single rubber strap—an excellent arrangement—but the lid for the combined battery and tool-box compartment was difficult to refit.

Owing to the angle of the oil-tank filler cap and the positioning of the internal return pipe, it was difficult to check the oil level. Fuel tank reserve supply is enough for about six miles.

With the red petrol tank offset by the large motif plates, chromium-plated mudguards, polished light-alloy castings and black frame, the Matchless G15 is as eyeable a model as one could wish to have. Its performance will see off most other roadsters. Yes, it's a *man's* mount.

SPECIFICATION and Performance Data

ENGINE: *Atlas 745 cc (73×89mm) overhead-valve twin. Crankshaft supported in ball and roller bearings; plain big-end bearings. Light-alloy cylinder head; compression ratio 7.5 to 1. Dry-sump lubrication; oil-tank capacity 4½ pints.*

CARBURETTORS: *Two Amal Monoblocs; air slides operated by handlebar lever.*

ELECTRICAL EQUIPMENT: *Lucas magneto with auto-advance. Lucas 12-volt lighting system with RM19 alternator. Rotor mounted on drive-side crankshaft, charging twin Lucas 6-volt, 13 amp-hour batteries through Zener diode. Lucas 7in-diameter pre-focus light unit with 50/40-watt main bulb.*

TRANSMISSION: *AMC four-speed, foot-change gear box. Gear ratios: bottom, 11.55; second, 7.68; third, 5.51; top, 4.51 to 1. Multi-plate clutch with bonded friction facings. Primary chain, ½×0.305in in light-alloy, oilbath case. Rear chain ⅝×⅜in with guard over top run. Engine rpm at 30 mph in top gear, 1,750.*

FUEL CAPACITY: 3¾ *gallons.*

TYRES: *Dunlop—front, 3.50×18in; rear, 4.00×18in. Both K70 Gold Seal.*

BRAKES: *8in-diameter front; 7in-diameter rear, both 1¼in wide. Finger adjusters, front and rear.*

SUSPENSION: *Roadholder front fork with hydraulic damping. Pivoted rear fork controlled by Girling spring-and-hydraulic units with three-position adjustment for load.*

DIMENSIONS: *Wheelbase, 56½in. Ground clearance, 6in. Seat height, 31½in. All unladen.*

WEIGHT: 410 *lb fully equipped, with full oil tank and approximately one gallon of petrol.*

PRICE: £355 4s 2d, *including British purchase tax. Revmeter, £7 7s extra.*

ROAD TAX: £4 10s *a year;* £1 13s *for four months.*

MAKERS: *Matchless Motor Cycles, Plumstead Road, London, SE18.*

PERFORMANCE FIGURES

(Obtained at the Motor Industry Research Association's proving ground at Lindley, Leicestershire).

MEAN MAXIMUM SPEEDS: *Bottom, 46 mph; *second, 68 mph; *third, 95 mph; top, 102 mph.
*Equivalent to peak power at 6,800 rpm.

HIGHEST ONE-WAY SPEED: 103 mph *(conditions: still air, 12½-stone rider wearing two-piece, waxed-cotton suit and overboots).*

MEAN ACCELERATION:

	10-30 mph	20-40 mph	30-50 mph
Bottom	2.2 sec	2.2 sec	—
Second	3 sec	2.6 sec	3.4 sec
Third	—	3.8 sec	4 sec
Top	—	5 sec	4.8 sec

Mean speed at end of quarter-mile from rest: 88 mph.
Mean time to cover standing quarter-mile: 14.3 sec.

PETROL CONSUMPTION: At 40 mph, 57 mph; at 50 mph, 52 mpg; at 60 mph, 44 mpg; at 70 mph, 40 mpg.

BRAKING: *From* 30 mph to rest, 31ft 6in *(surface, damp tarmac).*

TURNING CIRCLE: 18ft 9in.

MINIMUM NON-SNATCH SPEED: 18 mph *in top gear.*

WEIGHT PER CC: 0.55 *lb.*

Front wheel carries an 8" × 1.25" brake and is shod with 3.50" × 18" Dunlop tyre. The tread design is more suited to sidecar lots

Drive side of the Norton power-house. Norton gear box is coupled up by an A.M.C. alloy chain case. Note large engine breather

Ian Speller tests the new Atlas Matchless. The peppy, powerful

THE Norton Atlas-engined G.15 Matchless swiftly answers the critical enthusiasts who call it an "A.M.C. special"; 1,500 hard test miles have removed any doubt from my mind. The G.15 is a handsome bike with adequate brakes and roadholding, superb steering and tremendous bottom-end torque with mid-range acceleration through the gears almost up to sprint standards.

All this solo when it is intended as a sidecar bike! The out-and-out sports boys would probably prefer the G.15 C.S.R.—a machine I would love to get my hands on!

Standard, the G.15 comes with heavy duty tyres and a set of gear ratios more suited to chair work than solo. These factors do affect solo performance in that top speed is cut to just over a ton, the engine is really working hard in the mid-eighties, and that the handling is not quite up to normal Norton standards. Still, I repeat that these factors do not detract from a bike not really intended for sports use.

The 750 c.c. Atlas motor delivers almost frightening power right through the range. Top-end poke is such that gear changes on long runs are made out of habit rather than necessity and the big twin unit levels out gradients as effectively as a bull-dozer.

In case I have given the wrong impression, I stress that any handling imperfections only happen when the G.15 is being motored on the limit, and in any case the rider soon adjusts himself. On long, fast bends when the bike is cranked almost to the footrests, there is slight frame whip which transmits itself mainly through the left-side footrest. This is the only symp-

The rod operated rear brake is shown here; unfortunately, it was not smooth in operation and kept sticking, so brakes rubbed

Rear tyre is not meant to be oil-cooled but an enthusiastic breather makes it look that way, still the silencer won't go rusty!

Acceleration second to none came from the twin carbs and highly tuned motor. Note engine is 100% oil-tight at end of test

Although of a sporting appearance, the mudguards were well up to their job. Front brake required a lot of adjusting during test

750 G.15

tom, however, and both wheels stick religiously to their line. The time-honoured Norton forks soak up the bumps in their usual greedy manner and the steering is super-accurate.

One of the A.M.C. boys told me recently that the G.15 was the best sports sidecar machine ever, and although I haven't tried it with a chair I can well believe it. Although the ratios are too low for normal solo use they would be ideal for an outfit and I should imagine the peak speeds through the gears would be almost as easily available. Sidecar fittings are provided as standard which gives a clear indication as to the real function of the bike. Holding the revs to their maximum, top speeds through the gears were 35 in first; 58 in second, 80 in third and a speedo indicated 104 in top. Driven hard, with plenty of work in the gears, fuel consumption dropped to 40 m.p.g. but at 60 m.p.h. cruising it was a commendable 55 m.p.g.—pretty good for a 750 c.c. twin. I made one run with the bike very heavily loaded to roughly simulate performance with a sports chair, and under moderate throttle the G.15 returned an overall 50 m.p.g.

Logically, the bike is more Norton than Matchless. Mechanically Norton, with Norton forks and wheels, the Matchless influence is seen only in the frame and swinging arm—and, of course, the tank badges. Standard specification is excellent with centre stand and prop stand, sidecar lugs, pillion footrests and powerful electrics.

The finish, too, is good, with durable paintwork and plenty of alloy and good quality chrome. Speedo and rev counter are mounted on a solid bracket which bolts to the top of the forks. In the centre of this bracket is the " off, side, head "

Continued on page 150

PERFORMANCE POINTERS

MOTORCYCLE MECHANICS ROAD TEST No. 90

Vehicle G 15 Matchless **Price new** £355.4.2

Engine Norton ATLAS . 750 cc vertical twin

Gearbox 4 Speed foot change

Final drive Chain to Norton wheel

GENERAL INFORMATION

Weight	406 lbs
Saddle height	30 ins
Turning circle	16 ft.
Is toolbox lockable	No
Is steering lockable	Yes
Fuel tank capacity	4 Galls
Reserve capacity	½ Gall
Oil tank capacity	4 pts
Gearbox capacity	1 pt.
Fuel specified	Premium
Overall consumption	50 mpg.
Braking from 30 mph	28 ft
Acceleration 0-60 mph	6.2 secs

SPEEDS IN GEARS

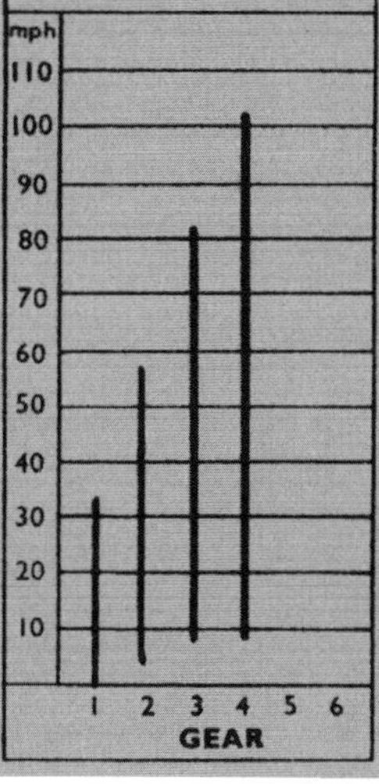

EQUIPMENT SUPPLIED

STANDARD FITTINGS Pillion rests, side car attachments

OPTIONAL EXTRAS Swept back pipes, semi-drop bars

SPARES PRICES

Engine gasket set	35s 6d
Set valves & guides	130s 0d
Piston with rings	76s 3d
Set of clutch plates	93s 8d
Silencer	118s 2d
Pr Exchange brake shoes	13s 8d

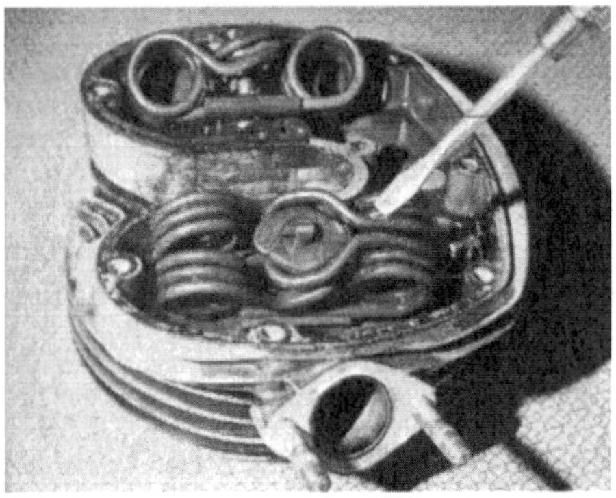

These models have hairpin valve springs and are removed by hooking fingers through coil and pulling them upwards

This screw controls oil to valve guide and can be adjusted by screwing right home and then backing out one turn

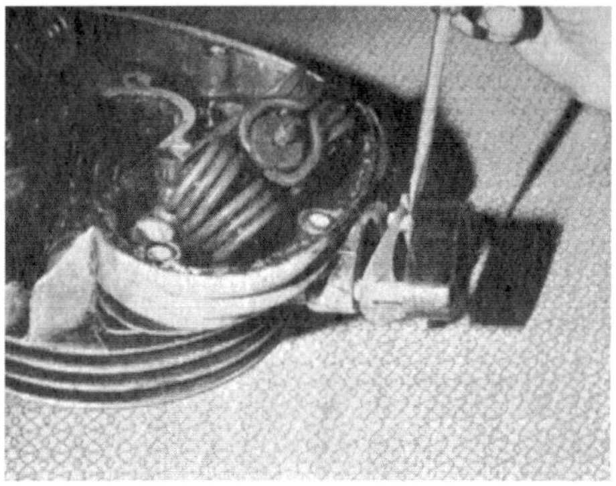

A fibre-compound heat-barrier block lies between carb and head. Make sure paper washers are on both sides

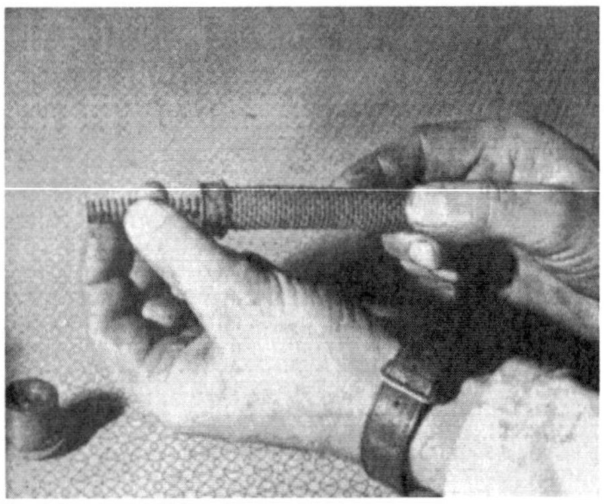

Don't forget spring which goes in base of the filter. This keeps filter tightly sealed against bottom of cap

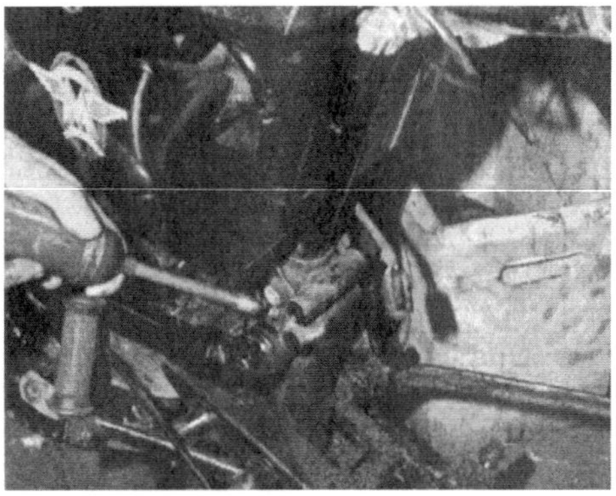

Every 2–3,000 miles, use a medium-heavy grease on swing-arm grease point to prevent wear in two pivot bushes

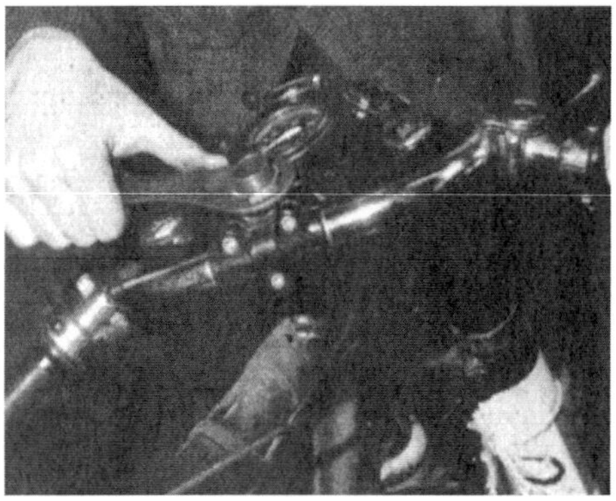

This model has no steering damper, so a close check must be kept on head bearings to make sure adjustment is OK

SPECIAL TIPS

AMC SPECIALIST, DAVE CURTIS, SHOWS HOW TO GET THE BEST FROM YOUR

Rocker-box inspection cover carries a rubber strip for oil sealing. New seal is cheap if there is any doubt

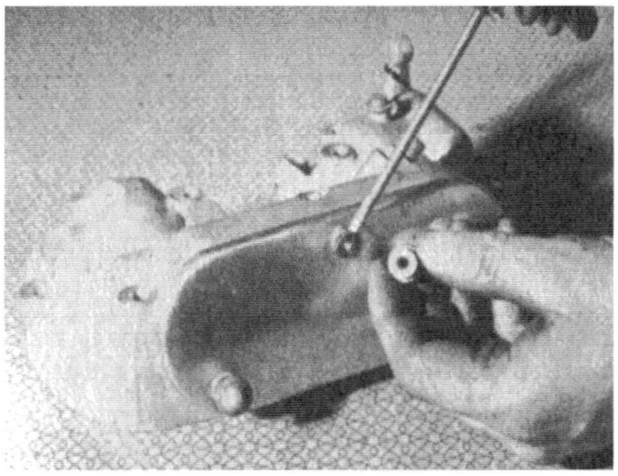

Always fit new fibre washers under rocker-cover holding nuts when you are replacing them to cut leak risks

Wash out wire cage and felt filter when you change the oil—and do this job regularly at least every 3,000 miles

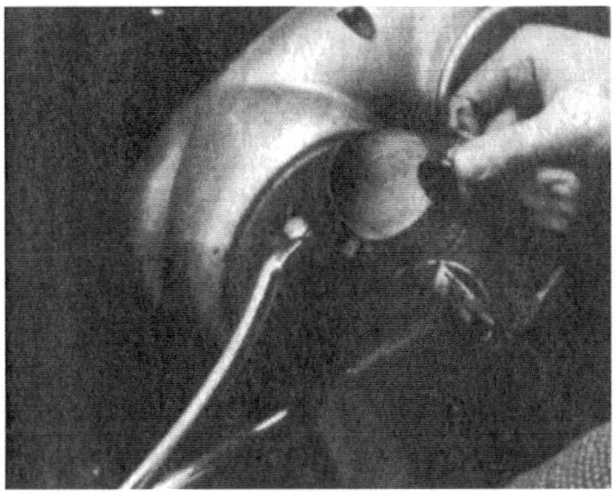

Gearbox level is critical and should be checked frequently. The box takes three pints of oil, up to lip of hole

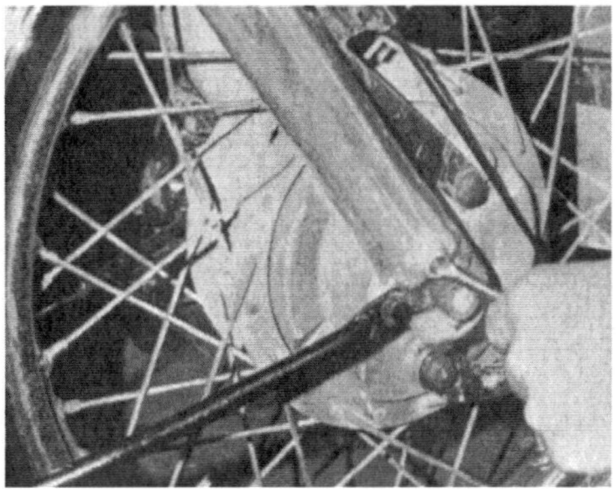

Every 5,000 miles front fork oil should be changed on both heavyweight and standard forks. Renew top washer

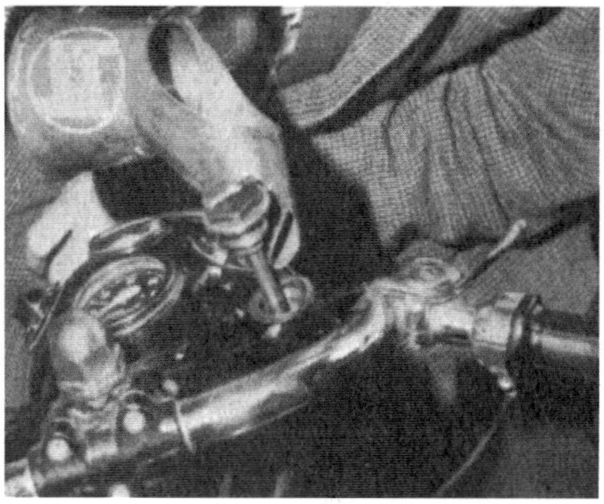

After draining fork, fill with a measured quantity. For heavyweight type, amount is 185 cc or 6½ fl. oz.

● Though it has been on the market for several years, the A J S or Matchless 250 really caught the public eye last summer when Peter Williams and Tony Woods won the lightweight class of the Thruxton 500 mile race. In the process they beat the cream of Continental and Japanese production machinery and gave a convincing display of British reliability and speed.

Obviously not every owner of one of these Associated Motor Cycles lightweights is capable of repeating this performance, but at least you can keep your machine running at its maximum efficiency by following the tips given by Dave Curtis, A M C agent of London Rd, Bicester, Oxon.

Big-end failure has been the bugbear of the A M C 250 in the past, but this can be avoided by regular oil changes, says Dave.

The normal oil level is 1 in. below the filler hole and the sump holds 2½ pints. Every 3,000 miles drain the motor, clean the oil filter and refill with fresh oil. The drain plug for the engine is near the bottom crankcase bolt. It is advisable to drain the oil when the engine is warm.

Gearbox oil level should also be carefully checked—and regularly. The gearbox takes three pints of oil and the level should be checked every 1,000 miles. The oil should be just up to the lip of the inspection hole.

Tappets should be adjusted when the engine is just warm and the clearances are nil. This means that the pushrods should be free to revolve but without any appreciable up-and-down movement.

Normally, tappets should be adjusted every 5,000 miles. Valve timing is as follows: inlet valve opens 40 deg. btdc and the exhaust valve closes 40 deg. atdc.

The front forks take 100 cc of oil if they are the lightweight type. When drained, 30 cc will remain in the damper plugs, so top up forks with 70 cc of S A E 30 oil to get the correct amount.

Abnormally lively action of the front end indicates that the forks need topping up. Road holding will suffer too. Rear swing-arm bushes should also be greased regularly or two pivot bearings will wear and give you dodgy steering.

The 250 is fitted with hairpin valve springs which are a bit unusual these days. They do not wear as rapidly as coil type, and can be checked by measuring between the collar portion and the centre point of the spring legs. Free length between these points should be 2 in. A decrease of more than ⅛ in. in this length means replacement.

MATCHLESS 250

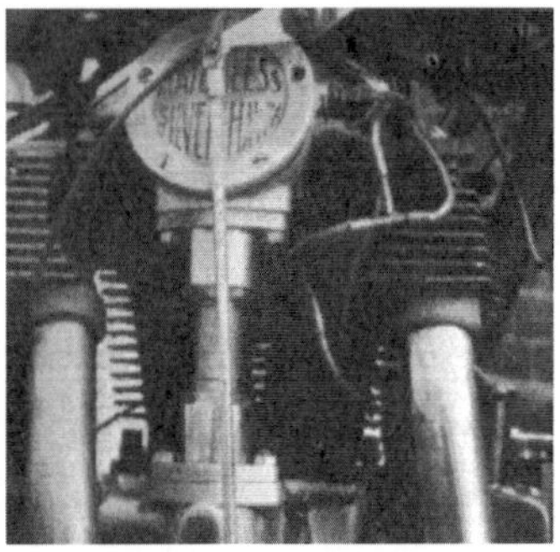

Massive camshaft-drive bevel is carried in alloy box. Narrow angle—26 degrees—of the cylinders can be seen between fins

Recent modification is the visible feed to the rocker box. Lower pipe is the drain return while front plugs show on right

Separate six-pint oil tank bolts direct to front of engine and feeds through oilways to pump. Note breather and filter

Exhaust manifolds are finned like block, and bolt on over the ports. H T distributor on left is similar to car type

Front girders had a massive "banana"-shaped central blade which gives a solid ride but cuts any tendency for twisting

Front fork suspension called for vigorous use of the grease gun. Spindles had fine thread for taking up any slack in them

Whole rear end of bike is a triangular spring unit, with damper knob and springs under saddle—same set-up as Vincents

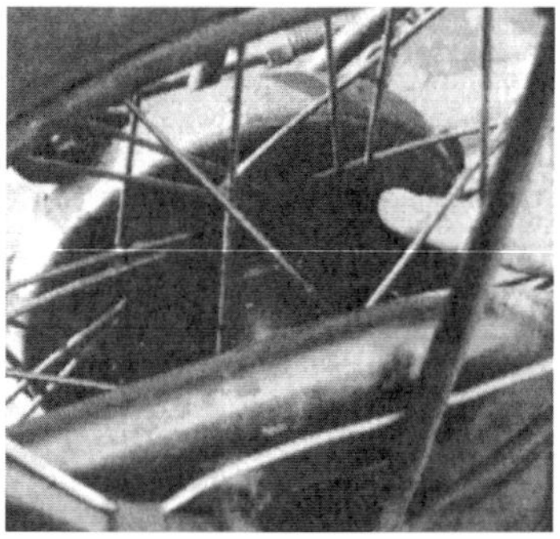

Brake drums, which carried eight-inch shoes, also fixed the spokes, which were laced up double on one side for chair work

SOMETHING SPECIAL No. 5

M M tries out a cammy vee-four—and samples the 85 mph thrills of grandfather's day

MIGHTY

●So you think that an overhead camshaft job is the modern lot? Especially if it is a four. Well, British motorcyclists now old enough to be your grandfathers were riding them around at 90 mph way back in the early 'thirties.

Proud owner David Bain of Pinner Road, Northwood, Middlesex, brought this 593 cc vee-four to M M after 15 months of using it for going to and from work and offered it for test.

Top speed of this 30 bhp ohc four was over 85, with exhilarating acceleration, although the road holding was a bit off, compared with present models.

Known as the Silver Hawk, this long-stroke (50·8 × 73·2 mm) four was first shown at Olympia in 1930 and was a successor to the side-valve twin Silver Arrow. It was in production for three years—and sold at £75.

Only recent modification of importance is the Monobloc carb which feeds a central inlet manifold through block centre

Lucas dynamo and coil are one unit, and the drive comes through a bevel and shaft damped by a flexible coupling

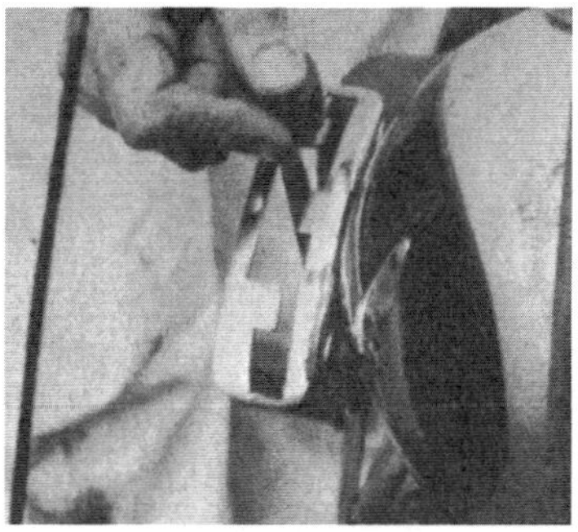

A high-class rarity in 1932 was the Matchless four-speed gearbox. Gate slots are designed to prevent jumping a gear

Neat instrument panel featured a central steering damper, speedo, and a sight glass for the original rocker oil-feed

Unique feature of the Hawk was the interconnected brake system, with finger adjuster on front for delicate balancing

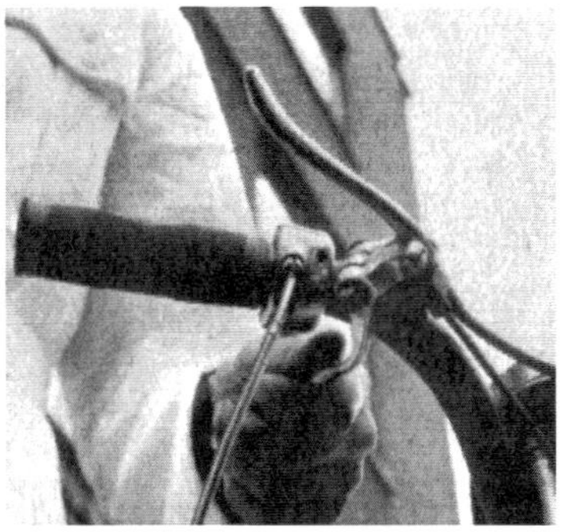

Twist-grip throttle was optional "extra" in those days! Underslung double lever was a neat control for advance/retard

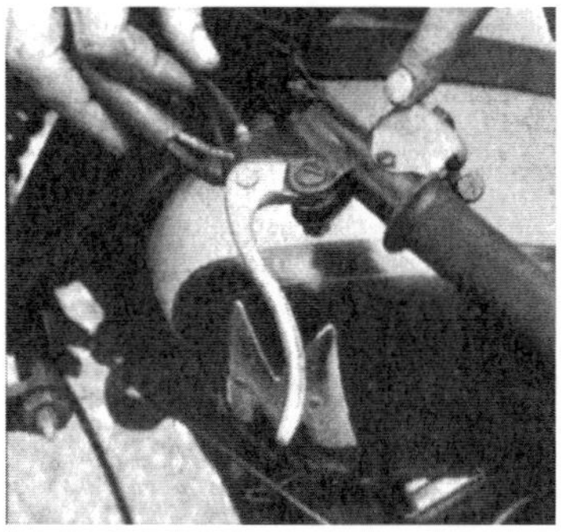

Double lever on other side was for mixture. Lever shape may be old fashioned but see how the fitting gave bars clean look

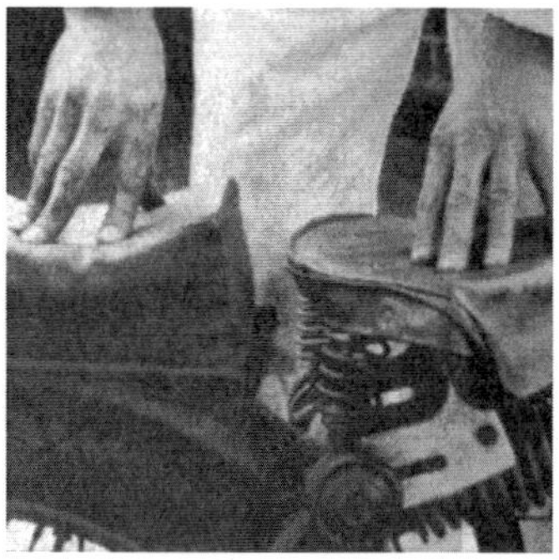

Pillion riding prewar was a penance! The typical passenger seat of the time was a small square stuck on the mudguard

MATCHLESS

ROAD TESTS OF NEW MODELS

Lavish use of chromium plating and highly polished light-alloy parts gives the G15 CSR eye-catching appeal. It is also one of the fastest sporting mounts available

Below: Twin six-volt batteries are housed in the forward section of the tool box; back of the box is cut away to accommodate the second battery. Right: Matching speedometer and revmeter dominate the handlebar layout; light switch is between the instruments

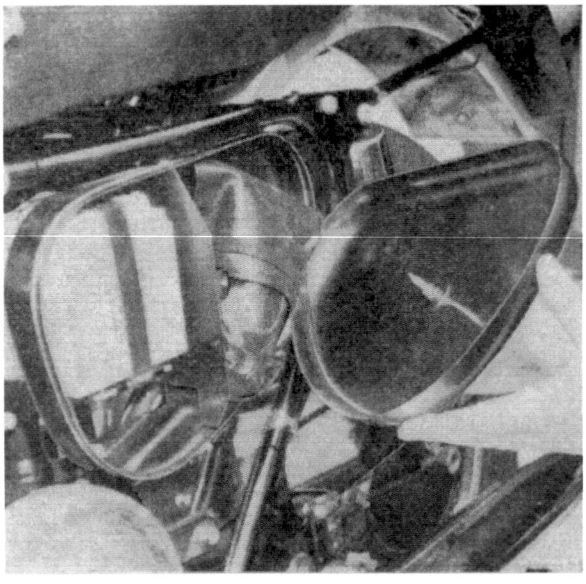

TO the true enthusiast, there is no greater joy than piloting a big, beefy solo with oodles of reserve and a wide spread of power that makes buzzing through the gear box unnecessary.

Power that takes 80 to 90 mph cruising in its stride, flattens the steepest of gradients as if they were molehills and annihilates a standing-start quarter-mile in a time fit to make an E type blink.

This is the sort of urge you get from the Matchless G15 CSR. Why CSR? Well, the standard G15, which we tested last October, is for the rider who wants an orthodox but ultra-powerful roadster.

The CSR version meets the demands of those who insist on rear-set footrests, narrow, sporty, light-alloy mudguards and racy exhaust system—a bike that looks every inch capable of its 114 mph maximum.

For those who prefer a low handlebar, this is also available.

Although the Atlas engine is identical to that used in the G15, the CSR proved appreciably faster than its stablemate, giving 12 mph more at maximum and knocking 0.5s off the quarter-mile time.

Apart from this, the main difference is in the riding position. The footrests are mounted about 5in forward of the normal pillion-rest position; pillion rests are not fitted. The rear-brake pedal is shortened to suit and the gear-change pedal is reversed.

As the gear-box mechanism has also been modified, gear selection is normal—one up, three down.

On the test machine, the

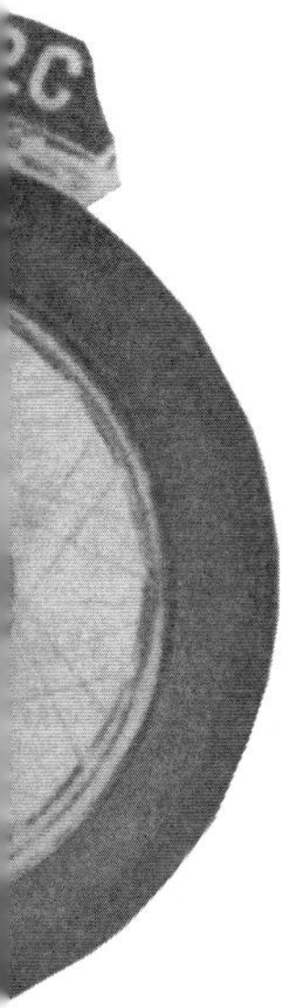

745 cc Matchless G15 CSR

Handling on tight corners is light, belying the size of the model

SPECIFICATION

ENGINE: Atlas 745 cc (73 × 89mm) overhead-valve twin. Crankshaft supported in ball and roller bearings; plain big-end bearings. Light-alloy cylinder head, compression ratio 7.5 to 1. Dry-sump lubrication; oil-tank capacity, 4½ pints.

CARBURETTORS: Twin Amal Monoblocs; air slides operated by handlebar lever.

ELECTRICAL EQUIPMENT: Lucas magneto with auto-advance. Lucas 12-volt, RM19 alternator, with rotor on drive-side crankshaft, charging twin Lucas six-volt, 13-amp-hour batteries, through rectifier; automatic voltage control. Lucas 7in-diameter prefocus light unit with 50/40-watt main bulb.

TRANSMISSION: AMC four-speed, foot-change gear box. Gear ratios: Bottom, 11.55; second, 7.68; third, 5.51; top, 4.51 to 1. Multi-plate clutch with bonded friction facings. Primary chain, ½ × 0.305in in light-alloy, oilbath case. Rear chain ⅝ × ⅜in with guard over top run. Engine rpm at 30 mph in top gear, 1,800.

FUEL CAPACITY: 3¾ gallons.

TYRES: Dunlop; front, 3.25 × 18in ribbed; rear, 3.50 × 18in K70 Gold Seal.

BRAKES: 8in-diameter front; 7in-diameter rear, both 1¼in wide and with finger adjusters.

SUSPENSION: Roadholder front fork with hydraulic damping. Pivoted rear fork controlled by Girling spring-and-hydraulic units with three-position adjustment for load.

DIMENSIONS; Wheelbase, 56½in. Ground clearance, 5¼in. Seat height, 33in. All unladen.

WEIGHT: 398 lb fully equipped, with full oil tank and approximately one gallon of petrol.

PRICE: £366 3s 7d, including British purchase tax.

ROAD TAX: £8 a year; £2 19s for four months.

MAKERS: Matchless Motor Cycles, Plumstead Road, London, SE18.

PERFORMANCE

(Obtained at the Motor Industry Research Association's proving ground at Lindley, Leicestershire).

MEAN MAXIMUM SPEEDS: *Bottom, 44 mph; *second, 66 mph; *third, 92 mph; top, 114 mph. *Equivalent to peak power at 6,800 rpm.

HIGHEST ONE-WAY SPEED: 115 mph (conditions: still air, 12½-stone rider wearing one-piece leathers).

MEAN ACCELERATION:

	10–30 mph	20–40 mph	30–50 mph
Bottom	2.4 sec	2.6 sec	—
Second	3.4 sec	3 sec	3.6 sec
Third	—	4.4 sec	4 sec
Top	—	—	5 sec

Mean speed at end of quarter-mile from rest: 95 mph.

Mean time to cover standing quarter-mile: 13.8s.

PETROL CONSUMPTION: At 40 mph, 64 mpg; at 50 mph, 59 mpg; at 60 mph, 55 mpg; at 70 mph, 52 mpg.

BRAKING: From 30 mph to rest, 36 ft (surface, dry tarmac).

TURNING CIRCLE: 18ft 9in.

MINIMUM NON-SNATCH SPEED: 24 mph in top gear.

WEIGHT PER CC: 0.53 lb.

flat, almost straight, 27in-wide handlebar was fitted. This gives the best compromise in riding position with a marked forward lean, which proved comfortable for a 6ft rider.

At 33in, the seat level is a little too high for short riders but, in combination with the handlebar and footrest layout, it was decidedly comfortable. Gear and rear-brake pedals could be operated without moving the feet from the footrests.

For kick-starting, the right footrest is hinged; to match up, the left rest also folds.

That the machine is essentially a man's mount was evident when starting. Two or three really hefty swings were required to bring the engine to life.

Because of the steeply inclined carburettors, flooding had to be avoided; for a cold start, one merely had to close the air slides and allow a minute or two for the float chambers to fill.

Except for slight piston slap when the engine was cold, mechanical noise was limited to no more than a slight rustle from the valve gear and the twin large-capacity silencers muted the exhaust note to a deep, well mannered throatiness, even at high revs.

For high-speed cruising, hard sparking plugs—KLG FE 220—were fitted and proved ideal.

At the bottom end of the scale, lumpy running made the tick-over relatively fast at 1,200 rpm and minimum non-snatch speed in top gear was rather high at 24 mph. Low-speed snap acceleration figures were also affected.

Bottom-end power was so beefy that upward gear changes could be made at 3,000 rpm, but occasionally the engine pinked when pulling hard in a high gear, in spite of 100-octane petrol.

On the open road, though, the model was in a class of its own.

Peak power is developed **at** 6,800 rpm, yet the spread is **so** wide that it is unnecessary to exceed 5,500 rpm in the gears when even the fastest motorway expresses were being left trailing far behind.

Acceleration to an indicated 100 mph (a true 92 mph) was shattering. And 13.8s for the standing quarter-mile makes this model one of the fastest production mounts yet tested by *Motor Cycle*.

The engine was at its best when cruising from 45 mph to twice that gait, and the semi-crouched riding position ensured that there was no undue drag on the arms at 90-mph-plus speeds.

Moderately smooth to 4,300 rpm, the engine then entered a spell of vibration, felt through the handlebar and seat, but became smooth again at 5,500 rpm.

An 80-mph range in top gear makes the choice of gear ratios uncritical. Nevertheless, those on the G15 CSR are well chosen.

Bottom is low enough for easy restarts on the 1-in-3 test hill at MIRA, while third is ideal for regaining 90-plus speeds.

Heavier in operation than on the G15, the clutch had a very slight tendency to drag. This took the edge off the pluperfect gear change normally associated with AMC boxes.

Only indication the clutch gave of six full-throttle standing starts was some extra play at the lever. Adjustment returned to normal when the plates cooled.

A big bike, the model belied its size when bend swinging. On only the slowest corners was there any noticeable top hamper.

Medium and fast bends called for slight effort when leaning. If the bike was banked over excessively on left-handers the stub of the centre stand grounded, but on similar right-side leans, nothing touched.

Front and rear suspensions are well matched and absorb road shocks satisfactorily at all speeds. Light at low speeds and therefore effortless when threading traffic, the steering became even lighter at high speeds and, above 85 mph, weaving was noticeable.

This was felt also when taking roughly surfaced bends at reasonably high speed.

On the test model, the brakes were not quite up to the extremely high standard usual on AMC models. Main fault was lack of bite in the front. Insufficient leverage made rear operation insensitive.

However, neither was affected by an hour-long, high-speed drenching on the M1.

Although headlamp diameter has remained constant at 7in, one big advantage of the 12-volt system is a more intense beam. On the CSR this allowed 70 mph cruising and the dip beam cut-off prevented dazzle.

The horn note, too, was above average but could still be improved with advantage.

Throughout the test, the engine remained commendably oil-tight. Overall petrol consumption, including very fast riding on motorways and performance testing, averaged 48 mpg. Petrol-tank reserve gave about six miles. One pint of engine oil was required in 400 miles.

Only maintenance needed was adjustment of the rear chain after a dash up the water-sodden M1. Normal spanner work is straightforward and the tool kit is adequate for routine jobs. However, the tool-box lid on the test model was a poor fit.

For normal parking the prop stand is convenient to use without danger of flooding the carburettors. Pulling the machine on to the centre stand involved a fair amount of muscle power.

Appreciated rider's features are the paired speedometer and revmeter—both well positioned and easy to read—the combined horn button and dip-switch convenient to the left thumb, a cut-out button by the twistgrip, and rubber handlebar grips instead of smooth plastic grips normally fitted.

Liberal use of chromium plating on the usual parts and on the headlamp shell, rear chainguard and instrument panel is nicely balanced by polished light-alloy mudguards.

Finished in an attractive cherry red; the tank top, tool box and oil tank blend well with the bright metal, making the model extremely eyeable.

And it performs even better than it looks.

Swept back exhaust pipes emphasize the sporty air of the model. The large-capacity silencers are really effective

Both footrests are hinged—the left one, here, also acts as a stop for the rear-brake pedal. The brake warning light switch is directly behind the arm of the pedal

The Model 16 has " Norton " brakes, which are first-class, and a mediocre riding position caused chiefly by the too-low dualseat

Impressions of a modern three-fifty: the AJS 16

THE three-fifty these days is the odd man out in motorcycling. Nobody wants a three-fifty.* This is not an entirely new situation. Thirteen years ago Vic Willoughby was writing in *The Motor Cycle :* " But while each type [of motor cycle] has its own particular appeal, there can be little doubt that many present-day motor cyclists have 'over bought'. In paying for the potential performance which they may never use, they are apt to sacrifice qualities which would be appropriate to their everyday needs. Is the modern three-fifty standard capable of meeting the requirements of a large proportion of motor cyclists . . . ? If so, does it offer any advantages over its contemporaries of larger engine capacity ? Has it any drawbacks ? " And concluding with the evidence (top speed of 75 m.p.h. and " distinctly snappy " acceleration) provided by a hard-working three-fifty Ariel that " . . . the requirements of large numbers of motor cyclists fall within scope of the three-fifty . . . a good modern three-fifty will meet the needs of those riders in an economical manner provided it incorporates such desirable features as an absence of excessive weight, correct gearing, good engine balance with adequate flywheel effect, tireless engine performance and robust engine bearings in the interest of long life." All true in 1952, when those words were written, but only true in part today.

The odd thing is that 13 years ago, when a considerable proportion of the motorcycling population rode far and reasonably hard, the three-fifty would have been adequate for only say 90 per cent of riders; and now when nobody apart from the old men of the Vintage Club covers more than 25 miles non-stop . . . the average distance between cafés, as revealed in an exhaustive survey conducted by this journal . . . the three-fifty would be eminently suitable for 99.9 per cent of us. Yet it has sunk to new depths of unpopularity. I am certain that an interview with any of the keen owners who congregate of a night at the hot dog establishment on the southern end of Chelsea Bridge would reveal a profound indifference to (possibly ignorance of ?) the desirability of correct gearing, good engine balance, adequate flywheel effect and robust engine bearings. They couldn't care less. Rather, I think, would they emphasize the desirability of a plenitude of chromium plating and polished light-alloy, a racing position, and a deafening exhaust note. Times change; motorcyclists change. Three-fifty motorcycles haven't changed. At least—not much.

Always a peculiarly British " in-between," the three-fifty had been made to feel unwelcome long before the 1960s. But in recent years its existence, never mind popularity, has been put in jeopardy by the 250 c.c. limit for learners; the vastly accelerated development of the two-fifty, both four and two-stroke, by British factories; the disinterest shown by the Japanese saviours of our sport and pastime in anything bigger than 250 c.c. (and, of course, the " 500 c.c." standards they have set with their delectable two-fifties); and the superior performance of the British six-fifties. . . . No manufacturer has displayed zeal in developing the three-fifty. With perhaps the exception of B.S.A., it has not been subjected to any major facelift in 15 years. In most cases the engine has remained basically unchanged since the early '50s, with the frame receiving the fruits of any rethinking in the design department—and any change in that direction has invariably meant a heavier (possibly stronger) frame to depress still further the already moderate performance of the conventional three-fifty.

This is not a fitting place for a diatribe on the three-fifty, its decline and possible fall. I make these comments merely to put the 350 c.c. class in perspective. The Model 16 A.J.S. I have been using in recent weeks has an engine outwardly very similar to that of the G3/L I rode in '49 and '50. The flywheels and the bottom half generally are, I understand, more robust than in those days, having now a close affinity with the scrambles engines, but in the main you can take it that the present-day A.M.C. single traces its pedigree in a very direct line through at least 25 years. Nothing wrong in having a long pedigree, of course. With a straightforward proposition such as the large-capacity, single-cylinder i.c. engine, it can be assumed that the really big

* This is not strictly true. Fewer people want five-hundreds, according to the latest returns.

strides in development were made over 25 years ago; the years since have been given to gilding this rumbustious lily. The G3/L, I remember, had a single downtube frame, no rear springing (though rear springing had recently become available), hairpin valve springs, light-alloy cylinder head, separate magneto and dynamo. The engine was quiet — much more so than contemporary B.S.A.s and Royal Enfields, for example— vibration was negligible and top speed nearly 75 m.p.h. The 1965 version (taking the A.J.S. as twin to a Matchless) is heavier, a little noisier, more vibratory—and, yes, a little faster.

Starting the 1965 Model 16 was simple. Turn the ignition switch to IGN, flood the Monobloc, close the air lever . . . kick. Usually one kick was enough. With the automatic ignition control at full retard the tickover was slow, even and reliable. The exhaust note was flat at tickover (because of the retarded spark) and remained so even when the A.J.S. was moving away through the gears, until decent r.p.m. were attained : it was the sort of note once referred to—in days when such antiques were manufactured and accorded some following—as typically " side-valve." Which is another way of saying the exhaust note was pretty unimpressive.

If I may return to talk of my " G3/L days," I feel it pertinent to mention that at that time the A.M.C. riding position was considered rather good. The saddle was high—about 31in from the ground—the footrests were wide apart but not too far forward, the handlebar was slightly raised and gave a relaxed position for the hands. Later A.M.C. productions, in the '50s, maintained a reasonable standard—possibly not quite so well balanced as before, because the twin seat was a little lower than the saddle and, riding solo, effort was called for not to slide back at speed . . . but, generally, the position was acceptable. Now there is a deterioration : at least, so far as the three-fifties are concerned. The seat is low at 28in, the footrests are 4in forward of the nose of the seat, and the handlebar is wide and flat. The result is that one sits with a decided curve to the spine, the arms at fair stretch to the handlebar, the feet high and forward. At 60 to 70 m.p.h., with wind pressure exerting its effect, this curved-back posture is pronounced and finally becomes uncomfortable, giving rise to what I believe are known as pains in the lumbar region. And slumped, with little balance between rests, seat and handlebar, it became an effort not to let one's knees flap wide from the unslender tank; and so keeping them in became and additional strain. These complaints may seem to have undue weight in this short report, but I believe that riding position is vitally important if there is any intention of covering more than 50 miles in the day. (Something of the sort goes for cars but to less degree because cars are not, in the same way, an extension of the driver's physique.) The A.J.S. encouraged what I used to hear described as a " sack of spuds " riding position : an inelegant way of putting it but remarkably descriptive. With relatively little of the rider's weight forward, the hard action of the Norton Roadholder fork was emphasized and the impression given was of light contact between front tyre and tarmac; which is not the most reassuring impression to have when you are out and about on wet and winding roads. Titch Allen, after a brief ride on the Model 16, described the steering as " remote," but in a not particularly denigratory tone; he also found the riding position " just right." Which proves something or other, I expect. The fork action, hard with plenty of damping control, was not responsive at moderate speeds and was obviously intended to give of its best at high speed; as the fork comes from the Norton side of the business its race heritage can be appreciated, but the value of that to a three-fifty ordinarily cruised at no more than 60 m.p.h. is open to doubt. The rear springing, conversely, provided plenty of movement and was sensitive to minor bumps and hollows in road surfaces.

Cruising speed was usually about 60 m.p.h. when the engine was working easily and the exhaust note though pronounced, was not deafening. Throttle opening at this speed was half to two-thirds; further opening did not produce much in the way of acceleration, which is to be expected with a relatively low-power engine, but increased exhaust noise to an irritating level. Near the end of the test, and after an hour's riding at almost full throttle, " blowing " began to occur between the cylinder head and barrel and worsened over the course of the following 100 (slow) miles to an audible bleep-bleep. Though the gasket appeared intact, from external examination, escaping compression could be clearly felt; possibly the cylinder head nuts had slackened. Cruising at above 60 m.p.h. showed up an unpleasing characteristic of the Model 16—vibration at the handlebar and footrests. A secondary effect of this vibration was a tendency for nuts and bolts and some other vital components to loosen—notably the retaining screw for the cover for the combined tool and battery compartment, which allowed the cover to fall off (and pass unnoticed by the rider, eardrums a-tingle from the exhaust's cacaphony), and the chrome-plated extensions in the front fork which unthreaded and danced free to expose the inner stanchions. A sympathetic resonance, not felt but quite audible, in the region of the headlamp gave clear indication of high revs.

The brakes were excellent. I am almost prepared to say they were better than in 1950 but recall that the front brake at that

Appearance of the engine, apart from enclosure of the push rod tubes, has altered little in recent years

Leak-free chaincase is a great improvement on the old pattern. Casing for the compartment vibrated loose at speed and was never seen again

IMPRESSIONS OF THE MODEL 16

time, just modified to take a torque arm in place of the bolt-through arrangement from fork leg to brake plate, was good enough to squeal the front tyre when applied hard. The 8in " Norton " front brake in the Model 16, bigger than in earlier years, would do no more—but might, I suspect, continue to do so for a longer period without " fading." The rear brake, of similar size and like the front contained in a full-width hub, was sensitive and powerful; all that an efficient rear brake should be. Using it, I realized that a one-time mentor, Torrens of *The Motor Cycle*, had been a little off beam in suggesting in one of his weekly columns that a tiny foot lever could be matched in sensitivity and ultimate power with the conventional, big lever. He made a case for two levers sized in strict proportion—one say 6in long with the vertical measuring 1½in, the other 18in × 4½in—being capable of equal performance. In theory, yes . . . but not so in practice, for I found that the A.J.S. brake lever, conventionally big, allowed much more sensitive control than the mini arrangement fitted to my fearless old Vincent.

All other controls of the Model 16 were similarly satisfying to use. The clutch was light, cut the transmission absolutely cleanly, and never slipped. The gearbox was quiet and the lever moved in a tiny arc with the lightest action in the world. The stop light gave up lighting early on, but the headlight gave a reasonable beam which was entirely adequate at 55 m.p.h. The horn, proud product of Joseph Lucas, looked as if it was designed to bleep unseen, being of miniature size, undistinguished styling, and having a note in harmony with its appearance.

Petrol consumption, with quite hard driving, at 60/65 m.p.h., and revving briskly in the gears, worked out almost precisely to 65 m.p.g., the figure we have long held to be inevitable with most motorcycles driven near their limit. (And if that sounds far fetched, let any Bantam owner try pushing his two-stroke along at 50 m.p.h. . . .) At 45/55 m.p.h. consumption improved to about 80 m.p.g. Maximum speed, with some help from the wind, was 85 m.p.h.—" against," just over 70; mean speed, 78 m.p.h.

A few months back I wrote : " These A.M.C. singles . . . are made for somebody who does not exist. He *was* alive, of course, just a few years back, and he liked a design with the time-honoured decencies . . . but he died in the late 1950s, unnoticed, apparently by Associated Motor Cycles. . . ." At the time I had not ridden one of the machines thus airily dismissed for a number of years; now that I have, I see no reason to change that judgement. *I* liked the Model 16 well enough because it was well made and nicely finished—if you overlooked (how ?) the execrable zig zag gold lining on the otherwise decently sombre tank—but nobody, surely, would advance it as any sort of champion to stem the Japanese tide ? To attract a sophisticated market such as exists in this country ? (It may of course be a wow in Zanzibar but we have only two readers in Zanzibar and are prepared to risk their displeasure.)

Incorporating the sum of modifications made to the basic product over many years, each change advertised as a significant improvement over last year's standards and faithfully echoed as such in Bowling Green Lane and other points further south, the latest A.J.S. is disappointingly akin to its forerunner of (let us say) 1950. In 1950 it was a beautiful motorcycle and second to none. Fifteen years on it shows its age too much for comfort. C.A.

One of the neatest power units to be found on any bike, the AMC 250 motor is a delight to clean. New swept-back exhaust pipe is a 1966 mod. Note how push-rod tunnels twist to follow offset cylinder

Sensible taps on Saphire are easy to get at and operate. Inclined Monobloc carb is already fitted with attractive trumpet bellmouth, which opens behind side panel removed in picture to show Wipac coil

Solid alloy British Hub Company front unit is same as used on racers—and really works, too. Air scoops are already cast into the brake plate, while adjustment screws are mounted on anchor arm

John Houslander tests the 1966 AJS 250, and finds power in a velvet glove

SUPER SAPPHIRE

MOTORCYCLE MECHANICS ROAD TEST No. 110

Vehicle AJS Model 14 CSR **Price new**

Engine 248 cc ohv single (69·8 bore x 64·8 stroke)

Gearbox 4 speed foot-change cylindrical unit

Final drive ½in single row chain drive.

GENERAL INFORMATION

Weight	330 lbs
Saddle height	30 ins
Turning circle	11 ft
Is toolbox lockable	No
Is steering lockable	No
Fuel tank capacity	3 galls
Reserve capacity	2½ pts
Oil tank capacity	2½ pts
Gearbox capacity	3 pts
Fuel specified	Premium
Overall consumption	80 mpg
Braking from 30 mph	28 ft
Acceleration 0-50 mph	11 secs

SPEEDS IN GEARS

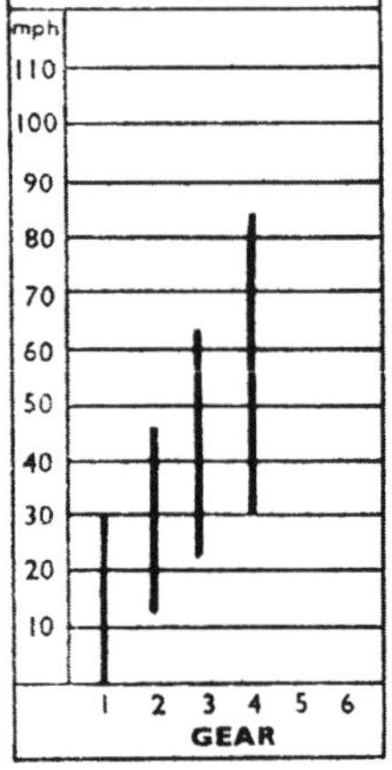

EQUIPMENT SUPPLIED

STANDARD FITTINGS

pillion rests

dual seat

OPTIONAL EXTRAS

SPARES PRICES

Engine gasket set	7s 9d
Set valves & guides	£2-0-5
Piston with rings	£2-17-0
Set of clutch plates	£2-2-8
Silencer	£7-7-1
Pr Exchange brake shoes	14s 9d

●A bike without any vices—and a whole vocabulary of virtues. That was the round-up of opinion by *MM* testers on the 1966 model of the 250 AJS Saphire, which is entering its fourth year of production.

A top favourite with the learning fraternity, the Saphire is probably one of the nicest bikes a learner can choose—yet it can still pack sufficient poke to warm the heart of some of *MM*'s hairy old testers.

Its virtues start the moment you put your leg across it, for it has one of the most comfortable seats I have had the pleasure of sitting upon. The motor springs into activity after two prods and a tickle whatever the weather, while the kickstart pedal is a proper man-sized affair, pitched well clear of the usual ankle-smashing ironmongery found on too many bikes.

In the pedal department, the gear change gets top marks too, for being precise, smooth and firm. Its travel is perhaps a little too long for some, but this again is ideal for the beginner.

Smooth as cream

The engine is as smooth and creamy as the top of the milk, with hardly a trace of vibration anywhere in the normal rev range. Coupled with a really well subdued exhaust, this smoothness has one surprising result—it can lead the rider into letting the speed drop below the snatch speed for a particular gear.

The smooth, clean lines of the engine and gearbox unit make clean-ups a pleasure—and can mislead you into thinking that engine and gearbox are in unit construction. The gearbox, however, is a cylindrical unit which nestles in a similar shaped recess behind the engine.

As the mainshaft is eccentrically placed, the box is rotated by adjusting screws to control primary chain tension.

The engine thrives on revs, as one would expect from the considerably oversquare unit (approx. 70 mm bore × 65 mm stroke). The quoted maximum given by AMC is 7,600 rpm, at which point it gives its maximum bhp of 22¼.

With a top whack of just over 80 mph, the Saphire puts itself in the same batting league as the rest of the hot 250s on the market today, even in its standard trim, so a touch of tune should make the magic 90 quite a reasonable possibility, and also sharpen the acceleration.

With such a powerful amount of "go" in the

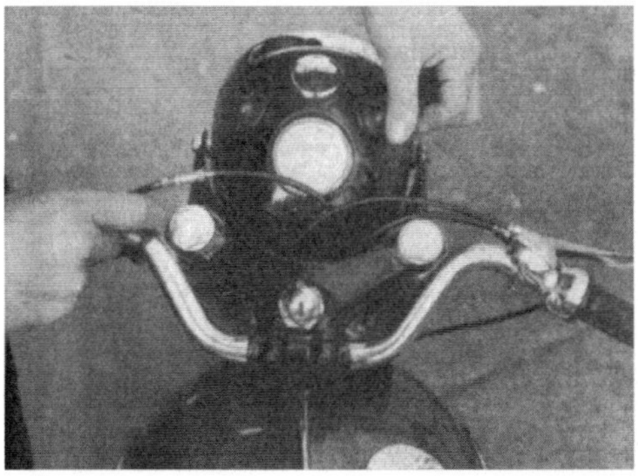

Wipac electrics are used throughout—and are really first class, with good headlamp beam. Ignition switch on right has emergency start position while centre ammeter is steady and accurate

Another 1966 modification on the AMC 250s is the light alloy mudguards, which are supported at front by neat trials-type loop stays. A heavy bridge plate between forks gives extra rigidity too

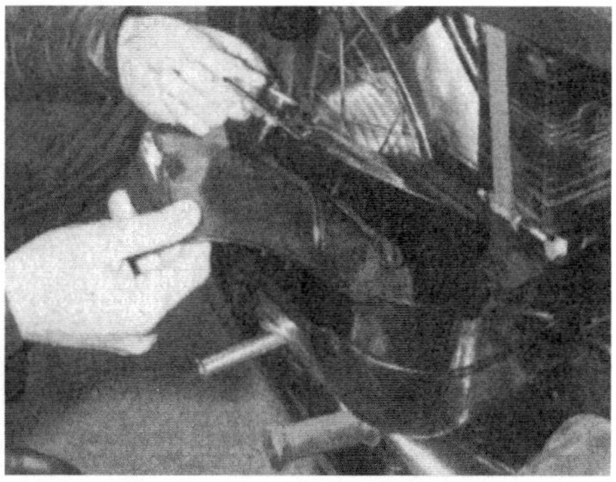

Right hand toolbox is neat and solid, with toolbag carried inside an inner locked compartment. Toolkit is quite comprehensive. Another good point is the rubber band fastening for the big battery

engine department, it was reassuring to find that there was plenty of "stop", too. The front brake now operates an all alloy British Hub Company unit identical to those fitted on some of the lightweight racers—and it really is a stopper. Unlike twin leading shoe front brakes, it has no sudden grab, and can be used to the point where the tyre starts to squeal without eye-popping exertion.

Electrics also are another Saphire strong point, from the neat, easy-to-use switches on the headlamp to the compact stop and tail lamp at the rear.

The headlamp beam is good, and the adjustable stoplamp operating cam is another useful modification. The only black spot on the whole machine was the horn which sounded rather like a strangled starling, and I will be charitable enough to suppose that this too might have been something that could be adjusted.

AJS and MATCHLESS TWO-FIFTY SINGLES

" SIX-FIFTIES fear it, two-fifties envy it—and girls who find their boyfriends having a mechanical affair with it, hate it. This is the Sapphire, probably the most perfect machine ever to come into existence. How many other sports machines have shared durability with performance?"

This charming quotation comes from the pen of 18-year-old Chris Pritchard of Shrewsbury. Naturally, it is a little colourful; but it reflects the basic satisfaction many owners feel as they sit astride their AJS or Matchless two-fifties. They are, as Anthony Head says, good bikes, especially for learners, with only a few bad points. " AMC could certainly improve this bike to make it compete with the best of our foreign competitors."

" A good, fast, cheap machine to run," adds 17-year-old Edward Yarde, a farm worker of Taunton, Somerset.

Want to know more about the average rider of these bikes? Predictably, he is about 18 years. A couple of middle-age reporters bump the average up to 20, but that would be unrepresentative. By and large, the Ajay or Matchless is the first machine—and it tends to be a trifle elderly. 1961 is average vintage. Experience of the riders is about two years in each case and they have, together, covered nearly 700,000 miles on their machines.

All two-fifties built since they were introduced in 1958 were eligible. Model designations are 14 and 14CSR (sports) for AJS and G2 and G2CSR for Matchless.

Performance

"ACCELERATION is good, especially in second and third gears, although in my opinion bottom gear is too low. I have had the speedometer needle at 85 mph on several occasions. On open roads, I can usually maintain a cruising speed of 60 or 65 mph almost indefinitely.

" My engine (with a 7.8 to 1 compression ratio) is very tractable."

Those words come from 18-year-old Anthony Head of London—he runs a 1960 Model 14 AJS.

Adds CSR owner David Maddison of Grantham: " The engine has good staying power over long distances. Provided it is well buzzed it will tackle anything."

Averaging readers' claims shows that the sports models give about 85 mph while the standard models are only 5 mph slower.

The engine unit likes revs and it is necessary to keep it on the boil for best results.

Says Chris Ball of Wellington, Salop: " Cruise along at 55 mph and you'll swear you're still in third. Oh for a two-fifty Ajay with a five-speed box! This high-revving motor thrives on hard work; keep it spinning and there will be no complaints."

Performance can be increased very easily, according to Barry Page, an interior decorator of Leigh-on-Sea: " The engine lends itself to modification readily. The valve timings are already good. With care the bike will show a clean pair of heels to almost any two-fifty." He doesn't say what modifications.

Starting

" STARTING couldn't be better. Just free the clutch plates, switch on petrol and ignition and give one kick, and she fires immediately, even on really cold mornings."

COLLATED BY MIKE EVANS

RIDERS' REPORT NUMBER FOURTEEN

Above left: This is the AJS Model 14 in 1961 guise; a standard model, it is the fore-runner of today's potent CSR machines. Above: The 1966 AMC sports engine with swept-back exhaust pipe. Below: Peter Fraser samples a 1965 job—the AJS Sapphire Ninety

That is from the owner of a standard two-fifty AJS.

Confirms David Maddison: "Starting is very good. One tickle and one good prod is enough in all weathers—provided all settings are at the makers' recommendations."

Fuel Consumption

BOTH sports and standard models return an overall average of 78 mpg, according to our reporters.

These two-fifties are obviously economical little mounts capable of up to 100 mpg under sedate riding.

Handling

THE FEW reporters who complain of bad handling are very definitely outweighed by the opinions of their colleagues.

"This is where the Model 14 comes into a class of its own," enthuses John Willson, 19, of Southampton.

"Hurl the bike round any bend, brake, accelerate and she won't stir a hair. With two-up she seems even happier—thanks, partly, to the adjustable rear-suspension units."

Braking

THE BRAKES come in for severe comments from many reporters, although there are, as ever, the enthusiastic ones.

Like Phillip Barton of Kettering who thinks his brakes are excellent. "No sign of fade when braking from high speed. At low speeds the front brake is a real stopper."

The majority are dissatisfied with the front brake—on standard machines. This is not surprising; from 1962 a much improved front brake and heavier fork were fitted to the sports models.

Constructive criticism comes from 48-year-old Walter Mardell, of Aintree, Liverpool: "After 3,000 miles from new on my 1958 Matchless, I decided that the front brake was not up to the standard of the machine.

"Since the factory considered that the 6in brake unit should be adequate, I set about improving it myself. I modified the shoe-plate mounting, to allow it to float, and relined with Ferodo MS3 linings.

"The immediate result was a fractured mudguard anchor because of the fork leg on the brake side flexing under braking load!

"The brake has always been adequate since, but I can never keep a decent front mudguard on the bike."

Transmission

OPINION is that, on the whole, this department is well up to scratch. Several have had trouble with clutch slip and have had difficulty remedying it. Also, nearly everyone has experienced a broken gear selector spring at some time or other.

Nevertheless, the gear box, in particular, receives praise aplenty for smoothness of operation and, basically, dependability.

"Gear selection is very smooth, but the pedal travel is a little long," reports Martyn

Jordan, 18, of Filey. "Juggling with the gears," he continues, "is one of the most enjoyable facets of motor cycling—and it is especially so with this machine."

Electrics

REPORTS on this subject are so conflicting—and temperatures run so high—that we must turn to the percentage figures for an accurate opinion. Lights, 57 per cent; horn, 8; other electrics, 52.

It would be unfair to average these three figures since the horn is obviously so bad that it tips the scales violently.

Nevertheless, the other two figures average to 54.5 per cent and certainly show the electrical system is not up to the mark.

Says David Maddison: "Very poor electrics. Bulbs blow regularly, the battery boils and the whole system invites water!"

Another opinion comes from 18-year-old Jonathan Fox of Edenbridge, Kent. "Electrics generally are unreliable and are not good enough for the standard of the machine. However, the actual lights are a different story—the headlamp allows comfortable speeds in the 60-mph region."

SPECIFICATION

ENGINE: AMC 248 cc (70 x 65mm) overhead valve single-cylinder. Double row roller big-end bearing; crankshaft supported in a ball and a roller bearing on the drive side and a plain bearing on the timing side. Light-alloy cylinder head, cast-iron cylinder barrel. Compression ratio, 9.5 to 1. Dry-sump lubrication; oil capacity, $2\frac{1}{2}$ pints.

CARBURETTOR: Amal Monobloc, $1\frac{1}{8}$in choke; air slide operated by handlebar lever.

ELECTRICAL EQUIPMENT: Coil ignition. Wipac 54-watt, ac generator mounted on drive-side crankshaft, charging Exide 11-amp-hour battery through rectifier. Wipac 7in-diameter headlamp with pre-focus light unit and 30/24-watt main bulb.

TRANSMISSION: AMC four-speed gear box clamped to rear of engine. Gear ratios: bottom, 17.97; second, 11.68; third, 8.05; top, 6.51. Multi-plate clutch with bonded friction facings. Primary chain $\frac{3}{8}$in duplex in cast-aluminium case. Rear chain $\frac{1}{2}$ x 0.305in with guard over top run. Engine rpm at 30 mph in top gear, 2,800.

FUEL CAPACITY: $3\frac{1}{4}$ gallons.

TYRES: 3.25 x 17in ribbed front; studded rear.

BRAKES: 6in-diameter, front and rear; finger adjusters.

SUSPENSION: AMC telescopic front fork with hydraulic damping. Pivoted rear fork controlled by Girling three-position, spring-and-hydraulic units.

DIMENSIONS: Wheelbase, 53in. Ground clearance, $5\frac{1}{2}$in. Seat height, 30in. All unladen.

WEIGHT: 325 lb without fuel.

PRICE: £248 including British purchase tax.

ROAD TAX: £4 a year.

MAKERS: Matchless Motor Cycles, Ltd, Plumstead Road, London, SE18.

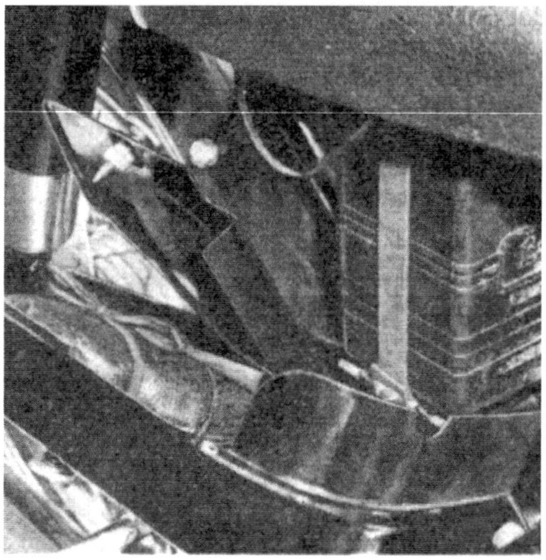

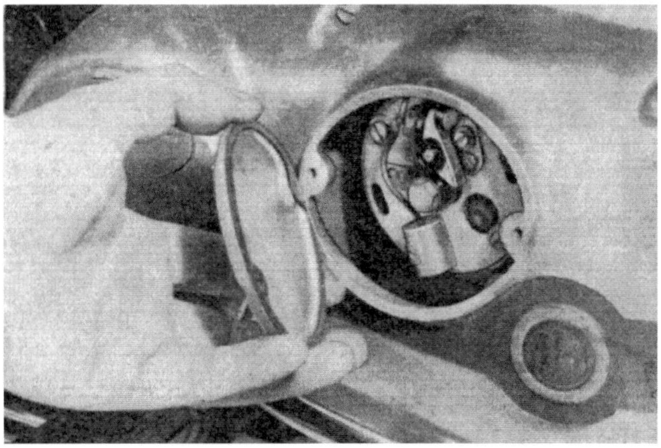

Left: Access to the contact breaker is excellent, say reporters. Top: Neat arrangement of horn and ignition coil behind the left-side panel; this is a 1961 standard version. Above: Tool roll and battery fit snugly behind the right-side panel of this 1965 model

OUR TEST VIEW	AJS 14 (April 1961)	M'less G2CSR (Aug 1962)	AJS 14CSR (Feb 1965)
Highest One-Way Speed	77 mph	75 mph	83 mph
Mean Maximum Speed	75 mph	73 mph	82 mph
Standing Quarter-mile (speed)	69 mph	67 mph	70 mph
(time)	19.6s	19.6s	19.4s
Fuel Consumption (at 30 mph)	104 mpg	112 mpg	112 mpg
(at 40 mph)	102 mpg	88 mpg	84 mpg
(at 50 mph)	81 mpg	72 mpg	66 mpg
(at 60 mph)	64 mpg	64 mpg	58 mpg
Braking (from 30 mph to rest)	36ft	30ft	30ft

Surprisingly, Chris Hall says: "The whole electrical system is first class. The main beam is good for 60 mph at night. But my horn is neither of use nor an ornament!"

Detail Finish

AVERAGE would be a fair estimation here. Says 21-year-old Michael Weaver, of Reading: "Paintwork is good all round. Chrome plating is of excellent quality as well, especially on the wheel rims."

While another reporter, Michael Leavesley, 19, of Nottingham, points out that the paintwork is prone to chipping, the majority of owners are satisfied.

However, many mention one detail defect—the stands. Listen to Colin Gascoyne, 17, of Southall: "The centre stand was really too narrow to support the bike on other than a perfectly level road. For this reason I purchased a prop stand.

"It had excellent fittings but the stand itself was a joke. It bent into a semi-circle after about six weeks' use."

Walter Mardell reports that he has never had a really reliable speedometer. "They are reliable enough at lowish speeds, but when cruising the bike in the upper sixties, the needle tends to swing and the mileage recorder goes haywire. I have had about five speedo heads in six years."

Riding Comfort

RIDING position gets a mark of 83 per cent in the table. And there are certainly not more than a couple of owners dissatisfied on this score.

As Jonathan Fox says, comfort is a key point with the AMC two-fifties.

Reliability

THESE quarter-litre four strokes are not outstandingly reliable; but, on the other hand, they are not notoriously *un*reliable. In fact, many readers sing their praises in no uncertain manner.

"During the months from May to November last year my bike did 450 miles a week. It took me about an hour to cover the 37 miles to my job. I was late only three times; once when the contact breaker stuck, once with a puncture and once with a snapped control cable." Jonathan Fox speaking.

Another similar opinion comes from John Willson, a motor-cycle stores assistant: "In 16,000 miles the bike has never let me down. If oil changes are carried out regularly and the engine is treated with respect, no troubles will occur."

Nevertheless, several of the letters relate stories of woe and vicissitude. David Waldron, a toolmaker, age 22, of Cannock, Staffs, gives his views: "The bike was not as reliable as I had expected. In 18,000 miles I had four new spark plugs, two sets of points, three gear-pedal return springs, one kick-starter spring, one new piston at 5,000 miles and a new exhaust valve at 7,000 miles.

"The engine was rebored at 13,000 miles and new piston, big end, main bearings and small end fitted. The big end started to rumble again at 18,000 miles so I sold the bike."

His machine was a Matchless G2 which he purchased new in May, 1961.

Offset David's experience with that of Walter Mardell. "My 1958 G2 gave me reliable, economical and fast transport until I took it off the road, in November, 1964, when it had covered all of 70,000 miles. It was still returning an mpg of about 80.

"My only major trouble was a broken crankpin at 20,000 miles—and the factory replaced it without charge."

It is hard to say why one owner should have endless trouble while another can be impressed by reliability.

John Willson probably puts his finger on the crux of the matter when he recommends regular oil changes. This is important with any four-stroke —and the AMC two-fifties are likely to benefit more than most as they are fitted with comparatively small oil tanks.

Too many people make light of oil changes. They can have only themselves to blame if a big end pegs out after 6,000 miles on the same oil! Some, I know, *never* change their oil.

This isn't to say, of course, that troubles can invariably be laid at the door of bad lubrication. There is little doubt that some of those who maintain their machines conscientiously have suffered considerable bad luck.

Below left: This front brake fitted to current models is an improvement on the criticized pre-1962 type. Below right: Rear-brake layout of a 1965 model. In spite of improvements, however, both brakes receive a lowish mark in the table

Service

THE spares situation from the factory—and, consequently, from dealers—is poor. You need only look at the extremely bad figures in the table. Respectively 44 and 43 per cent for factory and dealer spares is hardly a creditable situation.

The alarming fact is that it is probably easier to obtain over-the-counter spares for some foreign machines than it is for these two-fifties, from Britain's second-largest manufacturer.

"The factory doesn't seem to exist as far as spares and service is concerned. I had to wait three months for the return of a big-end assembly, despite repeated letters." From Chris Gilbert, 19, of Plymouth.

Confirms David Maddison: "Factory service and interest is a negative quantity. I am still waiting for a spares list ordered three years ago."

However, in the main, David's criticism of lack of interest is unfounded. There is always a genuine willingness to help down at Plumstead. But it is regrettable that they are unable to combine this with an ability to make spares 100 per cent available.

It is only fair to mention, however, that many of reporters' views are based on experiences of a year or more ago.

According to the factory, the spares position is much improved recently.

Lots of chromium plate, an ultra-clean power unit, comfortable riding position and sturdy construction—these are the features of this 1965 Matchless

Accessibility

REASONABLY good, report readers. Everything is easy to find and attend to.

One or two have a moan about the necessity of removing the tank in order to change throttle cables. But access to the contact breaker and rocker gear is good.

Overall View

" THE machine is very good for a learner or someone who just wants to potter to work. But for a more serious motor cyclist like myself it is ruined by small design faults which could easily have been rectified by the makers.

" For long distances, adjustable handlebars and controls would make an incredible difference. However, this particular fault has at last been attended to."

That quotation is from Chris Gilbert. And it is fairly representative of the opinions of all reporters.

Confirmation comes from 19-year-old machinist Barry Heywood, of Coventry: " One of the best two-fifties on the market. You get what you expect for your money. And now that they have fitted new forks and brakes and given it a face lift, the Matchless Sports is better than ever. If I was going back to a quarter-litre job, I would have another CSR any day."

" The acceleration, braking and road holding will match any British two-fifty. It must be the most beautiful motor cycle in the world." Ian Spooner of Chesterfield speaking.

Postscript is added by Michael Weaver: " A good little bike—nothing elaborate or fancy."

There you have it. The AJS and Matchless two-fifties are fast, comfortable and fun to ride. Unless you happen to be very unlucky, they are also reliable.

The only noticeable cloud in the sky is factory spares and service. Judging from the overwhelming condemnation by reporters—and by a similar result when we reported on AMC big twins—there is considerable room for improvement down at the Woolwich works.

Below: "As good as most" is the verdict on the standard tool kit supplied with the AMC two-fifties. Above: David Waldron's 1961 standard Matchless on which he covered 18,000 miles

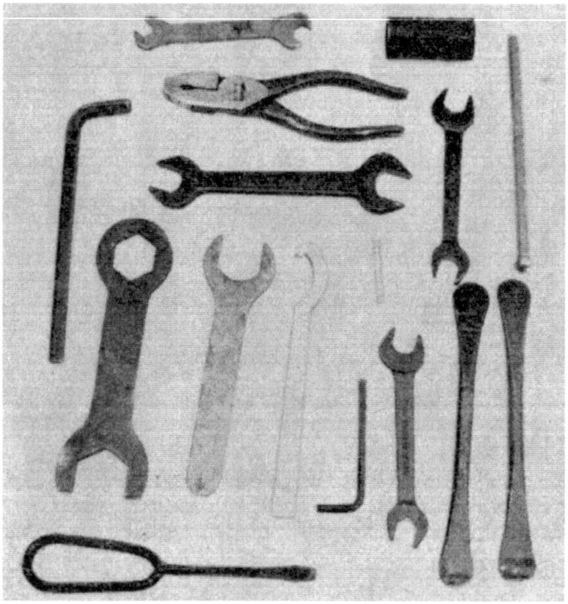

PERCENTAGE VOTE

After sending in their reports, readers were asked to complete a questionnaire in which they answered specific questions according to the formula good, middling or poor.

In calculating these figures we have allowed two points for good and one point for middling. Poor got nothing.

The marks below are given as percentages of the total possible marks.

Acceleration	77	Accessibility	65	Workmanship	70
Flexibility	78	Handling	87	Quality of Finish	73
Smoothness	73	Front Suspension	75	Lighting	57
Starting	88	Rear Suspension	80	Horn	8
Oil Tightness	74	Smoothness of Controls	74	Other Electrics	52
Reliability	74	Riding Position	83	Tool Kit	60
Clutch	74	Brakes	61	Spares from Manufacturer	44
Gear Box	83	Mudguarding	82	Spares from Dealer	43
Delivery Tune	80				

OVERALL MARK: 72 per cent.
GOOD BUY? 81 per cent said yes.
WOULD YOU BUY ANOTHER AJS OR MATCHLESS? 75 per cent said yes.

MATCHLESS TEST

Continued from page 133

light switch with a dipswitch and horn button conveniently placed on the left handlebar. A " kill-button " is placed on the right bar where it can't be confused with the horn.

For all normal purposes, handling was excellent, but I would have liked to see a steering damper. They are useful for solo work and under certain conditions almost invaluable for an outfit. The big dual seat was very comfortable and the riding position with the slightly upswept bars first class.

The electrics are immensely powerful, thanks to the 12-volt crank-shaft-mounted alternator and two six-volt batteries mounted in series. Headlight diameter is seven inches and the unit holds a 50w/40w bulb. The rear light is combined with the stoplight and an integral reflector.

To sum up, it's a pity that A.M.C. didn't offer the G.15 for road test coupled to a sports chair. As a solo it's a good honest performer with sensational acceleration and tremendous top-gear flexibility. As a sidecar bike it will certainly take a lot of beating.

Superbox, Son of Matchbox

One cannot apply the same set of standards to a competition machine that one applies to a street scrambler. In some areas, one is tougher. In other areas, one may be more forgiving. Thus, to clear up any misunderstanding, it should be known at the outset that the Matchless G85CS 500cc single is "strictly for serious." Otherwise, at $1,400, it would have to be considered the most expensive play bike in production . . . and one of the worst. Play bikes are for carefree fun, and a good one doesn't require much fiddling. By way of contrast, the G85 is somewhat of a thoroughbred and needs a lot of combing through to keep her happy.

For perspective, we should harken back to the previous Matchless dirt single, the G80CS, which, for the dashing (and not too particular) amateur-go-a-racing, was ready to romp off-the-floor. Take off the lights and away you'd go at the drop of a smokebomb. Or, if you were in Europe, it was at the twang of the starting gate. We know one American chap who used his G80CS for 6,000 miles of reliable continental touring before he stripped for a summer of international motocross, where he acquitted himself with honor if not with laurel. His only complaint about the bike was that it was "awfully heavy" (360 odd pounds stripped) and he theorized that it had acquired this rather anvil-like quality because the makers had a habit of adding three more pounds of metal to whatever part of the frame broke, until finally it stopped breaking.

All this brings us back to the new 500, which weighs a relatively feather-like 318 pounds with about half a tank of gas. In other words, Matchless, in the face of the successful onslaught of lightweight 500cc class machinery, has decided to fight fire with fire.

It takes a lot of slenderizing to bring a long-stroke single like the Matchless down to such *svelte* dimensions and still have a structure that resists the destructive effects of dirt racing. To do so, the makers took what has become the standard route these days by beginning with a frame of lightweight chrome-moly. Of conventional design, this frame is an improvement over its predecessor in that the main unit, consisting of double down tubes and cradle and single top tube, is of all-weld construction rather than being bolted in several spots as was that of the G80CS. The robust-appearing rear section is pieced together in like fashion so there should be no question that the set-up offers a great deal of rigidity.

Also lighter are the new Matchless forks, which, unfortunately, are springy in a way they were never intended to be — laterally. While this will not be noticeable on improved tracks, it will be a problem in rough scrambles, where sideways tweaking forces are in considerable abun-

MATCHLESS G85CS SCRAMBLER

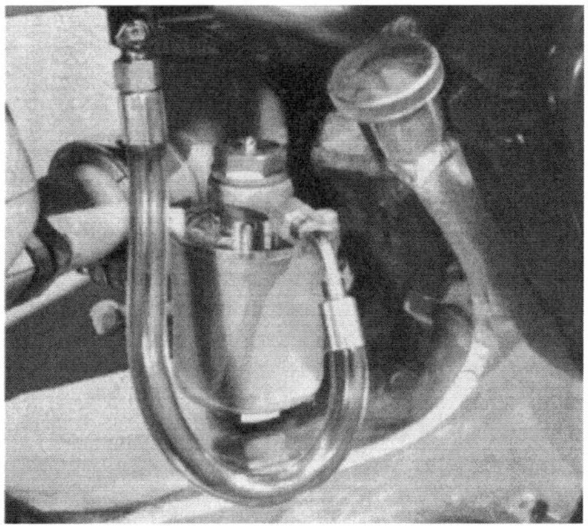

dance. A hefty U-brace is recommended to remedy this flexing.

The weight saving process has been extended to the fenders, where light alloy is used instead of steel; the air box, which is plastic; the handsomely styled polished aluminum oil tank; and the engine, where primary case covers are of much lighter construction than those of the G80 engine. While the lightness of these casings may cause a bit of worry, it should be noted that any mishap capable of punching levers or rocks through them would most likely do the same if they were of normal thickness. This chaincase, by the way, is quite oiltight, thanks to a clamp band seal.

As there is not the slightest pretense that the G85CS is for anything else but competition, the makers have also been able to cut weight by eliminating the flywheel alternator and battery that were on the G80. The reader may note that the rear brake is reminiscent of a road racer's; small wonder, for the brake is none other than from the famous AJS7R, lightweight magnesium hub, shoes and all! Combine that with the turned down G80CS front brake and there is more stopping power than this scrambler will ever need.

In the performance department, we found that the G85 is very competitive. It comes out of the hole fast and will hit 71 mph after only 8.9 seconds and 1/8 mile. Some of this increased acceleration is due to the machine's lighter weight, but most of it comes from minor engine modification.

The power unit is basically G80, and, in fact, has been available as an option to G80 owners. The only differences are that the G85 has a 12:1 compression ratio instead of 8.7:1, and a 1-3/8-inch Amal GP carburetor is mounted instead of the 1-3/16-inch Monobloc. Cam timing, valve size and the longish stroke-to-bore ratio remain the same.

The raised compression — which one would suspect to be rather high for scrambles — proved to be no problem, save to one staffer who got "bit" by all 38 degrees of advance while trying to start the beastie. No harm done, both the limp and the hurt feelings will pass with time. Understandably, pinging occurs at low rpm under full load, but disappears quickly as the engine reaches middle range. As with many a single, it took a bit of time to find the correct procedure for starting, which was to flood the float chamber completely, grab three-quarter throttle and heave ho. Ah yes, don't forget to retard the spark.

The GP carb was perfect for high-speed running, but we question its use for anything else but TT scrambles, where speeds rarely drop below 35 miles per hour. In the rough — through extremely tight turns, up twisting climbs or in high-gear, low rpm sand sections — the GP carb will prove a nuisance, for it loads up and power drops off until the clutch is slipped and the revs return. For riders who don't want the responsibility of keeping this instrument buzzing and "cleaned out," we would suggest a switch back to the prosaic but tractable Monobloc.

The G85 is delivered in European motocross rigging, with thin 21-inch front tire and straight, bridged bars. This will please those riders hailing from the loamier states who stand up a lot when they ride, assuming they brace the forks as suggested. In deference to the rearward crowd, who like to sit back with a set of Cooper bends and steer their machines across the sand like motorboats, we also tried the G85 with Western bars, and 4.00x19 tire up front. We found this much to our liking, as the bike is a beautiful slider, steering exactly with very little twitch or yaw, even when proceeding sideways over small ruts in a high-gear slide. No matter how the bike is set up, however, one should not expect a lightweight; even with its weight down to 318 pounds ready to go, it still feels heavy, although not uncomfortably so.

About halfway through the first day's testing, a small crack appeared in the rear fender around the front fixing bolt. Vibration. Then, after a half hour of fire road travel, the nut for the swing arm through-bolt fell off. This was followed by other bits and pieces in the next few days, which leads us to believe that a thorough safety wiring job would be good insurance before undertaking even the shortest race.

One should expect special touches which show a certain cognizance of the problems encountered in racing, and they are present in the G85. For instance, the front brake operating lever is mounted on the top half of the backing plate and therefore is given some protection by the fork leg. While Matchless' separate gearbox unit is "old hat," at least the makers had the grace to put the adjusters in easy reach just above the primary chaincase; we have absolutely no complaints about either the smoothness of the changes or the choice of ratios. While the front fender is rather short, an aluminum mud stopper is mounted on the down tubes and will help keep engine fins unclogged.

The foot rests are a picture of strength and simplicity and should work out well in motocross. However, AMA riders, depending on how their district competition committee feels, may have to switch to the folding variety.

It is hard to make an overall pronouncement on a motorcycle like the G85CS, because the needs of the American racing crowd are so diverse. We feel, however, that in as-delivered form, it is more ready for closed-circuit rough scrambles than for any other type of competition. Both desert riding and TT scrambles will require changing the front wheel, not to mention a host of other modifications unique to these sports, which grow more specialized each day.

The bike has much potential but, no matter the sport, there is need of much refinement by the owner-rider before he may circulate with great aplomb. ■

MATCHLESS G85CS SCRAMBLER

SPECIFICATIONS

List Price	$1,400
Suspension, front	telescopic
Suspension, rear	swing arm
Tire, front	3.25-21 (4.00-19)
Tire, rear	4.00-18
Brake, front	7.0 x .88
Brake, rear	8.0 x 1.2
Total brake swept area, sq.-in.	50.5
Brake loading (test weight/swept area) lb/sq-in.	9.4
Engine type	single-cyl., 4-stroke
Bore and stroke (inches-millimeters)	3.39 x 3.36, 87 x 86
Displacement (inches³-centimeters³)	30.4,
Compression ratio	12:1
Carburetion	1⅜″ Amal GP
Ignition	Lucas magneto
Bhp @ rpm	42
Oil system	dry sump
Oil capacity, pts.	12
Fuel capacity, gal.	1.7
Starting system	kick, folding crank
Lighting system	none
Air filtration	paper element
Clutch	multi-disc, wet-plate
Primary drive	single-row chain
Final drive	single-row chain
Gear ratios, overall:1	
5th	none
4th	7.4
3rd	8.5
2nd	11.9
1st	18.0
Wheelbase	57.0
Seat height	32.6
Seat width	9.7
Foot-peg height	13.0
Ground clearance	8.6
Curb weight (w/half-tank fuel)	318
Test weight (fuel and rider)	478

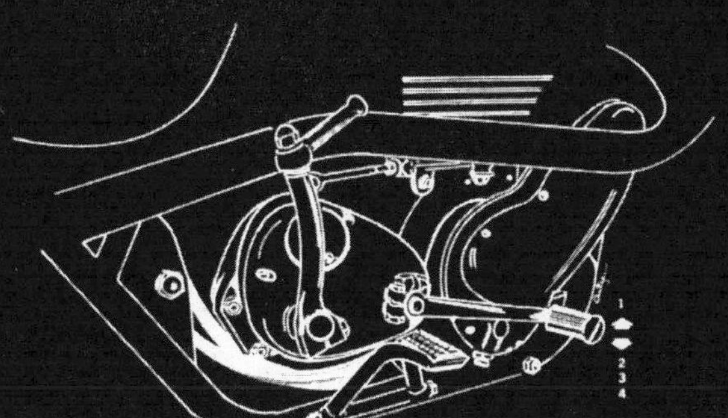

PERFORMANCE

Top speed	73
Maximum speed in gears (@ 7000 rpm)	
5th	none
4th	73
3rd	64
2nd	45
1st	29
Speedometer error	
30 mph indicated, actually	No speedo
50	
70	
Acceleration, zero to —	
30 mph, sec.	2.8
40	3.8
50	5.0
60	5.9
70	8.5
Standing ¼-mile, sec.	16.2
terminal speed	73
Standing ⅛-mile	8.9
terminal speed	71

ACCELERATION AND ENGINE / ROAD SPEED

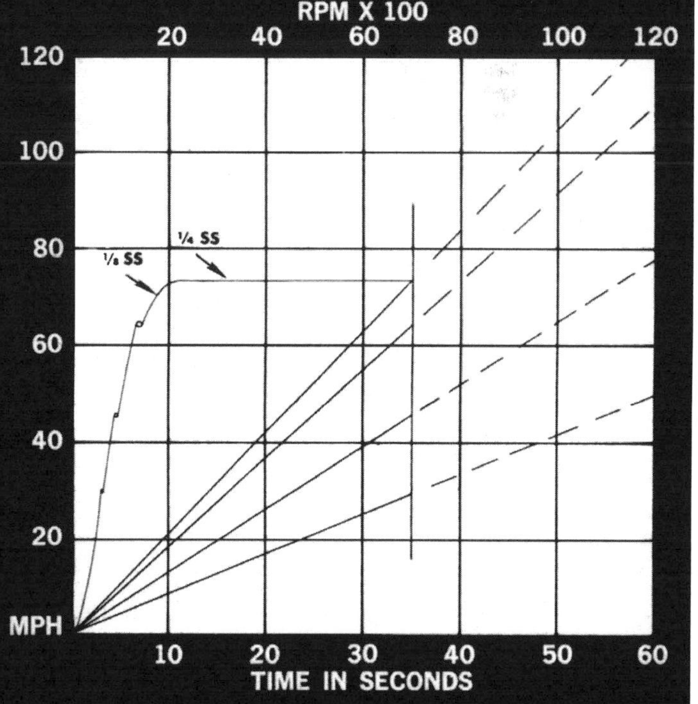

ROAD TESTS OF NEW MODELS

This picture shows the sturdy lines of the AJS. The raised handlebar may be specified at no additional cost. The luggage grid, of course, is an optional extra

646 cc AJS 31 CSR and Watsonian Monza Sidecar

LAST summer, we borrowed a six-fifty Matchless double-adult sidecar outfit from George Clarke Motors of Brixton, later returning it with a posy of congratulations on a splendid chair job. "What this bike really needs for fun," we said, "is a light, sports sidecar." "Keep in touch, then," replied Henry Smallbone, manager of the sidecar department. "I'm thinking about an AJS 31 CSR with Watsonian Monza chair."

The thinking became reality and we've given the outfit a 1,700-mile towsing, the sort of brisk life any keen enthusiast wants to get out of a sporty machine.

MIRA-timed performance figures tell a part of the story. Mean maximum speed was 75 mph; the standing quarter-mile was covered in 18.6s; petrol consumption was 60 mpg at a steady 50 mph; braking distance from 30 mph was 35ft 6in.

When you bear in mind the Ajay outfit is offered at an all-in price of £369, its performance-per-£ value is extraordinarily good.

The machine is adapted for sidecar use by fitting front-fork springs of 50 lb rate, instead of the standard 36½ lb; sidecar-type fork yokes to reduce trail from 3½ to 1¼in; 150 lb-rated Girling telescopic legs at the rear instead of the usual 110 lb solo units; and a 22-tooth engine sprocket (standard is 23-tooth).

Alignment could not be faulted. In traffic or on the open road the outfit handled superbly. There was exactly the right feeling of steering tautness, without any pull to the left, and perfect response to throttle work and braking through corners.

However, when returning from the Dragon Rally with camping gear and a pillionist as well as a sidecar passenger, the back end of the bike felt lively as we drove round those high-speed bends down to Capel Curig.

It could be that fitting even stronger rear springs would pay off; adjusting the 150 lb Girlings to the maximum-load position made an improvement, but did not cure the slight wallowing entirely when the outfit was really heavily laden.

All-round performance proved very good indeed. Gear-ratio spacings matched the power curve admirably: there was no need to fuss around in the rev band for the best output. The outfit could, and did, cruise for mile upon mile with the speedometer needle registering 75 mph—68 genuine.

ON TAP

There always seemed to be ample power in hand to stay with the fastest vehicles encountered, except solos enjoying their advantage of small bulk.

This despite the fact that top gear was much too high. Its 4.71-to-1 ratio meant that the test model proved faster in third than top. And unless conditions were favourable it was too high to be regarded as a motorway overdrive.

First gear also was too uppity. A really fast getaway from rest demanded conscious co-ordination of twistgrip and clutch lever.

Happily, the clutch nobly withstood all demands made on it. After a series of standing quarter-mile bids at MIRA, the clutch took up the drive and disengaged just as sweetly as it did initially.

Perhaps because of that over-high top gear, the outfit put up some outstandingly good constant-speed petrol-consumption figures.

SURPRISE

Naturally, though, hard driving on the open road with generous use of the lower gear ratios increased the consumption, and around 45 mpg was usual.

Reserve capacity guaranteed some five miles, a rather small allowance if petrol stations were widely spaced; and it was disappointing that the tank could not be refilled with a round four gallons even from apparently bone dry.

The braking department was beyond criticism. In fact, the 35½ft test figure was a surprise, so smoothly did the brakes do their work. Furthermore, the outfit remained easy to control, even under the most strenuous deceleration.

The engine turned out to be a sulky starter. When warm, more than one swing on the kickstarter was usually needed. Cold, it took five and more prods before firing.

A well-flooded carburettor, the air lever closed and half retard on the ignition lever offered the best prospects for a cold start.

Because the Ajay is equipped with magneto ignition, its lighting system enjoys full benefit from the Lucas 12-volt, 100-watt alternator.

Lights, indeed, were somewhat better than average, but

Below: John and Eileen Ebbrell out for a spin through the lanes of Essex. The six-fifty AJS-Monza makes a first-rate touring outfit. Right: The heat shield on the oil tank protects the rider's leg. Far right: Access to the two six-volt batteries is restricted by the sidecar connection. The space alongside the battery compartment is for the well-above average quality tool kit

the horn could have been more powerful to advantage.

Driver-comfort is excellent on all counts. The broad dualseat might have been expressly designed for sidecar men. It has plenty of room, besides giving good support to the thighs.

The footrests, too, are nicely placed for an average-size driver, though the folding starter pedal could foul the right leg in a rather annoying way. The wide handlebar gives ample leverage.

A lower-level, standard bar can be supplied if preferred.

Although the engine was not notably smooth, its vibrations are mild and more of the thumpety-thump variety reminiscent of a big single.

Mechanical noise was limited to a patter from the valve gear. The exhaust combined a sporty tone with a pleasing lack of decibels.

Maintenance proved to be straightforward if not invariably speedy. It is necessary to take off the fuel tank to adjust the rockers. The rear wheel detaches easily enough, but the three fixing bolts have to be withdrawn and the speedometer drive must be disconnected.

Above: Beefy six-fifty power unit with separate gear box. The engine was exceptionally oil-tight

The Watsonian Monza with the hood fitted. A tonneau is a neat optional extra

The tool kit, incidentally, is of good quality, though it was annoying that the test machine had no C-spanner to adjust the rear springing, or a socket key to fit the rocker covers.

The engine remained totally oil-tight throughout the test. There was a slight weep from the primary chaincase and an oil-pipe union at the tank remained obstinately leaky.

Because of the angle of the oil-tank filler cap, checking the level was awkward. Overall oil consumption worked out at about 400 miles to the pint.

Chromium plating and paintwork on the machine and sidecar body withstood the ravages of winter salt-spray very well—which was more than could be said for the sidecar connections. The exhaust pipes yellowed in the vicinity of the ports.

The speedometer was some 10 per cent inaccurate throughout its range; the mileage recorder did better—some 5 per cent optimistic.

These defects came to light during the test: the rear-brake cam tended to stick in the on position; one of the batteries suddenly refused to hold the charge.

The only involuntary roadside halt occurred when a spot of oil found its way to the contact-breaker points during a prolonged traffic stop.

Finally, the sidecar. This is the "austerity" Monza Standard, normally supplied without tonneau, luggage grid or brake, with a straightforward mudguard instead of the streamlined type fitted to the de luxe model.

The tonneau was much missed; an owner should buy it as an extra.

By sports-sidecar standards, the Monza is adequately comfortable. Springing is well chosen and there's plenty of leg-room even for a tall passenger. The semi-reclining seat affords good support to the small of the back.

Siamezed exhaust pipes and a mechanically quiet engine helped to make passengering a pleasant experience. Vibration was felt only at peak revs.

This AJS 31 CSR-Monza sidecar outfit is a first-class job; its few faults are emphasized only by the excellence of the whole. Offered at George Clarke's keen price, the value must be unrivalled today.

ACCELERATION

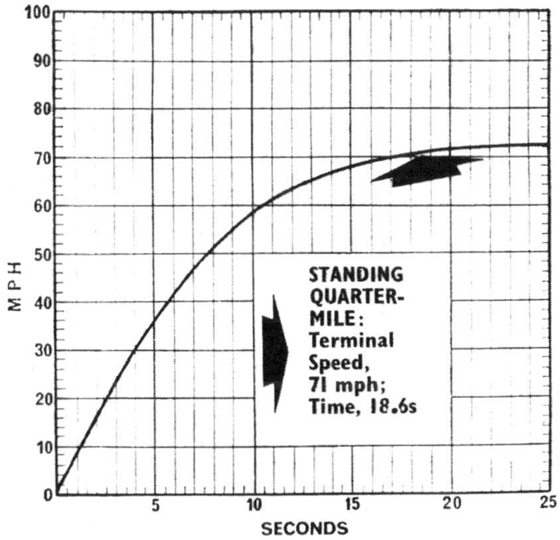

FUEL CONSUMPTION

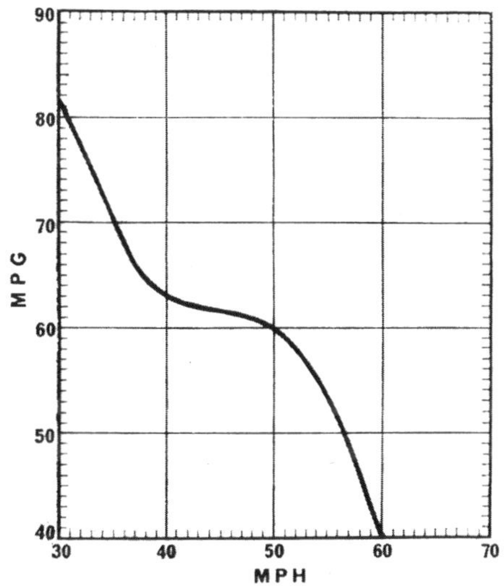

Motor Cycle ROAD TESTS

646 cc AJS 31 CSR and WATSONIAN MONZA SIDECAR

Bottom-, second- and third-gear speeds represent maximum-power revs, 6,200

Below: This Watsonian Monza is a standard model—plain mudguard and no grab rail or tonneau. The sidecar luggage grid is an extra. The siamezed exhaust pipe gives a welcome reduction in noise for the sidecar passenger

SPECIFICATION

ENGINE

Capacity and type: 646 cc (72 x 79mm) overhead-valve parallel twin.

Bearings: One plain and two roller mains; plain big-ends.

Lubrication: Dry sump; capacity, 4½ pints.

Compression ratio: 8.5 to 1.

Carburettor: Amal Monobloc, type 389/92 with air slide. No air filter.

Claimed output: 42 bhp at 6,200 rpm.

TRANSMISSION

Primary: ½ x 0.305in chain in oilbath case.

Secondary: ⅝ x ⅜in chain.

Clutch: Multi-plate.

Gear ratios: 12.04, 8.01, 5.74 and 4.71 to 1.

Engine rpm at 30 mph in top gear, 1,830.

ELECTRICAL EQUIPMENT

Ignition: Lucas type K2F magneto with manual advance and retard.

Charging: Lucas 100-watt, 12-volt RM19 alternator to two Lucas six-volt, eight-amp-hour batteries through rectifier.

Headlamp: Lucas 7in-diameter with 50/40-watt main bulb.

FUEL CAPACITY: 3⅞ gallons.

TYRES: Dunlop 3.50 × 18in rear; 3.25 × 18in ribbed front.

BRAKES: 8in-dia front; 7in-dia rear; both with finger adjusters.

SUSPENSION: Norton telescopic front fork with two-way hydraulic damping. Pivoted rear fork controlled by 150 lb Girling spring-and-hydraulic units; three-position adjustment for load.

DIMENSIONS: Motor cycle.—wheelbase, 56in; ground clearance 5½in; seat height, 29½in. Outfit.—overall length, 90in; width, 57½in; track, 43in. All unladen.

WEIGHT OF OUTFIT: 616 lb, including one gallon of petrol.

SIDECAR

Model: Watsonian Monza Standard.

Chassis: Watsonian Super Silk tubular loop, with brazed lugs and four-point attachment. Bell-crank wheel suspension controlled by Armstrong spring-and-hydraulic unit. Disc wheel with 4.00 × 10in tyre. No brake. Ground clearance, 7in.

Body: Moulded in reinforced glass-fibre. Perspex screen; collapsible hood secured by press studs. Luggage compartment behind seat.

Dimensions: Overall length, 83in; seat back to nose, 56in; seat squab to hood, 29in; back rest, 24 × 17in; seat squab, 17 × 13in; luggage compartment (irregular shape), 28 × 17 × 10in.

Makers: Glass Fibre (Birmingham), Ltd, Albion Road, Birmingham, 11.

PRICE: £369, including British purchase tax, as supplied by George Clarke Motors, Ltd, 1 to 13, Morrish Road, Brixton, London, SW2.

EXTRAS: Sidecar luggage grid, £1 5s; tonneau, £1 5s.

ROAD TAX: £8 a year; £2 19s for four months.

PERFORMANCE

(Obtained by "Motor Cycle" staff at the Motor Industry Research Association's proving ground, Lindley, Leics.)

MEAN MAXIMUM AND HIGHEST ONE-WAY SPEEDS: 75 mph (13-stone rider wearing two-piece oversuit: 8½-stone sidecar passenger; no wind).

BRAKING: From 30 mph to rest on dry tarmac, 35ft 6in.

TURNING CIRCLE: left, 14ft 8in; right, 23ft 2in, between kerbs.

MINIMUM NON-SNATCH SPEED: 20 mph in top gear.

WEIGHT PER CC: 0.95 lb.

RELIABLE is the word which sticks resolutely in my mind after reading through the comments of our 50-odd reporters. Time and time again this word has projected itself from the page; in every case it is allied with a touching story of faithful service that has certainly not been equalled in any previous Riders' Report.

"These bangers, with a design heritage dating back to well before the war, are old-fashioned by today's high-revving standards. But they represent a stage in British motor-cycle development and it will be a very sad day when they are no longer produced in their present form." These sentiments, from Charles Mitchell of Bristol, convey the feeling of genuine affection in which all owners seem to hold their AMC singles.

I must confess that I am more than a little touched in this respect. After owning two of these machines—my second was an ancient 1949 five-hundred—I couldn't help shedding a little tear at the thought that these eminently pleasant, though of late rather unsuccessful, motor cycles are now reaching the end of their production life.

This is confirmed by 20-year-old Tom Waterer of Surrey: "An excellent machine—but made for a now virtually dead market."

It is hard to say why this is so. You will see from this report that the AJS and Matchless big singles have everything to recommend them—except, perhaps, out-and-out go-ability of the modern kind.

The reporters are all experienced motor cyclists, with about seven years on wheels. Average age is 25; between them they have covered over 700,000 miles on the bikes; that works out at about 15,000 miles each.

They are a special brand of enthusiast. They stick to the old bikes because of the utter reliability and ease of maintenance. These AMC jobs are made for riding—and they don't object too strongly if they are a mite neglected.

Many of our readers certainly don't stick at owning one. Some have a fleet, in varying stages of repair. Perhaps they use the older specimens for spares (see the paragraph on service). Adrian Baldwin, a 19-year-old Bristol student, has owned four and still prides himself on his stable of three.

As he says: "The AMC singles are the most underrated motor cycles on the market. It is a wicked shame that they are not more popular, for here are excellent all-round machines—not soggy big bangers as most people think. They are all good motor cycles should be—fast, economical, reliable, safe and predictable.

Perhaps they are too predictable for modern enthusiasts? There is so little discrepancy in accounts—and remember we're dealing with bikes up to ten years old—that it appears to be an unexciting business to own one. You know what's going to happen and you get none of the surprises which may come your way with more modern designs.

AJS and MATCHLESS BIG SINGLES–1956 to 1966

RIDERS' REPORT number seventeen

COLLATED BY MIKE EVANS

One of the earlier models dealt with in this report—the 1958 Matchless three-fifty G3. Unlike the previous years' models, this machine was equipped with alternator and coil ignition

Performance

REPORTERS' average maximum speeds work out to 79 mph for the three-fifties and 84 mph for the five-hundreds. By modern standards this is quite paltry—and no doubt accounts to a large degree for the failing popularity of these 16s and 18s, and G3s and G80s.

However, as Bob Todd, 20, points out, the engine has a very wide power band. Whether revving hard through the gears or plodding along at

Below: This cutaway drawing of the 498 cc engine shows it in its 1958 guise—the AC generator installation can be clearly seen. A feature of these extremely simple power units is the hairpin valve springs, also obvious from the drawing. Right: Peter Fraser samples our 1962 road-test 498 cc Ajay

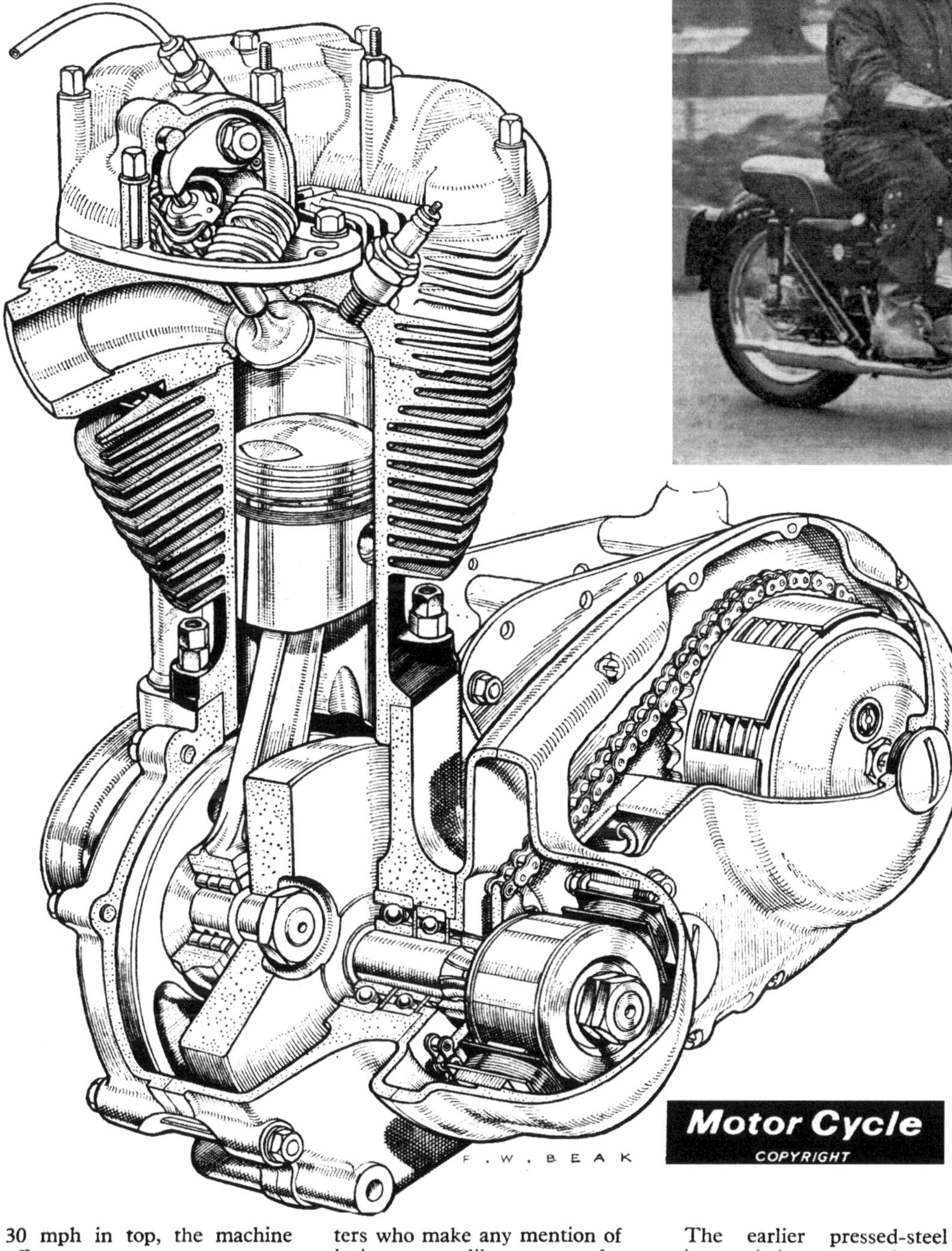

30 mph in top, the machine offers no protest.

He's a good judge since he regularly competes in trials and scrambles—though not on the same bike.

Maximum cruising speed appears to be about 70 mph on the smaller models, with another 5 mph or so available from the bigger bangers.

The characteristic thump of a big single is obvious to the rider, particularly so at slow speeds; only after about 60 mph does vibration as such become noticeable.

According to the few reporters who make any mention of it, it gets steadily worse as the speed builds up. On the other hand, many don't consider vibration a point worth commenting on.

Acceleration, of course, is nothing to write to the National Sprint Association about. Yet it is adequate—and surprises many riders of more sporty machinery.

Oil-tightness? A few moans here. Says Bob Todd: " Oil finds its way from the pushrod tubes, the alloy chaincase and, to a small extent, from the rocker box."

The earlier pressed-steel primary chaincase was, in the opinion of most, an abortion. Tom Waterer tells us that it can be sealed successfully by using a sandwich of outer cover, rubber band and inner cover. The metal band, he says, can be left off.

Starting

OH YES! For those as 'as got the knack, easy. For those as 'asn't, Pygmalion difficult. More than anything else, it's a question of knack. This goes for most single-cylinder bigsters, and the AMC versions are no different.

American reader Jimmy Christodoulou of San Diego wrote to complain of starting on his 1966 G80 Matchless. So I gave him the old rigmarole of on with petrol, lightly flood, ease over compression with valve lifter and then one almighty swinging kick.

Back came his reply the following week: " My Matchless has kicked off just perfectly."

Just shows, doesn't it? Really, these bikes are simplicity itself to start. They aren't even temperamental. As long as you use the right procedure she'll start. If you don't you can prod for hours!

Handling

THESE models have always been regarded as good handlers. Latest developments—the Norton fork, for instance—will have helped, if anything.

Moreover, since the introduction of Girling suspension units in the mid-fifties, there have been very few complaints about the rear end.

The earlier, " jam-pot " AMC rear suspension units were not so good.

There are a few conflicting reports, nonetheless. While Peter Jopson can " scrape the stand with no fears since it handles like a Norton," Dutch enthusiast Marcelis van Dongen, age 20, reports that handling is good—but not excellent " because in fast corners it

Technicalities

BOTH the three-fifty and five-hundred models have essentially simple overhead-valve, single-cylinder engines. The capacity of the smaller unit is 348 cc (74 x 81mm) and that of the larger, 498 cc (82.5 x 93mm).

For 1962 the three-fifty engines were redesigned with the shorter, 81mm stroke; previously the capacity had been 347 cc, with a bore and stroke of 69 x 93mm.

Basically, however, both the five-hundred and three-fifty models have retained a similar specification over the ten years covered by this report.

The major changes in this time have been the introduction of the duplex frame for 1961 and the change from the old-fashioned magneto-cum-dynamo ignition and lighting system to ac generator and coil ignition in 1958.

While the electrical modifications certainly made for a cleaner engine unit, it is apparent from readers' opinions that the old system is still preferred.

Dry-sump lubrication with an external four-pint oil tank follows orthodox practice.

Compression ratio of the later models is 8.5 to 1 (348 cc) and 7.3 to 1 (498 cc). The crankshaft is supported in plain and ball bearings, while a roller big-end bearing is employed.

The AMC four-speed gear box is mounted separately from the engine; drive to the gear box is by a single-row chain running in an oil bath. Chain tension is adjusted by moving the gear box longitudinally in its mounting brackets.

Specimen gear ratios, as quoted in our 1962 road tests, are 14.85, 9.85, 7.08 and 5.8 to 1 for the three-fifty, and 12.86, 8.53, 6.13 and 5.02 to 1 for the five-hundred.

Suspension is by Girling spring-and-hydraulic units controlling the pivoted fork at the rear, and by AMC Teledraulic front fork. On the latest models this has been superseded by the Norton Roadholder fork.

Models of both capacities are identical in other respects—although the three-fifty version has 3.25 x 19in tyres while its bigger sister sports a larger, 3.50in-section rear cover.

Throughout the long history of these singles, they have retained the same basic model designations.

In the AJS range the three-fifty is the Model 16 and the five-hundred the Model 18. Equivalents in the identical Matchless range are the G3 and G80.

Manufacturers are Matchless Motor Cycles, Ltd, Plumstead Road, London, SE18.

waggles its tail, often very much."

Final word to Ian Hamilton of East Kilbride: "Handling and steering are fine."

Braking

"**AS** near perfection as possible," reports enthusiastic Steve Bessant. Perhaps he is a little *too* enthusiastic—yet there are very few complaints about the brakes.

No need for detail here; there is little doubt that the brakes are to be considered excellent.

Consumption

NO surprises here. A big single is normally economical. Peter Jopson, a decorator from Maidstone, averages 60 mpg on his 1966 five-hundred.

A three-fifty owner, 22-year-old Anthony Bennett of Westgate-on-Sea, says that he frequently gets 80 to 85 mpg on long runs. If the speed is kept down to 30 or 35 mph, the bike will return over 100 mpg.

Averages? For the five-hundred, 61 mpg; for the three-fifty, 72 mpg.

Transmission

"FIRST class apart from one small point—the gear lever return spring lasts only about six months. Fitting a replacement is not a difficult or expensive job but the spring usually breaks at a most inconvenient time."

Roger Bradshaw's experience is typical. Most reporters complain of this, although Roger does seem to suffer more frequently than most.

Otherwise, as he says, everything is okay. The gear box is excellent and the clutch is trouble-free.

Electrics

AS usual, this is one of the most difficult categories to collate. Words like disgraceful are bandied gaily around, while other readers enthuse in no uncertain terms.

For the only clear picture it is necessary to turn to the table.

You will see that lighting gets an average mark of 60 per cent, while the electrics in general are only slightly better at 65 per cent. The horn, as usual, is beneath contempt.

It appears, therefore, that reporters are barely satisfied with the performance and reliability of the electrical systems.

What does emerge clearly, however, is the indisputable preference for the old set up of dynamo and magneto, superseded in 1958 by the "modern" coil ignition.

Detail Finish

"WORKMANSHIP is in keeping with the finish. There seems nothing on the bike that is skimped or can be classed as insubstantial." That view comes from one of our many Australian readers, John Bryant, of Melbourne.

English owners are no less generous with their praise:

"Although the bike is seven years old," writes 19-year-old Tony Trevail, of Perranporth, "the paintwork and chrome are still smart. When it is cleaned up, the bike is very handsome."

Alan Richardson, an engineering student from Beverley, Yorkshire, confirms this view: "The finish is one of the outstanding features.

"The seven-year-old paintwork is still in fine condition—much in contrast to my previous mounts."

Riding Comfort

SINCE these bikes are certainly not cut-and-thrust sportsters, it isn't surprising that they are comfortable.

"My two Matchlesses can be ridden all day without discomfort," reports Sidney Singfield, 18, of Liverpool.

Roger Bradshaw, who is 6ft 4in tall, found the riding position a little too cramped, but solved the problem by turning the original handlebar upside down.

On the other hand, fellow six-foot-fourer Mick Seaby, of Wimbledon, considers the

Above: This power unit, in a 1966 three-fifty, differs only in small details from those of earlier machines. Owners praise it for its accessibility, such as the way in which the clutch cable can be replaced in minutes after removing the small inspection cap on the gear-box cover.

standard position to be quite comfortable.

One of the charms of the "traditional" design is the way in which such things as the riding position can be readily altered to suit an individual rider.

Reliability

"IN 18 years of motor cycling I have owned three AMC singles—two three-fifties and one five-hundred. The 1952 498 cc Matchless gave me 108,000 miles of faithful service in its ten years on the road. My 1957 three-fifty (bought with a genuine 21,000 on the clock in 1962) has covered 55,000 miles, although it is now ready for an overhaul."

That is, perhaps, the most impressive of the many recommendations from readers. It comes from 34-year-old Peter Wallace of Manchester.

Peter's experience is, however, not far above average. Norman Crate, 49, of New Ferry, Cheshire, has clocked 27,000 miles on his 1961 Matchless G3 and he is "still waiting for the performance to drop off and show the need for a decoke."

It seems that these old bangers would give a few pointers to some latter-day twins.

Here are a few more words of praise—you won't find many voices raised in disagreement:

"In the seven years I have owned it, my 1957 G3 has proved thoroughly sound. It is far and away the most reliable vehicle I have possessed." (John Allison, Dudley, Worcs.)

"Both my 1957 three-fifty and 1956 five-hundred are old bangers. The one cost £20 and the other only £1! Yet both run well and reliably." (David Jowers, North Harrow, Middlesex.)

"The bike never let me down in 26,000 miles. The only replacements were one rear chain, one front tyre and a headlight unit." (John Bryant, Australia.)

"These bikes are the most reliable machines made in the world since the war." (Paul Ellis, Wellingborough, Northants.)

'Nuff said? There is little doubt that, on reliability, the AMC singles come out of this riders' report as good as, if not better than, any other machine so far reported on.

Service

AFTER previous AMC Riders' Reports I was expecting broadsides from all directions on the spares situation. Surprisingly enough, however, the picture isn't quite so bad as it has previously appeared.

Perhaps this is because many of the owners of these machines are able to ferret around and get secondhand spares.

The bike has been in production so long that most dealers have reasonable stocks and there are lots of secondhand bits available.

Nevertheless, there is sufficient criticism of the Plumstead spares set-up to warrant a paltry 56 per cent mark. Dealers come out slightly worse, at 54 per cent.

Service at the factory is reported to be efficient. But it is incredibly long-winded, according to Roy Poynting of Salisbury.

"Although I find spares relatively easy to come by, the factory service department does not warrant similar praise.

"In a rash moment I decided to return my crankshaft to Woolwich to have the big-end renewed. It took two months and three letters to get it back. The excuse, as usual, was shortage of fitters. However, the charge was reasonable."

Roy isn't without supporters in his criticism.

On the other hand, Michael Bee of East Dulwich, took his

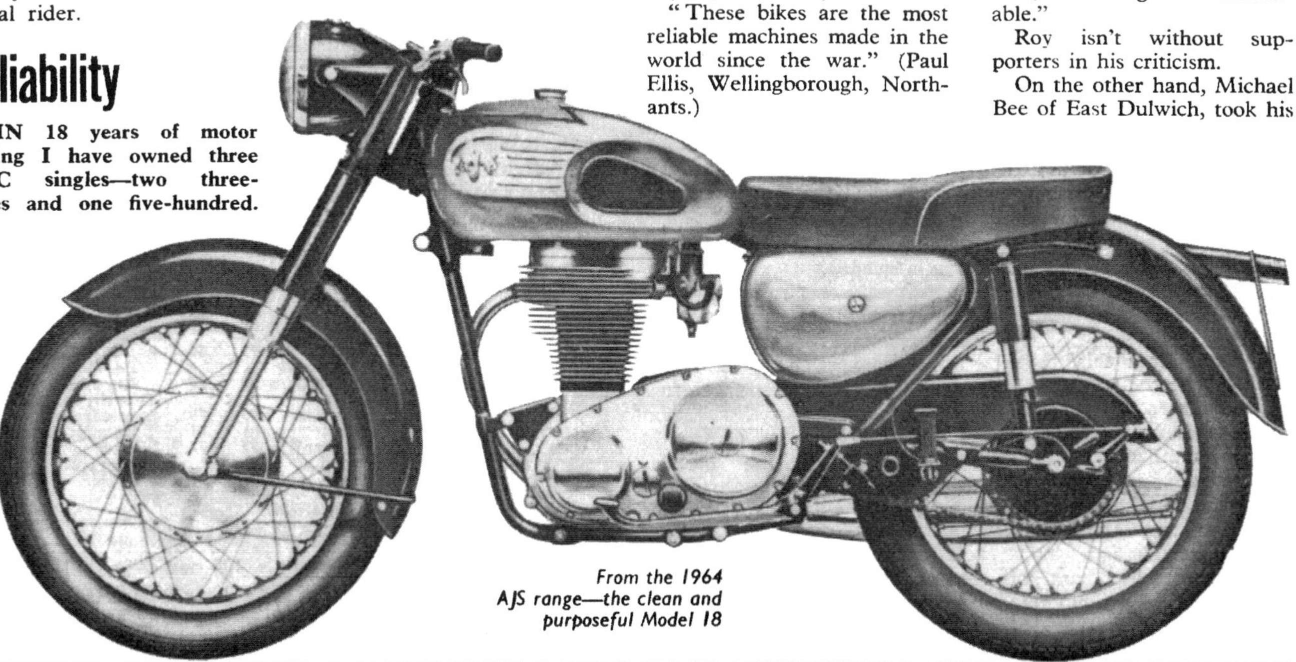

From the 1964 AJS range—the clean and purposeful Model 18

cylinder head to the service department and they fitted a new valve guide and recut the valve seats in 45 minutes. And the charge was only 10s!

Accessibility

"PERHAPS the designer of this machine had actually owned a bike," jokes Tom Waterer. In this he is probably nearer the truth than he suspects.

The basic layout of the AJS and Matchless singles was laid down so many years ago that the designer probably did use one of his creations for his day-to-day transport.

You could bet, then, that if he spent, say, three hours changing a clutch cable while on his way to the drawing board one morning, he jolly soon designed a simpler way of doing it.

There is ample evidence from readers' letters to prove that these models are very accessible.

The engine is particularly easy to work on and there are no snags.

That traditional AMC fuel-tank fixing arrangement (you know, the four bolts that strip their threads, the myriad rubber washers and metal washers and the bits of connecting wire) are the cause of frequent frayed tempers. Yet this is the only point for criticism.

Overall View

YOU want either out-and-out speed or utmost reliability. Anyone with long experience of motor cycling will tell you that you can very seldom have both. You *can* get a combination—a happy medium, if you like—but for utter reliability it would appear that you have to settle for sedate performance.

That is, of course, if you call 70 to 80 mph sedate. There is a good cause, in these days of the 70-mph limit, for having a bike like the AJS or Matchless single.

But apart from this, there will always be the select band of riders who go for reliability and ease of maintenance. That is why they will stick to their "old-fashioned" singles as long as the MoT test station will continue to accept them as roadworthy.

As Andrew Merchant of Upminster says, it is a pity that demand for this type of machine has virtually ceased.

"It offers such a pleasant mode of travel. Perhaps this opinion will condemn it in the eyes of younger riders, but I have come to regard my Model 16 as a permanency in my stable.

"Although no great demands are made on it, it has shown durability and reliability. Long live the AMC singles."

Adds John Allison, a teacher: "Ideal for someone like me who, because of many outside interests, can afford time for only minimum maintenance."

A secondhand Ajay or Matchbox is an extremely good bet. For one thing, it can be picked up fairly cheaply—and models have changed little over the years. Since the engine is easy and cheap to work on, it doesn't really matter if, say, a new big-end or main bearing is needed.

Further, either three-fifty or five-hundred is ideal as the in-between stage for the comparatively inexperienced rider—after the two-fifty and before the six-fifty.

He can learn a lot from stripping down this basic, single-cylinder four-stroke, which isn't half so daunting as the modern vertical twin. What's more, provided he takes care in reassembly, he is almost guaranteed faithful service for many miles.

That is the view of almost all reporters.

The famous heritage of the AMC singles is well illustrated by this 1941 shot of WD three-fifty Matchlesses in a wartime inter-services exercise. Note the girder front fork. In that year the Teledraulic fork made its appearance; it was the first of the modern generation of telescopic front forks

Most of our road tests (and some of the Riders' Reports) deal with new machines. They are bikes in their first flush of infancy.

As a type, the bikes in this report are different. They have had a long and distinguished history—including faithful war service—and many are now reaching the end of their useful life.

The AMC big singles are the last of the many—Spitfires among F111s. But, like the Spitfire, they will leave behind them the sort of memory that dies very hard in the minds of those who knew them.

ROAD TESTS	AJS 498 cc Model 18 March 1962	Matchless 348 cc G3 May 1962	Matchless 348 cc G3/Garrard GP July 1962
HIGHEST ONE-WAY SPEED	87 mph	78 mph	62 mph
MEAN MAXIMUM SPEED ..	82 mph	77 mph	60 mph
STANDING QUARTER-MILE			
(speed)	68 mph	67 mph	52 mph
(time)	18.4s	19.6s	22.8s
FUEL CONSUMPTION			
(at 30 mph)	84 mpg	93 mpg	74 mpg
(at 40 mph)	72 mpg	88 mpg	68 mpg
(at 50 mph)	64 mpg	80 mpg	58 mpg
(at 60 mph)	52 mpg	67 mpg	—
BRAKING (from 30 mph to rest)	29½ft	30ft	36ft

PERCENTAGE VOTE

■ Readers were asked to complete a questionnaire in which they answered specific questions according to the formula good, middling or poor.

In calculating these figures we have allowed two points for good and one for middling. Poor got nothing.

The marks below are given as percentages of the total possible marks.

	350s	500s		350s	500s		350s	500s
Acceleration	66	61	Front Suspension	84	83	Quality of Finish	84	84
Flexibility	91	83	Rear Suspension	79	56	Lighting	60	60
Smoothness	72	89	Smoothness of Controls	86	86	Horn	25	25
Starting	92	83	Riding Position	85	85	Other Electrics	65	65
Oil-tightness	55	50	Brakes	81	81	Tool Kit	60	60
Reliability	94	89	Mudguarding	77	77	Spares from Manufacturer	56	56
Clutch	77	78	Workmanship	87	78	Spares from Dealer	54	54
Gear Box	87	72						
Delivery Tune*	83	83						
Accessibility	73	89						
Handling	78	78						

**Secondhand machines not taken into account.*

OVERALL MARK. 74 (350s), 80 (500s)

GOOD BUY? For both models, 100 per cent said yes.

WOULD YOU BUY ANOTHER AMC MACHINE? 85 per cent of three-fifty owners and 89 per cent of five-hundred owners said yes.

AJS

A history rich in competition and design progress . . .

BY GEOFFREY WOOD

The 1937 AJS Single was available in both 350- and 500-cc sizes. A peppy ohv engine, four-speed footshift gearbox, rigid frame, and girder front fork were standard practice.

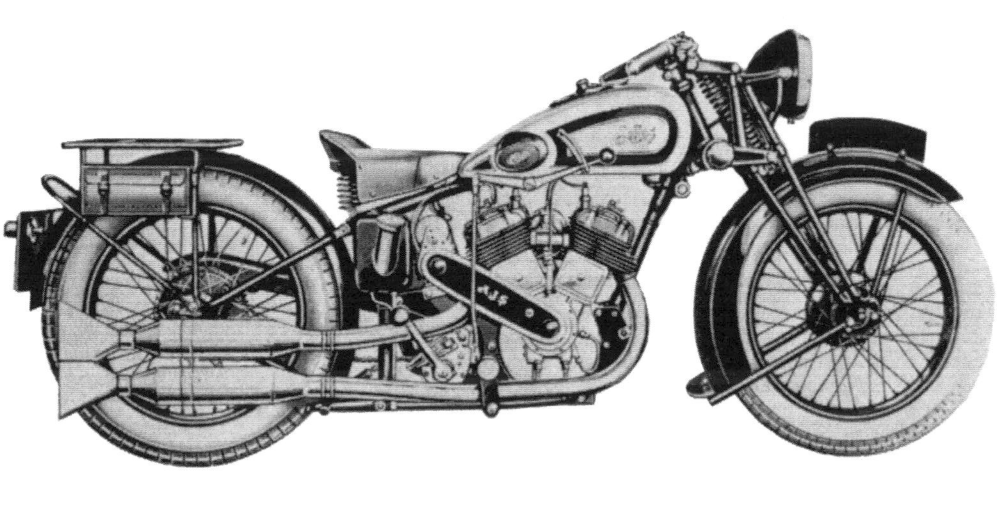

The 1932 AJS model T-2 was a smooth-running side-valve V-Twin in either 800- or 1000-cc sizes. A hand shift three-speed gearbox was used. Reliability was legend.

The 1938 model R7 featured a chain-driven ohc alloy engine and unusual pivoted-fork rear suspension with plunger spring boxes. The R7 was not as fast or reliable as the Velocettes.

IN THE colorful annals of motorcycling, there are a few manufacturers which have been pioneers of design progress as well as creators of legend in competition. One such company is AJS—a British marque that possesses a great legacy of success in trials, motocross and road racing, and which has exerted a driving force for better motorcycle design during the first half of the 20th Century.

The story of this progressive concern began in 1897 when an engineering blacksmith named Joseph Stevens gathered his five sons about him to design and produce an internal combustion engine. The new gasoline burning device was soon put into production, and within a short time it was powering such early day motorbikes as the Werner and the Wolf.

This engine proved to be a reliable little chugger, and the reputation of the Stevens family was established. Thus encouraged, the business expanded and production of motorcycle frames and other proprietary parts was initiated. By 1900, the 2.25- and 3.25-bhp single-cylinder engines were available in both air and water cooled versions, and the little company was on a firm financial base.

By 1909, the Stevens family was eager to try its own hand at producing a complete motorcycle, so the plans were drawn up and production was started. Named the AJS after the A.J. Stevens name, the new motorbike featured a 2.75-bhp single-cylinder engine mounted in a bicycle frame; belt drive was used. The model proved to be as reliable as any during those early days.

By 1911, the infant company was producing a genuine motorcycle with a heavier frame, a 298-cc (70 by 77.5 mm) side-valve engine, belt drive, and a three-speed gearbox. This use of a separate gearbox was a most unusual feature, as the majority of manufacturers still were using either a single-speed belt drive, a two-speed rear hub gear, or an expanding-pulley belt drive arrangement. Two of these AJS machines were entered in the 1911 Isle of Man TT, but 15th and 16th places were the best obtained.

In 1913, the factory made a really noteworthy design advance when it produced the 350-cc model with bore and stroke measurements of 74 by 81 mm. The engine was a side-valve unit that produced 2.75 snarling bhp, and this power was transmitted by an all-chain drive. The earlier three-speed gearbox was dropped in favor of a two-speed unit. Another progressive idea was the use of an internal expanding brake on the rear wheel. A caliper type brake still was used on the front wheel. One of these thumpers was entered by the factory in the famous Isle of Man TT, and it garnered a creditable 9th place.

The following year, the 350 was improved by the fitting of a double chain and sprocket arrangement from the engine to the clutch, with a lever and cable sliding one or the other set into engagement. The two chains ran on engine sprockets with an odd number of teeth, and this provided four speeds when used with the hand-shifted two-speed gearbox.

The side-valve engine also was developing greater power, with the remarkable engine speed of 5000 rpm possible on the road. The engine, incidentally, had a cast iron cylinder, and a steel piston with a slim clearance of only 0.001 in. between them. The lubrication system was still rather archaic. A hand pump and the total loss method were used.

The 1914 Junior TT saw five new AJS machines entered, and they all finished–1st, 2nd, 4th, 6th and 29th. Eric Williams was the victor, and he won at the record speed of 45.72 mph with a record lap at 47.57 mph. This magnificent showing greatly enhanced the company's reputation and boosted sales. A move soon was made to larger quarters at Wolverhampton. Then came the Kaiser's war, and the AJS folk were plunged into war contract work.

The war itself taught the Stevens brothers a great deal about metalurgy in general, and the value of aluminum and other light alloys in particular. When peace returned, much of this knowledge was put into practice in motorcycle production. This provided a great leap forward in design progress.

The first post-war TT race, in 1920, provided the marque with another victory in the Junior class. AJS was rather lucky, however, because six of the eight works entries broke down, and winner Cyril Williams pushed his bike in from Keppel Gate after it had ceased to run. Eric Williams did increase the lap record to 51.36 mph, though, which well demonstrated the speed of his Single.

The new AJS actually was a very advanced design for its day, and it can be said to have been the true father of the high performance single-cylinder design. The 350-cc engine had a bore and stroke of 74 by 81 mm, and it featured an overhead valve design. The valves were set at a 90-degree angle to each other, thus providing the now common hemispherical combustion chamber design. The advantages of the hemispherical combustion chamber were better fuel flow characteristics and greater head space for larger valves. Today this is a standard practice for most high performance ohv engines. The head and cylinder were cast iron; a steel piston was used.

The belt drive, by then, was just about dead, and AJS was progressive with its all-chain drive. An interesting feature was the six-speed gear setup (and present-day riders think six-speeds are something new!) which had a three-speed gearbox with the two-speed engine-clutch arrangement. An internal expanding brake was used on the rear, but a caliper brake was still featured at the front.

During 1921, the factory worked the bugs out of their new ohv design, and reliability took a big jump. The number of gear ratios was reduced to only three by doing away with the dual engine-clutch chain setup, and the gear ratios were 4.9:1, 6.0:1, and 9.3:1. A new frame was used in conjunction with 2.25-26 tires. Dry weight was only 190 lb. Maximum speed was 70-75 mph, which was teriffically fast for those days.

In the Junior TT, the marque scored a smashing 1st, 2nd, 3rd, 4th and 5th, with Eric Williams the winner at 52.11 mph. Howard Davies upped the record lap to 55.15 mph, then went on to win the Senior TT at the record speed of 54.49 mph–the only time a 350 ever has won the 500 event!

By then, the reputation and sales of the AJS were at a very high level. Profits from production were turned back into research and additional race results. The 1922 models were again improved, with the famous 350-cc Big Port Single heading the range of machines. This thumper ran on a compression ratio of 5.75:1 and had a huge 2.625-in. diameter exhaust pipe. Internal expanding brakes were used front and rear. The model again took a 1st and 2nd in the Junior TT with Tom Sheard winning at 54.75 mph.

The following year, the racers were modified by use of aluminum alloy pistons. In 1924, a large inlet valve was fitted to better charge the cylinder. The racing fortunes of the marque declined after 1922, as a result of some peculiarly rotten luck, rather than through deficiencies in design and engineering. A few places were gained in the TT races, plus a win or two in Continental grands prix, but for the time the coveted No. 1 was to elude AJS.

The 1939 V-4 was a fierce looking beast with its supercharger and water cooling. Bob Foster sits ahead of Brooklands mufflers. Top speed was 150 mph.

The last of the Porcupines–the 1954 model–this Twin produced 54 bhp at 7500 rpm, good enough for 145 mph. The engine was mounted at 45 degrees. The bike weighed 335 lb.

In the meantime, the production range was going strong with sales climbing to previously unheard of heights. By 1925, the range of machines offered was comprehensive and included models for just about every type of riding. For the budget minded there were 250-, 350- and 500-cc side-valve Singles that offered reliable day-to-day transportation at rock bottom prices. For the more sporting minded there were 350- and 500-cc ohv models that provided spirited performance. For the connoisseur of motorcycles there were the powerful, smooth-running 800- and 1000-cc side-valve V-twin models.

Another 1925 success was the 500-cc Single, built for a bash at the Senior TT. Ridden by Frank Longman into 2nd place, the ohv engine had a bore and stroke of 84 by 90 mm, and weighed only 270 lb. After winning the French Grand Prix, the factory decided to put the model into production for the 1926 season. By then, AJS had won renown throughout Europe for their well designed Singles.

Motorcycle racing in Europe became extremely competitive during the late 1920s. Greater horsepower was required to stay in contention. In response to this, the Wolverhampton firm developed new ohc works racers for the 1927 season. The Single camshaft was driven by chain, and the primitive foot-operated oil pump (total loss) was at last replaced with a double-gear pump and a dry sump system. A new four-speed gearbox also was used, with gear ratios of 4.6:1, 5.58:1, 7.5:1, and 10.29:1 on the 350-cc model, and 4.12:1, 4.99:1, 6.7:1, and 9.19:1 on the 500-cc model. An unusual feature was the foot operation of both brakes. The idea was to leave the right hand free to operate the throttle and gearshift lever.

The new camshaft engines proved reliable, but slow. In the TT, the best the AJS machines could do was a 3rd in the Junior. Jim Guthrie, however, went on to win the 350-cc class in the grands prix of Belgium, Switzerland, Austria, and Germany. Nevertheless, the company decided that the cammer needed more development, so AJS reverted to its pushrod design for the 1928 season—with a fair turn of success in Continental events.

In 1929, the factory reintroduced the ohc engine with small internal improvements, plus a stronger girder fork at the front. Large brakes also were used, with an 8-in. unit on the front and a 9-in. size on the rear. The 350 proved much faster that year with a 2nd place in the Junior TT, plus victories in the Ulster, Austrian, German, and European grands prix.

The 500-cc model generally proved unreliable, but it did set up a new one-hour record at 104.51 mph, plus a 500-cc one-kilometer record of 118.98 mph. The 350-cc model also set a kilometer record at 107.37 mph. Altogether, this marque set over 100 speed/endurance records that year.

During 1930, the standard production range was enhanced by the introduction of a new side-valve V-twin that was aimed at the rider who wished something smooth, powerful, and a bit different. The unusual thing about his new Twin was that the engine was mounted transversely in the frame. The model stayed in production only a few years and was then discarded as being too expensive to produce.

Another model that was introduced in 1930 was the R7 racer. This 350-cc model featured a chain-driven ohc 350-cc engine patterned after the works racer, and it had a rigid frame and girder front fork. The R7 proved to be popular with the private owners, but it never did achieve speeds to match the KTT Velocette or Manx Norton.

The AJS factory also had a bash at the world speed record with a 1000-cc V-twin in 1930, but 130 mph was the best obtained. The engine was a 50-degree V-twin with bore and stroke of 79 by 101 mm. It had chain-driven overhead camshafts. There were two carburetors, with both inlet ports facing rearward and exhaust ports facing forward. Cylinders were of steel and heads of light alloy. The V-twin appeared again in 1933, fitted with a supercharger, but 136 mph was the best speed achieved. The last attempt was made at Tat, Hungary, with 145 mph recorded, but this was 6 mph below world record.

During the early 1930s, the marque scored a few grand prix wins, plus Jim Guthrie's 1930 Lightweight TT victory. As time passed, AJS faded away from serious contention. The main cause was financial, because the economic depression severly damaged the financial standing of the company, By 1931, the concern was on the rocks, and the doors were closed, but, happily, the name and tradition were kept alive when the Matchless Co. purchased AJS and moved it to Woolwich, London.

This move to London, plus the depression, naturally caused a curtailment of AJS racing efforts, and increased emphasis was put on the standard production models. The side-valve and ohv Singles continued to be the most popular models in the range, and their reliability and performance endeared them to owners throughout Europe. The big side-valve V-twin also was continued, and this long wheelbase model offered about as much rider comfort as could be expected in those days of rigid frames and girder front forks.

There was an AJS reentry into international racing in 1934. From then, until 1939, the 350-cc machines followed the general trend of design. The engines remained basically the same chain-driven ohc units, but they gradually included more light alloy in their construction. Aluminum bronze heads were used in 1934, a bi-metal head with a bronze skull in 1935, and aluminum alloy with screwed-in valve seats in 1937.

Aluminum alloys also were used extensively for the cycle parts. Magnesium alloys were employed for gearbox and engine castings. This all helped keep the weight down to a light 270 lb. on the 1935 model, but the weight climbed to over 300 lb. in 1938 when rear suspension was added. The suspension was by an unusual pivoted-fork that attached to plunger spring boxes at the rear of the frame. Another improvement was the huge alloy front hub-brake unit, and nearly all of these modifications were adopted on the R7 model which was produced in limited number for the private owner.

By the middle 1930s, the world was on its way to economic recovery, and the AJS higher echelon decided that the expanding sales justified a more intensive racing campaign. Those also were the days when everyone else had the same idea, and international road racing became terribly competitive. Matt Wright, who was AJS' chief design engineer, thought really big. When his pet project was laid out on the drawing boards it caused even the strongest to gasp.

The basic plot was a 50-degree V-4 with bore and stroke of 50 by 63 mm. The light alloy cylinders and heads were very similar to the R7 design. Overhead cams were driven by a chain in the center of the engine. The valves were controlled by exposed hairpin valve springs, and two carburetors fed the cylinders. Eccentrically mounted rocker arms were used to set valve clearances. The compression ratio was 8:1. The hollow, two-throw crankshaft was supported by five roller and ball bearings. A vertical shaft and bevel gear setup drove the oil pumps and two magnetos.

Two of these Fours were ridden in the 1936 Senior TT by Harold Daniell and George Rowley, but they both retired halfway through the race. The machines were not terribly speedy, and they obviously needed a great deal of development work to make them raceworthy. This Four was put on display at the 1935 Olympia Show. It was to be sold either with or without supercharger; the price was very high, of course. History books make little mention of how many of these super/-sport road models were sold.

During 1937, the Four was under development and was not raced. In 1938, it reappeared with a supercharger mounted in front of the crankcase. The new spring frame also was used, as was a huge 8-in. by 1.75-in. front brake that was housed in a magnesium alloy hub with deep fins. The blown Four proved to be very fast, but overheating and poor steering plagued the model. Little success was obtained.

Meanwhile, the production models were faring exceptionally well. The trend away from side-valve engines to ohv models had been wholeheartedly accepted by the factory, and the 350- and 500-cc Singles were very popular. With Burman four-speed footshift gearboxes the performance was good, even if the rigid frames and girder front forks provided a far from smooth ride. These fine handling Singles were particularly successful in trials competition, and their fame continued to spread the world over.

The final year of international motorsport before the war was 1939; AJS contributed greatly to that classical year of speed. The V-4 model had been converted to water cooling during the winter, and it was ready to do battle with the best. In the Senior TT, the 405-lb. Four was handicapped by the twisty nature of the circuit, plus a need to take on more water, and 11th and 13th places were the best obtained.

Late in the season, in the fast Ulster GP, the great Walter Rusk finally was able to show the speed of the Four, as the 7-mile Clady Straight allowed him lots of room to unleash the blown engine's fury. Walter and Dario Serefini, on a blown Gilera Four, were locked in a terrific battle, and on Lap 3 Rusk pulled clear by 19 sec. and recorded the first ever 100 mph lap. Then came disaster when a front fork link broke, thus ending a colorful era for both motorcycle sport and the AJS V-4.

After the war, supercharging was abolished by the FIM for international racing, and, at any rate, the factories could not have afforded such an expensive luxury. In war-torn England, the most pressing need was to rebuild factories and start production again. To this goal AJS dedicated itself, and a whole new range of machines was introduced.

The basis of the new range was 350- and 500-cc ohv Singles with bore and stroke measurements of 69 by 93 mm, and 82.5 by 93 mm. Both models featured the new Teledraulic front fork, which provided a much smoother ride. Frames still were rigid. These Singles were of an exceptionally sound and rugged design, and riders the world over came to know and love them for their ease of operation and dependability.

These Singles were used by the factory in both trials and the newly founded motocross sport, in which they achieved tremendous success. Hugh Viney, for example, won the coveted trophy in the Scottish Six Days Trial, and other riders kept the marque well to the fore in all types of dirt competition.

The factory also made a return to road racing with a new 500-cc Twin. The Twin was rather unusual, with its engine mounted horizontally in the frame, and spike fins on the head that led to the nickname of "Porcupine." The bore and stroke were 68 by 68.25 mm. The one-piece crankshaft featured two roller bearings, plus a third plain bearing in the middle. Rods of RR-56 alloy were used with plain big-end bearings, and the engine ran "backward."

The drive to the double overhead camshafts was by spur gears. Exhaust valves were sodium filled to facilitate cooling. Oil was circulated at the amazing rate of 45 gph at 7000 rpm. The engine was mounted in a wide duplex cradle frame that featured air-oil shock absorbers on a swinging arm suspension. Front suspension was by the Teledraulic fork with 3 in. of travel. Huge brakes with air scoops were used.

In the 1947 Senior TT, the Porcupine showed a fair turn of speed in the hands of Jock West and Les Graham, but minor teething troubles kept them well down on the finishing list. A few other places were obtained in grand prix events, but additional development obviously was needed to make the Twin a winner.

The following year, the marque intensified its racing campaign with the introduction of the 350-cc 7R model. Produced both as a works and production racer, the 74 by 81-mm Single had the traditional chain drive to a single overhead cam. The "Boy Racer," as it was affectionately called, had a duplex cradle frame, Teledraulic front fork, and huge conical hubs cast from magnesium alloy.

The alloy engine was fitted with a 1.156-in. TT carburetor. A Lucas magneto provided spark. An 8.45:1 compression ratio was used. The engine produced about 30 bhp at 7000 rpm. The four-speed Burman gearbox had ratios of 5.24:1, 5.95:1, 7.07:1, and 10.14:1, which produced a speed of 110 mph in top cog. Other specifications of this sleek new racer were a 56-in. wheelbase, 298-lb. dry weight, 4.75-gal. fuel tank, large megaphone exhaust, a 3.00-21 front tire, and a 3.25-20 rear tire.

In classical racing that year, the team fared well with both the 350- and 500-cc models, but they still lacked the speed of the all-conquering 350 Velocette and 500-cc Norton. Many good places were obtained, and private owners did exceptionally well on the 7R model. In dirt competition, the AJS figured prominently. The great Hugh Viney again won the Scottish Six Days Trial.

The year 1949 generally is regarded as the greatest in the AJS history. The production range was expanded to include a new 500-cc (66 by 72.8 mm) Twin. All the models could be had with the swinging-arm frame that made the AJS one of the most comfortable bikes in the world. The marque also produced special 350- and 500-cc models in trials and motocross trim, and these alloy-engined bikes achieved tremendous eminence in their fields.

In grand prix racing, the works team experienced a fabulous year, although Les Graham did lose the Senior TT when his magneto drive shaft broke, just a few miles from victory while he led by a wide margin. The Porcupine was a very speedy model that year, and on the Continent AJS engaged in some epic struggles with the equally speedy Gilera Fours. Les won the Ulster and Swiss Grands Prix, plus taking a 2nd in the Dutch event, and this gave him enough points to win the coveted 500-cc World Championship. Teammate Bill Doran won the Belgian GP and took a 4th in the Ulster and a 3rd in the Monza classic. This enabled the marque to also win the 500-cc manufacturers' championship. In the Junior class the 350 failed to score any wins, but it did provide honest opposition to the terrifically fast dohc Velocette.

In trials competition, Hugh Viney etched his name indelibly in the history books with his third straight victory in the Scottish Six Days Trial. This remains the lone "hat trick" victory string in the famous trial, and it is perhaps one of the all-time greats in motorcycle competition. In motocross, the AJS recorded another great year, and in America the Singles won many of the most important dirt competitions that were held.

In 1950, the AJS did not fare so well in the Senior class, as the new Norton Featherbed was a little too fast. The team placed well in many races, and Les Graham did win the Swiss GP. In the Junior class the Velo was again too fast for the 7R, but Les did score a popular win in the Swiss meeting on wet roads.

During 1951 and 1952, grand prix racing became more competitive, and AJS responded with many technical improvements to its race ware. The works 7R model had the valve angle reduced from 79 to 70 degrees, and the intake port was steepened 12 degrees. The engine also was mounted farther forward in the frame, the wheelbase was shortened to 55 in., a narrower crankcase fitted, and clip-on handlebars were used. Then the height was lowered by using 3.00-19 front and 3.25-19 rear tires, and the front fork was shortened 1 in. These fork and wheel modifications also were made on the Porcupine. Then in 1952, the engine was raised to 45 degrees, orthodox engine finning was used, and the weight was lowered by mounting the oil sump beneath the engine.

For 1952, the factory also designed a unique new works 350 model with three cams and three valves. Two exhaust valves were used in conjunction with two ports and two exhaust pipes. The bore and stroke also were changed to 75.5 by 78 mm, and these measurements were later used on the production 7R in the late 1950s. The three-cam model fared well on the racetracks, and it also set a one-hour record of 115.66 mph for the 350-cc class late in the year at Montlehery, France.

During the next few years, the line of models continued the same with road, trials, scrambles, and the racing 7R models being offered. Improvements were mainly of small technical nature to the existing sound design, and the sales and reputation of the AJS continued to expand around the world. By 1954, the 7R model was producing 37 bhp at 7200 rpm, and it was by then the most popular mount in the world for 350-cc class road racing.

The year of 1954 also was a memorable one for the marque, as AJS won the Junior TT with a three-cam Single in the hands of Rod Coleman of New Zealand. Both the 350- and 500-cc racers were weird in appearance with their huge low-slung 6.5-gal. alloy fuel tanks that required a fuel pump to get the fuel up to the carburetor. The idea was a bit of streamlining, with a non-stop ride for fuel and a lower center of gravity. The 500 Twin by then weighed 335 lb., was breathing through a pair of 1.125-in. GP carburetors, and developed 54 bhp at 7500 rpm–good enough for 145 mph.

After 1954, the marque lost interest in grand prix racing, and AJS discontinued active participation to spend time on production model development. In 1956, a new 600-cc Twin was produced, along with the traditional 350- and 500-cc Singles in road, trials, and motocross trim. In 1958, a clean looking 250-cc ohv model was added to the range, and, in 1959, the 600 Twin became a powerful 650-cc model.

Meanwhile, the 7R model had been under constant development, and by 1961 it was developing 41.5 bhp at 7800 rpm. With a wide torque range that produced 33 bhp as low as 6000 rpm, the 285-lb. Single delivered terrific acceleration. The compression ratio, by then, had been upped to 12:1, and the Amal GP carburetor had a bore of 1.375-in. The four-speed gearbox had ratios of 4.87:1, 5.36:1, 6.48:1, and 8.67:1, which provided a top speed of 120 mph.

During the early 1960s, the company underwent a great change. The dynamic leadership that had characterized AJS history was lacking. Design progress and sales began to sag. The company continued to make a modest effort in trials and scrambles competition, and Gordon Jackson's magnificent one-point victory in the 1961 Scottish Trial (his only point was for one dab of his foot) helped keep the name alive.

Things were not well, though, and, in 1963, the last of the illustrious 7R models rolled off the production lines. Next came the amalgamation with Norton, and the AJS Twin became a Norton with AJS on the fuel tank. Today the Singles still are produced, but lack of design improvement has hurt their image and competitiveness. Just at the darkest hour, when it appeared as though the doors were going to close again, a company, Manganese Bronze Holdings, came forward, and out of this has arisen a new company called Norton-Villiers that holds promise of great things to come. Already, the name AJS has been restored to racing by a 250-cc two-stroke of Villiers ancestry, and the company has given a hint that more racing activity is to follow.

So this is the story of the AJS–a tale of a make that has a great legacy in both motorcycle sport and production. From the early trend-setting with gearboxes and overhead valve engines to the wonderful trials, motocross, and road racing models, it is a legend that is rich in tradition and engineering achievement. ■

MATCHLESS

Proud Pioneer of British Motorcycling

BY GEOFFREY WOOD

THERE HAVE BEEN many hundreds of motorcycle manufacturers since the turn of the century. Some of these companies have carried on their business in an ostentatious manner—winning all sorts of competitions or producing fabulous models of great classical stature. Others have been more modest, producing well designed bikes that perform reliably and gain many friends, but which fail to make the big headlines.

One of these marques that has been rather quietly successful is Matchless—a British concern whose history is woven ever so closely with the basic history of the motorcycle. It is not that the Matchless has failed to produce many good designs or successes in international competition, for the firm's history is spiced with honorable achievements in both areas. Rather, it is a case of the marque never having managed a string of truly magnificent successes that make headlines the world over. The philosophy of this legendary British concern has been to produce a good, sound motorcycle for the common man. In this, Matchless has succeeded admirably.

The story of this Anglo-Saxon pioneer began in 1878 when H.H. Collier began to produce his Matchless brand of bicycle. Collier senior soon was joined by his two elder sons, Harry and Charles, and these enterprising men tried to fit a Continental-made engine over the front wheel of one of their bicycles. This setup proved not to be the best, so they next tried mounting the engine behind the seat tube. This didn't work very well either, so then they mounted the 2.75-bhp De Dion Bouten single-cylinder powerplant centrally in the frame. These belt-driven and pedal-assisted models proved to be fairly reliable for their day, and both of the Collier boys won a goodly share of the races in the vicinity of London.

The Collier brothers proved to be good inventors, and they had some remarkably progressive ideas for those early days. In 1903, they added a pillion seat to their Single, so that a passenger could be carried and, in 1904, they produced a Tricar which was fitted with a chain drive in 1905. In that same year, the company also produced a powerful V-twin model that had a JAP engine. This twin-cylinder model featured a leading-link front fork and an experimental swinging-fork rear suspension—and all this in 1905.

The Collier brothers were great competitors. In 1905, Harry represented his country in the International Cup Race that was contested in France. In 1906, the two brothers both qualified, with Charlie riding to a 3rd place—the only British machine to finish. The boys were dissatisfied with the rules that favored the Continental bikes—big engines in tiny frames, and a 110-lb. maximum weight limit. So, they joined with others in organizing a race on the Isle of Man for touring or production motorcycles. The Marquis de Mousilly of St. Mars, who was a friend of the Collier family, agreed to donate a trophy, and thus was born the glorious tradition of the "TT."

The first TT race was staged in 1907, and Charlie Collier won at 38.23 mph. The following year Charlie came in 3rd on his Single. Then, brother Harry won the trophy in 1909 at the fantastic speed of 49.0 mph. That same year, Harry set up a new 24-hour record at 32.3 mph to prove how reliable the company's products were. All this publicity helped establish the reputation of the tiny concern.

These early thumpers used a JAP pushrod-operated ohv engine, which returned no less than 94.5 mpg fuel consumption in the race. The frame was rigid, with a leading-link front fork that had a single coil spring, and a single-speed belt drive was used. By 1908, magneto ignition had been adopted, and the marque also produced a beastly racing model with a 120 by 120-mm JAP V-twin engine that punched out 24 bhp.

The Collier brothers continued to race with great success—C.R. winning the 1910 TT, with H.A. in 2nd place. Charlie also lapped Brooklands on a 976-cc JAP-powered model at 80.24 mph, and then he clocked a top speed of 84.92 mph. The following year, the Indian factory sent over some potent models with chain drive and two-speed

The 1932 Model CS was a 500-cc "sloper"—used a three-speed handshift gearbox. Rigid frame and girder fork were standard practice then.

The 1933 Silver Arrow was a luxurious mount for those days, with pivoted-fork rear suspension. Engine was a narrow 26-degree transverse V-twin of 400 cc.

gearboxes, and these models placed 1st, 2nd and 3rd in the Senior TT ahead of Harry. This led to the challenging by the Matchless Co. of Jake De Rosier for a match race at Brooklands to save England's honor, but the American won and the British began to switch to chain drive and two-speed gearboxes. Charlie did salvage some degree of honor when he clocked a sizzling 91.37 mph on his big Twin, which stole the world speed record away from De Rosier.

After those early days of glory, the name slowly faded away from the Isle of Man, with Harry's 3rd place and Charlie's 4th place in 1912 being somewhat of a swan song. The brothers had decided to concentrate on improvement of their production models, because those were the days when the motorcycle was undergoing very rapid development. This design work was manifest in several three- and six-speed drives that were tried, with rear hub gearboxes and adjustable belt pulleys in use. In 1912, the marque also started to produce its own engine—an 85.5- by 85-mm bore and stroke Single that delivered 3.5 bhp.

In 1913, the company went "foreign" for engines—using a Swiss Motosacoche V-twin that had a bore and stroke of 64 by 77 mm. This overhead valve engine featured the now common hemispherical combustion chamber, which historians regard as the first use of this type of design. A Motosacoche inlet-over-exhaust-valve V-twin was also used to power the Model 8B, which featured a three-speed gearbox and all-chain drive. This model also displayed a kickstarter and an internal expanding rear brake, instead of the previous caliper type brake on the wheelrim.

Then came World War I, and the factory concerned itself with producing war materials and aircraft parts. After the war, the company lost all interest in racing and concentrated on improvement of the standard production models. The whole range was comprised of V-twins, with both side-valve and ohv engines being used. In 1923, a 350-cc side-valve Single was added to the range, which had chain drive and hub brakes. These models were sound and reliable, but there was nothing about them that made headlines or won races.

This 350-cc Single underwent continued development in 1926 with installation of an ohc valve system, but its appearance in the Isle of Man Junior TT was doomed to failure. A 600-cc low compression ohv Single was then introduced for the burgeoning sidecar market, and this was followed by a 250-cc side-valve Single. By 1927, the range of models included a 250-cc side-valve Single, 500-cc ohv sports Single, 600-cc ohv Single, and 1000-cc side-valve and ohv V-twins.

In 1926, H.H. Collier Sr. died, and the concern was renamed Matchless Motorcycle Co. The elder Collier had been busy, before his death, in design of new models, and these were introduced in 1928. The main features were modern saddle type fuel tanks that went over, instead of between, the top frame tubes, two-exhaust-port single-cylinder engines, and a new standard of reliability. The gearboxes were still three-speed affairs with a clumsy hand shift and foot clutch. These models were very reliable, and they saw the company through the depression years that wiped out so many of their competitors.

By 1930, Charlie Collier was beginning to make felt his design ability. With introduction of the 400-cc "Silver Arrow," the company embarked on a new approach at catering to the more sophisticated rider who wanted a luxurious motorcycle. The Silver Arrow was knicknamed the "whispering wonder," by reason of its ultra-quiet, smooth-running, side-valve two-cylinder engine. The Twin looked almost like a vertical Twin, but it was, in fact, a narrow 26-degree V-twin set crosswise in the frame. The Silver Arrow was very comfortable for its day, with a pivoted-fork rear suspension, which was in the era when the rigid frame was still standard practice.

In 1931, the company introduced an even more sophisticated mount in the "Silver Hawk." This model featured a narrow V-four engine set transversally in

Designed in 1936, the "G" series Matchless engine has proven to be exceptionally reliable. The cylinder and head are alloy—note the hairpin valvesprings.

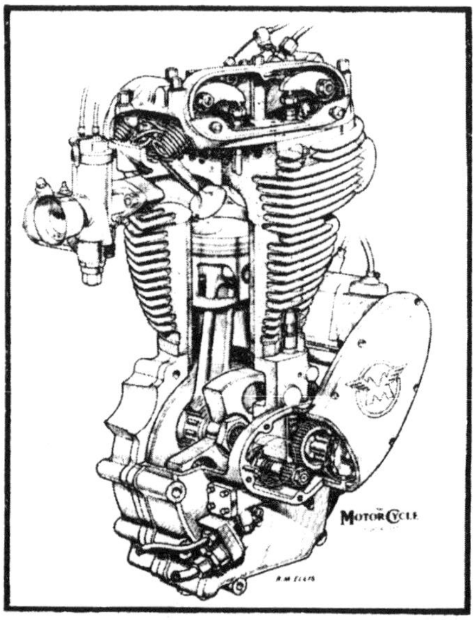

The 1937 side-valve 1000-cc V-twin was a popular mount for sidecar owners. These old flatheads were known for their low speed power, low maintenance expense.

The prototype racing Matchless Twin was unveiled in 1951—featured a modified roadster pushrod engine. The AJS fuel tank was from a 7R racer.

The early 1960s saw this 500-cc Single fall from favor with the world's riders. A low initial cost, reliability, and moderate performance were its virtues.

The 1960 Matchless Twin with its 500- or 600-cc AMC powerplant was not an especially good design and later was dropped in favor of the Norton engine.

the frame. The 593-cc powerplant featured ohc valve operation. The deluxe specifications included such things as a chromed fuel tank and an instrument panel. A fully sprung sidecar was available. This V-four sold at a price that was actually not terribly far above the ohv 500-cc Single's price, but the scarcity of pounds in those depression years prevented the sales figure from ever attaining a commercially profitable rate. The Four was dropped from production within a few years—the verdict being excellent engineering, but unsound business practice. Today, the Silver Hawk is prized as a collector's item, and as a great classic, but few are known to exist.

In 1931, the Matchless Co. bought the A.J. Stevens concern, and thereafter the AJS and Matchless began to acquire similar specifications. During the early 1930s, the fashion in England was for "sloping" Singles, and Matchless responded with 500-cc side-valve and ohv models that were quite light. With a rigid frame, girder front fork, and three-speed gearbox, the side-valve machine weighed only 220 lb.

The big side-valve V-twin had been redesigned in 1929, and this model, called the Model X, was produced until World War II. The market for the husky 1000-cc V-twins was substantial during the prewar days in England, since the sidecar owners loved the low-speed power, reliability, and low maintenance expense that these chuggers possesed.

In 1936, the amalgamation of the AJS and Matchless was apparent when the "G" series was introduced. These 350-cc and 500-cc Singles featured pushrod-operated ohv powerplant, a four-speed footshift gearbox, a rigid frame, and a girder front fork. About the only difference between the AJS and Matchless in later years was the location of the magneto—which was in front of the cylinder on the AJS and behind the cylinder on the big "M." These thumpers were known as exceptionally rugged bikes, and their handling and performance were up to the high standards that motorcyclists had come to expect during the late 1930s.

These G models were soon followed by the Clubman and Clubman Special models, which were more highly tuned Singles for the sporting minded riders. The Clubman Special model could be purchased in competition trim with a 21-in. front wheel, knobby tires, upswept exhaust, and special gear ratios. These Singles were used in trials and scrambles events with great success during the late 1930s, and this publicity helped further the marque's export sales.

During these years, the international road racing scene had taken on great stature, with many companies fielding works teams on everything from fine

A model that many old-timers will remember is the 1952 G-80CS. These great Singles were nearly unbeatable in California desert races.

The 1959 trials 350 was a cobby looking 306-lb. mount with 10 in. of ground clearance. Wide ratio gears were used. This model won many important trials in Europe.

The 1964 G-80CS offered a robust 500-cc engine, good handling, and moderate performance. The current edition still is known for its reliability.

handling ohc Singles to exotic water-cooled and supercharged Fours. The Matchless name (under the company name of Associated Motor Cycles, after 1938) was quietly absent from this grand prix game, preferring instead to develop a range of bikes for the man on the street. In all fairness to the marque, it should be stated that the firm conducted an intensive racing program under the AJS banner, which was a policy that was continued after World War II.

During the war, the factory once again was switched over to wartime production, only this time it was for military motorcycles. The model produced was the G3L—the ohv 350 thumper that was exceptionally reliable. Altogether, no less than 80,000 bikes were produced during the war, and even today one can occasionally see the trusty little Singles thumping down the country lanes of England.

After the war, the marque narrowed its range down to the G-3 (350-cc) model and G-80 (500-cc) model—both with "iron" engines and the new hydraulically dampened telescopic front fork. The design was cleaned up a little from the prewar days, but it remained the basic design that was laid down in 1936.

The company also produced some trials mounts, which were mainly 350s set up with 21-in. front wheels, knobby tires, upswept exhaust, wide ratio gears, and alloy fenders. These were superb little trials mounts, and men such as Hugh Viney and the Ratcliffe brothers established a splendid reputation with these machines. The 500-cc model was used extensively in motocross racing, and Basil Hall scored many wins on his factory model.

It was during these immediate post-war years that the company greatly expanded its export market, with the United States figuring prominently in this expansion. The fine Singles were especially loved by California desert riders, because handling, traction, and reliability were better than any of the others for cross-country and scrambles races of that era. Indeed, AJS and Matchless riders were the most successful of all the riders until the late 1950s, and it would take a story in itself just to cover all of the victories racked up by riders such as Dutch Sterner, Del Kuhn, Julius Kroeger, Guy Louis, Aub LeBard, Earl Flanders, Don Bishop, Rod Coates, Ralph Adams, Lee Carey, Dalton Holliday, Vern Hancock, Bud Ekins, Butsy Mueller, Dick Dean, and Fred Borgeson.

In 1949, the marque made another great step forward when the new "Super Clubman" model was introduced. The new Matchless was a 500-cc Twin with a bore and stroke of 66 and 72.8 mm. It featured a three-bearing crankshaft, alloy head, and a comfortable dual seat. The Twin displayed an additional characteristic—the new swinging arm rear suspension. This new frame was a major contributor to rider comfort and roadability, and it was available on all of the models.

During the next several years, the range of models expanded to include mounts for trials and motocross riders, and these fine Singles established a great record on both sides of the Atlantic. The 1952 catalog listed a total of nine models—one Twin and eight Singles. This also was the year that the AJS and Matchless Singles became virtually the same bike, with the magneto being mounted in front of the cylinder on both makes.

Serious touring riders loved the G-9 500-cc Twin for its comfort, 90-95 mph performance, and sporting appearance. The Twin had a four-speed gearbox with ratios of 13.35:1, 8.8:1, 6.4:1, and 5.0:1. A favorite with everyday riders was the 350-cc Single–a reliable model with brisk performance. The G-3L model was built around a rigid frame; the G-3LS had a spring frame. Next came the G-80 and G-80S models–both 500-cc Singles in rigid and spring frame trim. An improvement to the Singles that year was the alloy head, which helped the engines run cooler and allowed higher compression ratios to be used. These 85-90-mph thumpers were known for their reliability and low maintenance expense, and they proved popular with riders the world over.

The 1952 catalog listed four competition models–two rigid frame models (350 and 500 cc) for trials and two spring frame models (350 and 500 cc) for motocross. The engines were considerably modified on these competition mounts, with alloy cylinders and heads, special cams, high compression pistons on the two motocross models, and a heavier gearbox with wide ratio gears on the trials bikes and close ratio gears on the scrambler models. Both the trials and scrambles versions were delivered with 3.00 by 21-in. front tires and 4.00 by 19-in. rear tires. Both models were fitted with small 2.25-gal. fuel tanks. The wheelbase was a short 52.5 in. on the competition models, and alloy fenders, sports solo seats, and upswept exhaust systems were fitted.

The next significant chapter in the history of the marque began in 1951 when a racing department was established to develop the 500-cc Twin. The idea was to determine if the standard G-9 Twin showed potential as a road racing model, with the long range goal being to produce a pukka racing machine at a cost well below that of an ohc Single. The Twin first appeared in the 1951 Manx Grand Prix in the hands of Robin Sherry, who proceeded to garner a 4th place with the machine behind the famous Norton Manx Singles. Justifiably encouraged, the factory continued to develop the Twin in preparation for the 1952 Manx GP. Derek Farrant rode the Matchless prototype that year, and he led from start to finish–setting up race and lap records at 88.65 and 89.64 mph.

This sensation virtually required the factory to start production of the Twin, which appeared in 1953 as the G-45 model. The G-45 was produced for five years, during which time it underwent detailed improvements. By 1956, the 66 by 72.8-mm Twin was producing 48 bhp at 7200 rpm on a 10.0:1 compression ratio, which provided a level road speed of 120 mph without a fairing. The G-45 had a four-speed gearbox with ratios of 8.57:1, 6.19:1, 5.0:1, and 4.58:1; and the wheelbase was 55.25 in. A deep, knee-knotched 6-gal. fuel tank was used, and the dry weight was 320 lb. Tire sizes were 3.00 by 19-in. front and 3.25 by 19-in. rear, and both brakes were a huge 8.25-in. size with an air scoop and twin leading shoe arrangement being used on the front brake. This Twin was a beautiful thing with its massive alloy cylinder and head, plus the pair of 1.094-in. GP carbs, and it attained a splendid record in the hands of private owners the world over.

This range of models remained much the same until 1956, when a new 600-cc Twin was introduced that had a bore and stroke of 72 and 72.8 mm, respectively. Gone were the rigid frame models, with the catalog down to only seven machines–the 500- and 600-cc Twins, the 350- and 500-cc road Singles, the G-80CS scrambler, the G-45 road racer, and the new G-3LC trials model.

This new trials model was a direct result of the works mounts that had done so well in 1954 and 1955 (Artie Ratcliffe won the 1954 Scottish Six Days Trial), and it had the swinging arm frame for the first time. A trim and fine handling Single, the G-3LC ran on a 6.5:1 compression ratio and pumped out 18 bhp at 5750 rpm. The bike weighed a light 320 lb. and had gear ratios of 21.0:1, 16.0:1, 10.3:1, and 6.6:1. Very few of these superb trials models ever made their way to this country, but they achieved a great number of outstanding successes in British and Scottish trials.

The range of machines continued much the same until 1959, when the G-45 Twin was dropped and the G-50 ohc Single was introduced. The G-50 also was derived from a previous model–this one the 350-cc 7R AJS racer. The big Single had a bore and stroke of 88.5 by 76.7 mm, and it pumped out 47 bhp at 7200 rpm on a 10.6:1 compression ratio. The Single breathed through a huge 1.375-in. GP carburetor, and it was known for being able to power a very high gear ratio. The Matchless was slightly slower than the double-knocker Manx Norton, but its simple chain-driven single-cam engine was easier to tune and less expensive to maintain than the Manx. With a top speed of 125 mph, plus good handling, the G-50 proved to be especially popular with the private owners.

Other notable changes in 1959 included the increasing of the 600-cc Twin's stroke to 79.3 mm, thus making it a full blown 650. The trials 350 also was improved by chopping the weight to 306 lb. and increasing the ground clearance to a whopping 10 in. The trials mount continued to be the finest bike of its type in the world.

During the early 1960s, the international motorcycle scene underwent a great change with the arrival of the Japanese models on the market. This had a profound effect on the British scene, and especially upon the Matchless. In 1956, AMC purchased Norton Motors, Ltd., and within a few years the standardization began within the AJS-Matchless-Norton range of machines. The first to go were the Norton Singles in favor of the AMC design, and then the AMC Twin went by the board in favor of the big 650- and 750-cc Norton powerplants. The lusty G-50 racer also was dropped at the end of the 1963 season–a fact that saddened motorcycle sportsmen throughout the world.

During these years, the design progress seemed to stagnate at the factory, with the only new model being the G-85CS scrambler. This 500-cc Single was produced in 1966 and 1967 as a fire breathing motocross mount with a 1.375-in. GP carburetor, Metisse type frame with the central frame tube serving as an oil tank, light 300-lb. weight, and 42-bhp engine. The red hot two-strokes already had taken the play away, though, and this expensive scrambler faded from the picture.

During 1967, the company was known to be virtually bankrupt, and it appeared that the famous name would disappear. At the last hour, a new company was formed when Manganese Bronze Holdings bought out the Norton-Matchless Co., plus the old Villiers concern, and today the future looks brighter for the big M. At present, the range is comprised of the Norton Atlas-powered street and scrambles Twins, plus the G-80CS scrambles model. The bore and stroke of the trusty old thumper was changed to 86 by 86 mm a few years ago in order to obtain greater rpm and power, but the design is still much the same as it was in 1952, or even 1936, for that matter! These fine Singles are probably the most reliable scrambles type mounts going, though, and this, plus the good handling, ensures a devoted following of American riders.